Operator Theory: Advances and Applications
Vol. 156

Quadrature Domains and Their Applications

The Harold S. Shapiro Anniversary Volume

Peter Ebenfelt
Björn Gustafsson
Dmitry Khavinson
Mihai Putinar
Editors

Birkhäuser Verlag
Basel · Boston · Berlin

Editors:

Peter Ebenfelt
Department of Mathematics
University of California, San Diego
La Jolla, CA 92093
USA
pebenfel@math.ucsd.edu

Dmitry Khavinson
Department of Mathematical Sciences
University of Arkansas
Fayetteville, AR 72701
USA
dmitry@uark.edu

Björn Gustafsson
Department of Mathematics
Royal Institute of Technology (KTH)
100 44 Stockholm
Sweden
Email: gbjorn@math.kth.se

Mihai Putinar
Mathematics Department
University of California, Santa Barbara
Santa Barbara, CA 93106
USA
mputinar@math.ucsb.edu

2000 Mathematics Subject Classification 00B25, 00B30, 30-06, 31-06, 32-06, 35-06, 47-06, 76-06

A CIP catalogue record for this book is available from the
Library of Congress, Washington D.C., USA

Bibliographic information published by Die Deutsche Bibliothek
Die Deutsche Bibliothek lists this publication in the Deutsche Nationalbibliografie; detailed bibliographic data is available in the Internet at <http://dnb.ddb.de>.

ISBN 3-7643-7145-5 Birkhäuser Verlag, Basel – Boston – Berlin

Part of Springer Science+Business Media
Printed on acid-free paper produced from chlorine-free pulp. TCF ∞
Cover design: Heinz Hiltbrunner, Basel
Printed in Germany
ISBN-10: 3-7643-7145-5
ISBN-13: 978-3-7643-7145-6
9 8 7 6 5 4 3 2 1 www.birkhauser.ch

Contents

Preface

This book is an expanded version of talks and contributed papers presented at a conference held at the University of California at Santa Barbara in March 2003, to celebrate the 75th birthday of Professor Harold S. Shapiro. The main theme of the conference was Quadrature Domains and Their Applications.

Harold S. Shapiro

The idea of having Gaussian type quadratures, on a fixed domain and for the full class of integrable analytic or harmonic functions, independently originated in the early seventies in the works of D. Aharonov and H.S. Shapiro, Ph. Davis, S. Richardson and M. Sakai. However, later it was discovered that the potential theoretic equivalent of the concept of a quadrature domain was clearly formulated already in a seminal memoir of Herglotz in 1914 (*Über die analytische Fortsetzung des Potentials ins Innere der anziehenden Massen*, Gekrönte Preisschr. der Jablonowskischen Gesellsch. zu Leipzig, 56 pp.) and was developed further in the works of Schmidt and Wavre. The modern time authors were motivated by quite a diverse spectrum of problems in the theory of univalent functions of a complex variable, approximation theory, fluid mechanics and potential theory on Riemann surfaces. This interdisciplinary trend has continued, and has even been amplified over the decades. Today, we can add to the ramifications of the theory of quadrature domains chapters of partial differential equations in the complex domain, variational problems for PDE and free boundaries, completely integrable systems related to quantum physics, gravitational lensing and many new aspects of the modern theory of fluid dynamics.

The theme of quadrature domains has been largely cultivated for more than three decades by Professor Shapiro and his PhD students. His 1992 book "*The Schwarz function and its generalization to higher dimensions*", Univ. Arkansas Lecture Notes Math. vol. 9, Wiley, New York, well illustrates the state of the field at the time of its writing. The present collection of articles reflects some of the progress that has been made in the last decade. However, the subject is so much alive nowadays that even in the short time that has passed between the occasion of the conference and the publication of this volume, new striking appearances of quadrature domains in mathematical physics have emerged.

The book contains both original articles and survey papers covering quite a wide scope of ideas in classical and modern analysis and applications. The survey articles written by the leading experts in the field will help to orient the beginners in the vastly increasing literature on the subject. The monograph may also help young researchers and graduate students wanting to familiarize themselves with this active and beautiful area of analysis that thrives on techniques from potential theory, complex analysis, geometry and partial differential equations. The book concludes with a selection of some open problems that were discussed in a special session at the conference. We hope that the book will attract young analysts looking for interesting and easily formulated questions, though still deep and important for numerous applications. On the other hand, the experts may find it helpful as a reference for the current status of the subject as well.

We were all inspired by Harold Shapiro's contagious enthusiasm, energy and never ending interest in finding new paths and beautiful problems in all areas of mathematics. We wholeheartedly wish him, on this occasion, a very happy birthday and many fruitful, healthy, and productive years ahead.

We are grateful to the National Science Foundation, University of Arkansas and Universities of California at Santa Barbara and San Diego for financial support of the conference.

The editors

Selected Bibliography of Harold S. Shapiro

Books

1. Harold S. Shapiro, , *Smoothing and approximation of functions. Revised and expanded edition of mimiographed notes* (Matscience Report No. 55), Van Nostrand Reinhold Mathematical Studies, Van Nostrand Reinhold Co., New York-Toronto, Ont.-London, 1969, viii + 136 pp.
2. Harold S. Shapiro, *Topics in approximation theory, With appendices by Jan Boman and Torbjrn Hedberg,* Lecture Notes in Math., Vol. **187**, Springer-Verlag, Berlin-New York, 1971. viii + 275 pp.
3. Harold S. Shapiro, *The Schwarz function and its generalization to higher dimensions*, University of Arkansas Lecture Notes in the Mathematical Sciences, Vol. **9**, A Wiley-Interscience Publication, John Wiley and Sons, Inc., New York, 1992. xiv + 108 pp.
4. William T. Ross and Harold S. Shapiro, *Generalized analytic continuation*, University Lecture Series Vol. **25**, American Mathematical Society, Providence, RI, 2002.

Selected Research Articles

1. W.W. Rogosinski and Harold S. Shapiro, *On certain extremum problems for analytic functions*, Acta Math. **90** (1953), 287–318.
2. Harold S. Shapiro, *The expansion of mean-periodic functions in series of exponentials*, Comm. Pure Appl. Math. **11** (1958), 1–21.
3. Harold S. Shapiro, *A Tauberian theorem related to approximation theory*, Acta Math. **120** (1968), 279–292.
4. Dov Aharonov and Harold S. Shapiro, *Domains on which analytic functions satisfy quadrature identities,* J. Analyse Math. **30** (1976), 39–73.
5. Peter Duren, Dmitry Khavinson, Harold S. Shapiro and Carl Sundberg, *Contractive zero-divisors in Bergman spaces*, Pacific J. Math. **157** (1993), no. 1, 37–56.

6. Lowell J. Hansen and Harold S. Shapiro, *Graphs and functional equations*, Ann. Acad. Sci. Fenn. Ser. A I Math. **18** (1993), no. 1, 125–146.

7. Peter Ebenfelt and Harold S. Shapiro, *The mixed Cauchy problem for holomorphic partial differential operators,* J. Anal. Math. **65** (1995), 237–295.

8. Björn Gustafsson, Makoto Sakai and Harold S. Shapiro, *On domains in which harmonic functions satisfy generalized mean value properties,* Potential Anal. **7** (1997), no. 1, 467–484.

9. Dov Aharonov, Harold S. Shapiro and Alexander Yu. Solynin, *A minimal area problem in conformal mapping*, J. Anal. Math. **78** (1999), 157–176.

10. Dmitry Khavinson, John E. McCarthy and Harold S. Shapiro, *Best approximation in the mean by analytic and harmonic functions*, Indiana Univ. Math. J. **49** (2000), no. 4, 1481–1513.

11. Dmitry Khavinson and Harold S. Shapiro, *Best approximation in the supremum norm by analytic and harmonic functions*, Ark. Mat. **39** (2001), no. 2, 339–359.

12. Harold S. Shapiro, *Spectral aspects of a class of differential operators*, in Vol. Operator methods in ordinary and partial differential equations (Stockholm, 2000), 361–385, Oper. Theory Adv. Appl., **132**, Birkhäuser, Basel, 2002.

13. Björn Gustafsson, Mihai Putinar and Harold S. Shapiro, *Restriction operators, balayage and doubly orthogonal systems of analytic functions,* J. Funct. Anal. **199** (2003), no. 2, 332–378.

Open Problems Related to Quadrature Domains

1. Two questions concerning quadrature domains

DOV AHARONOV, Department of Mathematics, Technion – Israel Institute of Technology, Haifa 32000, Israel.

1.1. The a_3 problem

The a_2 problem was presented by Harold S. Shapiro in [4] and was finally solved in [2]. The similar "a_3 problem" (meaning replacing the condition $a_2(f) = \alpha,\ 1/2 < \alpha < 2$, by $a_3(f) = \alpha,\ 1/3 < \alpha < 3$) seems to be essentially harder. Indeed, not as for the "a_2 problem", here symmetrization does not seem to help and one needs to find other method to replace it.

Partial results. Any extremal f has a bounded derivative, namely $|f'| < M < \infty$ in the unit disk U, for some constant M. From this one easily deduces that all complex moments

$$\int_D w^n dA(w) = 0, \quad n \geq 3,$$

vanish, where D is the image domain $f(U)$ and dA stands for the area measure.

We note that uniqueness is not clear at all.

One would like to prove, like for the "a_2 problem" that $\tilde{D} = \mathrm{int}\overline{D}$ (i.e. the domain D with "erased slits") is a Jordan domain $\tilde{D}$.

If this could be done, then it would follow that there are absolute constants A, B, C with the property

$$\int_{\tilde{D}} g(w)dA(w) = Ag(0) + Bg'(0) + Cg''(0),$$

for any analytic, Lebesgue integrable function g in $\tilde{D}$. In other words, $\tilde{D}$ is a quadrature domain. From this point on, it should not be hard to find the complete solution.

1.2. The nature of the boundary of a generalized quadrature domain

It was shown in [1] that the boundary of a quadrature domain is algebraic. This was done by investigating the Schwarz function, based on a standard process of eliminating its poles. For generalized quadrature domains, the situation is more complicated.

To be more specific, we restrict ourselves to the case of a domain D arising in the recent article [3]:

$$\int_D g(w)dA(w) = Ag(0) + B\int_0^b g(u)du,$$

where g is an analytic, Lebesgue integrable function in D, the points $0, b$ belong to D and the line integral is taken along any curve inside D.

The case of a simply connected domain D was resolved in [3].

References

[1] D. Aharonov, H.S. Shapiro, *Domains in which analytic functions satisfy quadrature identities*, J. Analyse Math. **30** (1976), 39–73.

[2] D. Aharonov, H.S. Shapiro, A. Yu. Solynin, *A minimal area problem in conformal mapping*, J. Analyse Math. **78** (1999), 157–176.

[3] D. Aharonov, H.S. Shapiro, A. Yu. Solynin, *A minimal area problem in conformal mapping. II*, J. Analyse Math. **83** (2001), 259–288.

[4] W.K. Hayman, *Research Problems in Function Theory*, Athlone Press, University of London, 1967.

---

2. Quadrature domains of infinite order and infinite connectivity

DARREN CROWDY, Department of Mathematics, Massachusetts Institute of Technology, 2-392, 77 Massachusetts Avenue, Cambridge, MA 02139.

The simplest example of a bounded quadrature domain is a circular disc of radius r. Such a domain D satisfies the quadrature identity

$$\int\int_D h(z)d\sigma = \pi r^2 h(z_0)$$

where z_0 is the centre of the disc, $h(z)$ is a suitable analytic function integrable over D and $d\sigma$ denotes area measure.

Any *finite* collection of disconnected circular discs also represents a quadrature domain. Connected quadrature domains, of finite order and finite connectivity, can be constructed by "continuing" such collections of disconnected discs (see, for example, [Crowdy, *Proc. Roy. Soc. A*, **457**, (2001)] where such ideas have been used in applications to vortex equilibria).

Consider $2N + 1$ identical circular discs of radius $1/2$ centred at $z = -N$, $-(N-1), \ldots, -1, 0, 1, 2, \ldots, N$. The circular discs touch and the disconnected bounded domain D_N, say, satisfies the (order $2N + 1$) quadrature identity

$$\int\int_{D_N} h(z)d\sigma = \sum_{k=-N}^{N} \pi r^2 h(k)$$

with $r = 1/2$. For $r > 1/2$, the domain D_N "continues" to a simply-connected quadrature domain of order $2N + 1$. Conformal maps from a unit circle to such a domain are known to be rational functions.

It is natural to consider the following limit: let the number of circular discs centred at the integers tend to infinity, i.e., consider the domain of disconnected discs D_∞. Such a domain will be unbounded. Provided $h(z)$ decays sufficiently fast at infinity, it will also satisfy the "generalized" quadrature identity

$$\int\int_{D_\infty} h(z)d\sigma = \sum_{k=-\infty}^{\infty} \pi r^2 h(k),$$

which has *infinite* order. Yet, it can still be considered a quadrature domain in the sense that the set of quadrature data is *finite*.

The question arises if D_∞ can be similarly "continued" to a *connected* unbounded quadrature domain of infinite order when $r > 1/2$. This question can be answered by direct construction (see, Crowdy (unpublished notes)). A natural extended definition is to consider quadrature domains satisfying

$$\int\int_{D_\infty} h(z)d\sigma = \sum_{n=1}^{N} \sum_{k_n=-\infty}^{\infty} \pi r_n^2 h(z_n + k_n) \tag{1}$$

which, although of infinite order, depend on just a finite set of quadrature data $\{r_n, z_n \in \mathbb{C}\}$ where, by the expected 1-periodicity of the configuration, z_n can be taken in the interval $-1/2 < |z_n| < 1/2$.

In a similar way, it is possible to envisage infinite-order quadrature domains of infinite connectivity, but still with a finite set of quadrature data. This can be done by defining the bounded domain $D_{M,N}$ of identical touching disconnected discs in a square array centred on the points $z_n = m + in$ where m, n are integers $|m| \leq M, |n| \leq N$. Taking the limits $M \to \infty$, $N \to \infty$ and "continuing" to a connected domain produces $D_{\infty,\infty}$ — an unbounded quadrature domain of infinite order and infinite connectivity satisfying

$$\int\int_{D_{\infty,\infty}} h(z)d\sigma = \sum_{m=-\infty}^{\infty} \sum_{n=-\infty}^{\infty} \pi r^2 h(m+n)$$

which will be (infinitely) connected for r just greater than $1/2$. Generalized quadrature identities analogous to (1) can be envisaged.

The following questions arise given these "generalized" infinite-order quadrature domains:

- Can it be proven rigorously that such domains exist?
- If they do, what it the best way to parametrize/uniformize the boundaries? What can be said about the analytic structure of conformal mappings to, for example, a simple "unit cell" in these singly/doubly-periodic infinite-order quadrature domains?
- Bounded, finite-order quadrature domains have algebraic boundaries. What can be said about the boundaries of the generalized domains?
- What can be said about the Cauchy transforms of the generalized domains?

These classes of domains are potentially of great use in applications.

3. Problems on quadrature domains

BJÖRN GUSTAFSSON, Mathematics Department, Royal Institute of Technology, S-10044 Stockholm, Sweden.

3.1. Uniqueness of simply connected quadrature domains

For "classical" quadrature domains [1], [9], [22] in two dimensions the question may be stated as follows: Do there exist two different simply connected domains $\Omega \subset \mathbf{C}$ ($\Omega = \Omega_1, \Omega_2$ say) admitting one and the same quadrature identity

$$\iint_\Omega f(z)dxdy = \sum_{j=1}^{m}\sum_{k=0}^{n_j-1} c_{jk} f^{(k)}(z_j),$$

to hold for every analytic and integrable function f in Ω?

Variants of the question are obtained by replacing the right-hand side by $\int f d\mu$ for a measure μ with compact support and working with harmonic test functions instead of analytic ones. In higher dimension the assumption of simple connectivity should be replaced by the assumption that the domains are "solid", meaning that the complement of the closure of the domain in question has only one component (the unbounded one).

The above uniqueness question is essentially equivalent to the exterior inverse problem in potential theory: Does there exist two different solid domains such that their Newtonian potentials agree outside their union? (The domains are considered as bodies of density one.)

See e.g. [26], [16] for general information about this inverse problem and [10] for its relation to quadrature domain theory. See also the last section (on open problems) in [24].

If one allows suitable weights (in place of pure Lebesgue measure) there are counterexamples for the inverse problem of potential theory, see [18]. Similarly, if the potentials are asked to agree only in a neighbourhood of infinity [20], [26]. There are also counterexamples for the corresponding question when area measure (in the two dimensional case) is replaced by arclength measure on the boundary, [15], Proposition 6.2.

Therefore one might conjecture that there is no uniqueness in the above stated problems. Nevertheless it would be good to have a definite answer.

3.2. On algebraic boundaries

Given points $x_j \in \mathbf{R}^n$ and coefficients $c_j > 0$ ($j = 1, \ldots, m$ say) it is known [21], [10], [11] that there exists an open set (possibly disconnected) $\Omega \subset \mathbf{R}^n$ admitting the quadrature identity

$$\int_\Omega h dx = \sum_{j=1}^{m} c_j h(x_j)$$

for all integrable harmonic functions h in Ω. It is uniquely determined (up to nullsets) if the appropriate inequality ($\geq$) is required to hold for all subharmonic test functions.

It is known from general theory for free boundaries [7] that $\partial\Omega$ is a real analytic hypersurface, except possibly for a small set of singular points (see more precisely [2]). In two dimensions it is moreover known [1], [9], [22] that $\partial\Omega$ is algebraic. For example, if $r > 1$ in the two point quadrature identity (for $\Omega \subset \mathbf{C}$)

$$\int_\Omega h dx dy = \pi r^2 (h(-1) + h(1)),$$

then $\partial\Omega$ is given by

$$(x^2 + y^2)^2 - 2r^2(x^2 + y^2) - 2(x^2 - y^2) = 0.$$

See e.g. [22], Section 3.1, and for more examples [5], [4].

Now the question is whether anything similar is true in higher dimension? Is $\partial\Omega$ necessarily algebraic or belongs to some other special class of surfaces? Even for two point quadrature domains in higher dimensions no explicit description of the boundary is known.

There are known a few examples of quadrature domains in $\mathbf{R}^4$ which do have algebraic boundaries, see [17], [24].

3.3. Quadrature domains and Hele-Shaw flow with weights, or on Riemannian manifolds

Let $\rho > 0$ be a suitably smooth weight function in $\mathbf{R}^n$. Then for any $t > 0$ there exists a unique (up to nullsets) domain $\Omega_t \subset \mathbf{R}^n$ such that

$$th(0) \leq \int_{\Omega_t} h \rho dx$$

for every integrable subharmonic function h in Ω_t (if h is harmonic, the inequality will become equality).

The presence of the time parameter t allows for the interpretation that $\{\Omega_t : t > 0\}$ is the solution of a Hele-Shaw flow moving boundary problem with an injection point [19], [8], [11] and the weight ρ allows for the interpretation that everything takes place on a Riemannian manifold.

It was shown in [13] and [14] that if, in the case of two dimensions, the weight ρ is logarithmically subharmonic, i.e. $\Delta \log \rho \geq 0$, then all the Ω_t are simply connected. In the view of Riemannian manifolds the logarithmic subharmonicity means that the Gaussian curvature of the manifold is nonpositive (hyperbolic manifold). The result can be used to provide the hyperbolic manifold with global polar coordinates. The radii in these polar coordinates can be thought of as Hele-Shaw geodesics and with respect to these the fluid domains Ω_t are starlike. See [14] for further discussions.

It is natural, and probably rewarding, to search for corresponding results in higher dimensions.

A few more results on the geometry of Hele-Shaw flow on manifolds (this time surfaces in $\mathbf{R}^3$) can be found in [24], Section 7.

Thus we expect that there is a rich theory concerning the geometry of quadrature domains and Hele-Shaw flows on manifolds to be discovered.

3.4. Parabolic quadrature domains

Parabolic potential theory is already well-established (see e.g. [6]) and there are some examples and beginning of theory for quadrature domain type questions. See for example [23] and [12]. Probably a general theory can be developed.

3.5. Quadrature domain theory based on Einstein's theory of gravitation rather than Newton's theory ("hyperbolic quadrature domains")

This programme is very tentative and it may very well be that it makes no sense to look for quadrature domains in full space-time. However, restricting to space-like slices we are still within the range of (nonlinear) elliptic potential theory, and it is quite likely that quadrature domain theory could be useful for certain problems there, exactly as it has turned out to be useful in a variety of problems in fluid dynamics [3]. One example is the type of problem discussed in [25].

References

[1] D. Aharonov, H.S. Shapiro, *Domains in which analytic functions satisfy quadrature identities*, J. Analyse Math. **30** (1976), 39–73.

[2] L.A. Caffarelli, L. Karp, H. Shahgholian, *Regularity of a free boundary with application to the Pompeiu problem*, Ann. Math. **151** (2000), 269–292.

[3] D. Crowdy, *Quadrature domains and fluid dynamics*, preprint 2002.

[4] D. Crowdy, *Constructing multiply-connected quadrature domains I: algebraic curves*, preprint 2003.

[5] P.J. Davis, *The Schwarz Function and its Applications*, Carus Math. Mongraphs No. 17, Math. Assoc. Amer., 1974.

[6] J. Doob, *Classical Potential Theory and its Probabilistic Counterpart*, Springer-Verlag, Berlin, 1983.

[7] A. Friedman, *Variational Principles and Free Boundaries*, Wiley and Sons, 1982.

[8] K.A. Gillow and S.D. Howison, *A bibliography of free and moving boundary problems for Hele-Shaw and Stokes flow*, published electronically at URL http://www.maths.ox.ac.uk/~howison/Hele-Shaw.

[9] B. Gustafsson, *Quadrature identities and the Schottky double*, Acta Appl. Math. **1** (1983), 209–240.

[10] B. Gustafsson, *On quadrature domains and an inverse problem in potential theory*, J. Analyse Math. **55** (1990), 172–216.

[11] B. Gustafsson, *Lectures on Balayage*, preprint 2003 (available via http://www.math.kth.se/~gbjorn/).

[12] A. Hakobyan, H. Shahgholian, *Generalized mean value property for caloric functions*, preprint 2003 (available via http://www.math.kth.se/~henriksh/).

[13] H. Hedenmalm, S. Jakobsson, S. Shimorin, *A biharmonic maximum principle for hyperbolic surfaces*, J. Reine Angew. Math. **550** (2002), 25–75.

[14] H. Hedenmalm, S. Shimorin, *Hele-Shaw flow on hyperbolic surfaces*, J. Math Pures Appl. **81** (2002), 187–222.

[15] A. Henrot, *Subsolutions and supersolutions in free boundary problems* Ark. Mat. **32** (1994), 79–98.

[16] V. Isakov, *Inverse Source Problems*, AMS Math. Surveys and Monographs 34, Providence Rhode Island, 1990.

[17] L. Karp, *Construction of quadrature domains in* $\mathbf{R}^n$ *from quadrature domains in* $\mathbf{R}^2$, Complex variables **17** (1992), 179–188.

[18] N.S. Nadirashvili, *Universal classes of uniqueness of domains in the inverse problem of Newtonian potential theory*, Soviet Math. Dokl. **44** (1992), 287–290.

[19] S. Richardson, *Hele Shaw flows with a free boundary produced by the injection of fluid into a narrow channel*, J. Fluid Mech. **56** (1972), 609–618.

[20] M. Sakai, *A moment problem for Jordan domains*, Proc. Amer. Math. Soc. **70** (1978), 35–38.

[21] M. Sakai, *Quadrature Domains*, Lect. Notes Math. **934**, Springer-Verlag, Berlin-Heidelberg 1982.

[22] H.S. Shapiro, *The Schwarz function and its generalization to higher dimensions*, Uni. of Arkansas Lect. Notes Math. Vol. 9, Wiley, New York, 1992.

[23] N. Suzuki, N. Watson, *A characterization of heat balls by a mean value property for the temperature*, Proc. Amer. Math. Soc. **129** (2001), 2709–2713 (electronic).

[24] A.N. Varchenko, P.I. Etingof, *Why the Boundary of a Round Drop Becomes a Curve of Order Four*, AMS University Lecture Series, Volume 3, Providence, Rhode Island 1992.

[25] R. Wegmann, *Keplerian discs around magnetized neutron stars–a free boundary problem*, Direct and inverse boundary value problems (Oberwolfach 1989), 233–253, Methoden Verfahren Math.Phys., **37**, Lang, Frankfurt am Main, 1991.

[26] L. Zalcman, *Some inverse problems of potential theory*, Contemp. Math. **63** (1987), 337–350.

4. A question on the regularity of the boundary of a quadrature domain

LAVI KARP, Department of Mathematics, Ort Braude College, P.O. Box 78, Karmiel 21982, Israel.

This open problem deals with the regularity of the boundary of quadrature domains.

Let Ω be a bounded quadrature domain for a measure μ and for the class of harmonic integrable function in Ω. By this we mean that

$$U^{\Omega}(x) = U^{\mu}(x) \quad \text{and} \quad \nabla U^{\Omega}(x) = \nabla U^{\mu}(x) \quad \text{for } x \in \mathbf{R}^n \setminus \Omega,$$

where $\text{supp}(\mu) \subset \overline{\Omega}$, U^{μ} denotes the Newtonian potential of μ and U^{Ω} is the Newtonian potential of the characteristic function of Ω.

Theorem 1. *Let Ω be a quadrature domain for a measure μ, $x_0 \in \partial\Omega$ and assume $\{x : |x - x_0| < r\} \cap \text{supp}(\mu) = \emptyset$ for some positive r. If the complement of Ω, $\mathbf{R}^n \setminus \Omega$, is not "too thin" near x_0, then $\partial\Omega$ is a real analytic surface in a neighborhood of x_0.*

See Sakai [4], Gustafsson and Putinar [2] for $n = 2$, and Caffarelli, Karp and Shahgholian [1] for $n \geq 3$. The second result deals with the case where the support of the measure μ meets the boundary $\partial\Omega$ in a certain way (see Karp and Margulis [3]).

Theorem 2. *Let Ω be a quadrature domain for a measure μ, $0 \in \partial\Omega$ and assume*

$$\int_{\{|x|<r\}} \frac{d|\mu|(x)}{|x|^n} < \infty. \tag{2}$$

If there is a sequence $\{\varepsilon_k\} \searrow 0$ such that $\lim_k \chi_{\Omega}(\varepsilon_k x) = \chi_K(x)$, where K is a cone and $\text{supp}(\mu) \subset K$, then the cone K equals to a half-space.

In other words, if the measure μ hits the boundary in a "soft" way (i.e. (2) holds), then the boundary $\partial\Omega$ has, in a certain sense, a tangent plane at $0 \in \partial\Omega$.

Open Problem: *Prove that under the assumptions of Theorem 2 the boundary is a real analytic surface in a neighborhood of $\{0\}$.*

The solution of that problem will show that either the measure μ hits the boundary of a quadrature domain strongly such that

$$\int_{\{|x|<r\}} \frac{d|\mu|(x)}{|x|^n} = \infty, \tag{3}$$

or it does not hit the boundary at all.

References

[1] L.A. Caffarelli, L. Karp, H. Shahgholian, *Regularity of a Free Boundary Problem with Application to the Pompeiu Problem*, Ann. of Math., **151**, No.1 (2000), 269–292

[2] B. Gustafsson and M. Putinar, *An exponential transform and regularity of free boundaries in two dimensions,* Ann. Scuola Norm. Sup. Pisa Cl. Sci. (2) **26**, No. 3 (1998), 507–543.

[3] L. Karp and A. Margulis, *On the Newtonian potential theory for unbounded sources and its application to free boundary problems*, J. Analyse Math. **70** (1996), 1–63 .

[4] M. Sakai, *Regularity of a boundary having a Schwarz function*, Acta Math. **166**, No. 3-4 (1991), 263–297.

---

5. Two questions in potential theory

Dmitry Khavinson, Department of Mathematical Sciences, University of Arkansas, Fayetteville, AR 72701.

5.1. Bounded point evaluation for harmonic polynomials

Problem 1. *Suppose a measure μ supported in the unit ball B in $\mathbf{R}^n, n \geq 3$, annihilates harmonic polynomials. Does there exist a set E in B of positive volume such that for each point x in E one could find a measure $\nu_x, \nu_x \ll \mu$, which represents all harmonic polynomials at x, i.e.,*

$$u(x) = \int u d\nu_x$$

for all harmonic polynomials u.

As is shown in [KS1], the affirmative answer would immediately, by using the argument in [KS1], imply the uniqueness of the best uniform harmonic approximation to functions continuous on B, thus answering in the affirmative an old question of W. Hayman. In $\mathbf{R}^2$ the answer is "yes" and follows at once from the Cauchy transform techniques (cf. [KS1]). The difficulty in higher dimensions is that unlike analytic polynomials one cannot "divide" harmonic polynomials.

Along the same line of thought the following potential-theoretic question seems very attractive (cf. [KS1].)

Problem 2. *If a real measure μ (considered as a charge distribution on B) produces the same electric field outside B as does a positive point charge at some point w in B, then there is also a (positive) charge distribution m with $m \ll |\mu|$ producing the same field outside B. (The absolute continuity implies, in particular, that m is not permitted to place charges anywhere off supp μ.)*

This is again true in the plane [KS1], but the argument relies on Banach algebras techniques (harmonic polynomials are real parts of analytic polynomials there!) and does not extend to higher dimensions.

Yet, the result seems not unreasonable, on "physical" grounds also in $n > 2$ dimensions. Indeed, what we seem to be looking for is simply to ignore the negative charges of μ while at the same time redistributing the positive charges within the support of μ.

5.2. Global behaviour of the Schwarz potentials

Let Γ be a real analytic nonsingular surface in $\mathbf{R}^n$. Let U be the function defined locally near Γ, having its Laplacian equal to 1 and vanishing on Γ together with its first derivatives. (The existence of such U follows at once from the Cauchy–Kovalevskaya theorem.) This function U (cf. [KS2]), often called the *modified Schwarz potential* of Γ, is the basic tool in almost all questions associated with quadrature domains for harmonic functions(cf. [KS2]). It naturally extends to $\mathbf{R}^n$ the concept of the Schwarz function in the plane (cf. [Da]).

Conjecture. *If U is an entire real-analytic function in $\mathbf{R}^n$, then Γ must be a plane.*

The similar result due to Ph. Davis [Da] holds in the plane and is easy to prove. In $\mathbf{R}^n$ it was proved in [KS2] for the case when U is a polynomial. Later on Karp and Shahgholian extended the argument to a slightly more general situation, but the main bulk of the conjecture remains open. Interestingly, G. Johnsson constructed simple examples in $\mathbf{C}^n$ of analytic hypersurfaces for which the corresponding modified Schwarz potentials are entire, or even polynomials (cf. [K]). The feeling, why the Conjecture may still hold in $\mathbf{R}^n$ relies on a general (non-proven) consent that the singularities of U in $\mathbf{R}^n$ reflect the appearance of nonzero curvature on Γ. However, even in two dimensions, no proof of Davis' theorem based on calculating curvature via the Schwarz potential is known.

References

[Da] P.L. Davis, *The Schwarz Function and its applications,* Carus Math. Monographs, No. 17, MAA, 1974.

[K] D. Khavinson, *Singularities of harmonic functions in $\mathbf{C}^n$*, Proc. Symp. Pure and Applied Math., A.M.S., **52** (1991), Part 3, 207–217.

[KS1] D. Khavinson, H. S. Shapiro, *Best approximation by analytic and harmonic functions in the supremum norm*, Arkiv för Math., Vol. **39**, No. 2 (2001), 339–359.

[KS2] D. Khavinson, H. S. Shapiro, *The Schwarz potential in $\mathbf{R}^n$ and Cauchy's problem for the Laplace equation,* TRITA-MAT-1989-*36*, Royal Institute of Technology, Stockholm, 112 pp.

6. Linear data of a quadrature domain

MIHAI PUTINAR, Department of Mathematics, University of California, Santa Barbara, CA 93106.

Let $\Omega \subset \mathbf{C}$ be a bounded quadrature domain with d nodes, counting multiplicities. Let $p(z)$ be the monic polynomial of degree d vanishing at these quadrature nodes. Then it is known [GP] that the defining equation of $\partial\Omega$ has the form

$$P(z, \overline{z}) = |p(z)|^2 - \sum_{k=0}^{d-1} |p_k(z)|^2,$$

where $p_k(z)$ are polynomials of degree k. Moreover, simple linear algebra arguments show that there exists a $d \times d$ matrix A, with a cyclic vector ξ, such that

$$P(z, \overline{z}) = |p(z)|^2(1 - \|(A - z)^{-1}\xi\|^2). \tag{4}$$

In this case, the spectrum of A coincides with the zeros of p and the quadrature nodes of Ω. In this way, a remarkable interplay between the properties of A and those of Ω can be established.

An indirect characterization of the pairs (A, ξ) of square matrices with a distinguished cyclic vector corresponding to a quadrature domain is available: Let A be the restriction of a pure co-hyponormal operator T with rank-one self commutator $[T, T^*] = \xi \otimes \xi$ to the finite dimensional subspace $\vee_{j=0}^{\infty} T^j \xi$. Moreover, a block-diagonal matricial representation of T is then derivable by simple operations from (A, ξ), see [P1, P2]. This is similar to a Jacobi matrix decomposition of T.

However, an intrinsic characterization of the pairs (A, ξ) (attached to quadrature domains) among the variety of all matrices with a cyclic vector is not known. Examples of 2×2 such matrices can be found in [GP].

References

[GP] B. Gustafsson, M. Putinar *Linear analysis of quadrature domains. II*, Israel J. Math. **119** (2000), 187–216.

[P1] M. Putinar, *Linear analysis of quadrature domains*, Arkiv För Mat. **33** (1995), 357–376.

[P2] M. Putinar, *Linear analysis of quadrature domains. III*, J. Math. Analysis Appl. **239** (1999), 101–117.

7. Fischer's problem

STEPHANE RIGAT, Centre de Mathématiques et Informatique, 39 Rue F. Joliot Curie, F-13453 Marseille, Cedex 13, France.

Let $\mathbf{F}$ be the set of holomorphic functions on $\mathbf{C}^n$ such that

$$\int_{\mathbf{C}^n} |f|^2 e^{-|z|^2} d\lambda < +\infty$$

where $d\lambda$ is the Lebesgue measure on $\mathbf{C}^n$, and for $z = (z_1, \ldots, z_n) \in \mathbf{C}^n$, $|z|^2 = |z_1|^2 + \cdots + |z_n|^2$. We denote by $\mathrm{H}(\mathbf{C}^n)$ the space of holomorphic functions defined on $\mathbf{C}^n$, by $\mathrm{Exp}(\mathbf{C}^n)$ the space of holomorphic functions on $\mathbf{C}^n$ of exponential type which means that

$$\exists C > 0, \qquad \forall z \in \mathbf{C}^n, \qquad |f(z)| \leq Ce^{C|z|}.$$

Let P and Q be polynomials in $\mathbf{C}[z_1, \ldots, z_n]$. If P is of the form $\sum a_\alpha z^\alpha$ with the usual multi-index notation, we denote by $P(D)$ the holomorphic differential operator with constant coefficients $\sum a_\alpha \frac{\partial}{\partial z^\alpha}$.

$\mathbf{A}$ will denote one of the three functions spaces defined as above, which means that $\mathbf{A} = \mathbf{F}$ or $\mathbf{A} = \mathrm{H}(\mathbf{C}^n)$ or $\mathbf{A} = \mathrm{Exp}(\mathbf{C}^n)$.

The problem is the following: *given g and h in $\mathbf{A}$ find the necessary and sufficient conditions on P and Q in order that there is a unique $f \in \mathbf{A}$ such that*

$$\begin{cases} P(D)f = g \\ Q|(f-h). \end{cases}$$

The notation $\phi|\psi$ in $\mathbf{A}$ means that there exists $\theta \in \mathbf{A}$ such that $\psi = \phi\theta$. Moreover, we want to find an explicit expression of the solution f in terms of residue currents, g and h (*cf.* [Pa], [Ri]). This kind of problem is called by Hörmander a "Mixed Cauchy Problem" ([Eb-Sh1], [Eb-Sh2]). An obvious necessary condition is deg $P =$ deg Q ([Me-Yg]).

If we are interested in the case $Q = z_1^m$, the condition $Q|(f-h)$ means that the derivatives of f of order less or equal to $m-1$ are the the derivatives of h on the hyperplane $\{z_1 = 0\}$, and we are in the situation of the usual Cauchy Problem. In this case, we have explicit formulas.

If we are now interested in the case $Q = z_1^{m_1} z_2^{m_2}$, then the condition $Q|(f-h)$ means that the derivatives of f of order less or equal to $m_1 - 1$ with respect to z_1 and the derivatives of f of order less or equal to $m_2 - 1$ with respect to z_2 are the derivatives of h on the hyperplanes $\{z_1 = 0\}$ and $\{z_2 = 0\}$. We see in this case that we are not in the situation of the usual Cauchy Problem.

If $\mathbf{A} = \mathbf{F}$ and if $Q = P^*$, which means that $Q(z) = \overline{P(\bar{z})}$ then it is not too hard to see that the mixed Cauchy Problem above as a unique solution ([Me-Yg]). To see the history of Fischer's problem, more references and applications, one can also consult [An].

Here is (a non exhaustive !) bibliography:

References

[An] J. Aniansson, *Some intergal representations in real and complex analysis, Peano-Sards kernels and Fischer kernels*, Doctoral Thesis, (1999), Department of Math., Royal Institute of Technology, Stockholm.

[Eb-Sh1] P. Ebenfelt , H.S. Shapiro, *The mixed Cauchy Problem for holomorphic partial differential operators*, J. Analyse Math. **65**, 1995, 237–295.

[Eb-Sh2] P. Ebenfelt , H.S. Shapiro, *A quasi maximum principle for holomorphic solutions of partial differential equations in* $\mathbf{C}^n$, J. Funct. Anal. **146**, (1997), 27–61.

[Ma] A. Martineau, *Equations différentielles d'ordre infini*, Bull Soc. Math. France **95**, (1967), 109–154.

[Me-St] A. Méril , D.C. Struppa, *Equivalence of Cauchy Problems for entire and exponential types functions*, Bull London Math. Soc. **17**, 1985, 469–473.

[Me-Yg] A. Méril, A. Yger, *Problèmes de Cauchy Globaux*, Bull Soc. Math. France **120**, (1992), 87–111.

[Pa] M. Passare, *Residue solutions to holomorphic Cauchy problems*, Seminars in Complex Analysis and Geometry (Guenot, J. and Struppa, D., eds), 99–105, Editoria Elettronica, Rende, 1988.

[Ri] S. Rigat, *Application of the fundamental principle to Complex Cauchy Problem*, Ark. Mat. **38**, (2000), 355–380.

[Sh] H.S. Shapiro, *An algebraic theorem of E. Fischer, and the holomorphic Goursat problem*, Bull London Math Soc. **21**, (1989), 513–537.

8. Two problems on quadrature domains

Makoto Sakai, Department of Mathematics, Tokyo Metropolitan University, Minami-Ohsawa 1-1, Hachioji-shi, Tokyo 192-0397, Japan.

8.1. Estimates of the number of special points

This is a problem proposed by Harold S. Shapiro at the University of Maryland in 1985 when we were discussing quadrature domains, see [5].

Let Ω be a bounded domain in the complex plane surrounded by a finite number of nondegenerate curves and let $z_1, \ldots, z_m$ be distinct points in Ω. Assume that

$$\iint_\Omega f(z)dxdy = \sum_{j=1}^{m} \sum_{k=0}^{n_j-1} c_{jk} f^{(k)}(z_j) \qquad (z = x + iy)$$

holds for every holomorphic and integrable function f in Ω. Then the boundary $\partial\Omega$ of Ω is algebraic and has possibly a finite number of cusps and double points on it, see [1]. We assume that $c_{jn_j-1} \neq 0$ for every j. The Schwarz function S for $\partial\Omega$ has $n = \sum_{j=1}^{m} n_j$ poles in Ω. We are interested in points z in Ω satisfying

$$S(z) = \overline{z}$$

called "special points" by Shapiro. They play important roles in studying the quadrature domain Ω.

Let s be the number of special points, a the connectivity of Ω, c the number of cusps on $\partial\Omega$ and d the number of double points on $\partial\Omega$. The problem is to find the relations between n, s, a, c and d.

There have been several results since then:

$$s \leq (n-1)^2 + 1 - a - c - 2d$$

by Gustafsson [2], see also McCarthy and Yang [3], and

$$s \geq n - 2 + a - c$$

by myself [4], but it seems that we have not yet obtained good estimates of s.

8.2. Quadrature domains for holomorphic functions of several variables

Discuss quadrature domains for holomorphic functions of several complex variables and establish a theory of such domains.

References

[1] B. Gustafsson, *Quadrature identities and the Schottky double,* Acta Appl. Math. **1** (1983), 209–240.

[2] B. Gustafsson, *Singular and special points on quadrature domains from an algebraic geometric point of view,* J. Analyse. Math. **51** (1988), 91–117.

[3] J.E. McCarthy, Liming Yang, *Subnormal operators and quadrature domains,* Adv. Math. **127** (1997), 52–72.

[4] M. Sakai, *An index theorem on singular points and cusps of quadrature domains,* in vol. *Holomorphic functions and moduli, Vol. I (Berkeley, CA, 1986), D. Drasin(ed.),* pp. 119–131, Math. Sci. Res. Inst. Publ. **10**, Springer, New York, 1988.

[5] H.S. Shapiro, *Unbounded quadrature domains* in vol. *Complex Analysis, I (College Park, MD, 1985-86), C. A. Berenstein(ed.),* pp. 287–331, Lecture Notes in Math. **1275**, Springer, Berlin, 1987.

9. Generalized quadrature domains

Henrik Shahgholian, Mathematics Department, Royal Institute of Technology, S-10044 Stockholm, Sweden.

In this section I will present some new directions in the theory of quadrature domains (QD). Two possible ways of extending this notion will be discussed in more detail: *parabolic QD* and *QD with partially fixed boundary.* Both of these problems are related to some of my recent research.

9.1. The original problem

A quadrature domain is a domain $\Omega \in \mathbf{R}^n$ with the property that for a given measure μ (with support in $\overline{\Omega}$) and an appropriately defined class of functions $\mathcal{A}$ one has the integral identity:

$$\int_\Omega h(x)dx = \int h(x)d\mu \qquad \forall h \in \mathcal{A}.$$

The deep connection between QD and variational problems (and or complementary problems), discovered by M. Sakai, has made it possible to study such problems from a variational inequality point of view. Although, not all QD can be studied within the framework of variational analysis it is still one of the main tools for deep analysis of such problems.

My objective here is to present two (these are partially new) directions of study in free boundary problems and give a QD-formulation of them. I hope that in this way one can find some new tools in solving certain problems that I will pose below.

9.2. Parabolic QD

In this part I will discuss a generalized form of QD. So let f, and μ, be non-negative functions (measures), then find a bounded $\Omega \subset \mathbf{R}^{n+1}$ with $\text{supp}(\mu) \subset \overline{\Omega}$ and such that

$$\int\int_\Omega f(x,t)h(x,t)\ dxdt \ = \int\int h(x,t)\mu(x,t)\ dxdt\ , \tag{5}$$

for all integrable caloric functions h in Ω. The particular case of μ being the Dirac measure at (x^0, t^0) gives the mean-value property for caloric functions with respect to the point (x^0, t^0). One also observes immediately that, by (9.1) $\Omega \subset \mathbf{R}^n \times (-\infty, t^0)$.

A, probably, well-known identity of the type (9.1) with $f = |x|^2/t^2$ and μ a multiple of the Dirac mass is given by the so-called heat balls, which are the level sets of the fundamental solution to the adjoint heat operator $\Delta + D_t$. For a nice presentation of the mean value property for heat balls see the reader-friendly book by L.C. Evans [E].

9.3. Reformulation

Reflect the domain Ω in $\{t = 0\}$ so that the identity will now hold for anti-caloric functions. Therefore we assume that

$$\int\int_\Omega h(x,t)f(x,t)\ dxdt \ = \int\int h(x,t)\mu(x,t)\ dxdt \tag{6}$$

for all integrable anti-caloric functions h in Ω; here we used the same notation Ω for the reflected domain.

A couple of assumptions are in order now:

- $\Omega \subset \mathbf{R}^n \times (0, \infty)$.

- Set $\mu = g + g_0$ where
$$\mathrm{supp}(g) \subset \overline{\Omega} \cap \{t > 0\} \quad \text{and} \quad \mathrm{supp}(g_0) \subset \{t = 0\}.$$
In this way we separate the initial data g_0 and the heat source g.

Set
$$u(x,t) = \int\int_{\Omega} K(x-y, t-s)\mu(y,s)dyds - \int\int K(x-y,t-s)f(y,s)dyds,$$
where
$$K(z,\tau) = (4\pi\tau)^{-n/2} \exp\left(-|z|^2/(4\tau)\right), \qquad \tau > 0,$$
and $K(z,\tau) = 0$ for $\tau \leq 0$.

Now one readily verifies using identity (9.2) that
$$\begin{cases} \Delta u - D_t u = f\chi_\Omega - g & \text{in } \mathbf{R}^{n+1}_+, \\ u = 0 & \text{in } \mathbf{R}^{n+1}_+ \setminus \Omega, \\ u(x,0) = g_0, \\ \mathrm{supp} g \subset \overline{\Omega}, \end{cases} \tag{7}$$
where
$$\mathbf{R}^{n+1}_+ = \mathbf{R}^n \times (0,\infty),$$
and the differential equation is interpreted in the weak or distributional sense.

Hence we have a connection between parabolic-QD and the free boundary formulation above. For some details see [ASU]. The above setting appears in several problems in mathematical physics, e.g., in the Stefan problem.

9.4. QD with partially fixed boundary

Another way of generalizing the notion of QD, is to considering part of the boundary fixed. The easiest way is to start from the free boundary formulation and go back to QD.

For simplicity we work in the upper half space $\mathbf{R}^n_+ = \{x_n > 0\}$, where the plane $\{x_n = 0\}$ will be considered as the fixed part of the boundary.

As before for a given measure μ with support in $\mathbf{R}^n_+$, find a function u and a domain $\Omega \subset \mathbf{R}^n_+$ (usually the only way to find the domain is to have $\Omega = \{u > 0\}$) such that
$$\Delta u = \chi_\Omega - \mu \qquad \text{in } \mathbf{R}^n_+$$
and $u = \nabla u = 0$ in $\Omega^c \cap \mathbf{R}^n_+$. Now on the boundary $\{x_n = 0\}$ one needs to assign values for u
$$u(x', 0) = f(x'). \tag{8}$$
We simplify the problem even further by assuming $f \equiv 0$. Now one can reformulate the above problem by Green's identity as
$$\int_\Omega h dx = \int h d\mu$$
for all integrable, harmonic functions h (in Ω), vanishing on $x_n = 0$.

9.5. Open questions

Let us now present some open questions around the topic of QD. I will consider problems related to the standard QD and also the above variants.

Problem 1 (*Null QD/ Global solutions*)
A null quadrature domain Ω is defined through the identity

$$\int_\Omega h dx = 0$$

for all integrable harmonic functions in Ω. Reformulation of this in terms of global solutions to the pde

$$u(\Delta u - 1) = 0, \qquad |u(x)| \le C|x|^2.$$

For details see [Sh] and [CKS]. It was conjectured in [Sh] that global solutions are limit domains of the exterior of ellipsoids. The corresponding problem in two space dimensions was settled by M. Sakai. Later H.S. Shapiro gave another proof. Global solutions in half-spaces were completely classified in [SU]). They appear to be one dimensional.

So we repose the question whether one can prove the conjecture in [Sh]. It is noteworthy that by classifications of [CKS] (and earlier results on QD with bounded complements), it only remains to show that Null QD which are thin at infinity must be paraboloids.

Problem 2 (*Parabolic QD.*)
For parabolic QD it was recently shown that these solutions are convex (see [CPS]). There are many questions still open around the parabolic QD, and the subject has just been born!

- Existence question,
- Regularity issues (near the initial datum),
- Complete classification of global solutions/null QD, for the parabolic case,
- Asymptotic behavior of solutions, as time grows: finite time extinction.

Problem 3 (*QD with partially fixed boundary.*)
There are many problems related to QD with fixed boundaries. Most of these results concern regularity issues, the behavior of solutions and the free boundary. I refer to [SU], for the elliptic case and to [ASU] for parabolic case. (These papers treat the case $f \equiv 0$.)

Question of interest are the behavior of the solution/free boundary when f in (8) is nonzero. E.g., when f is $C^{1,1}$ one can relate the problem to that of the Dam problem of water reservoirs.

References

[ASU] D.E. Apushkinskaya, H. Shahgholian, N. Uraltseva, *On the global solutions of the parabolic obstacle problem,* Algebra i Analiz **14** (2002), no. 1, 3–25.

[CKS] L. Caffarelli and L. Karp, H. Shahgholian, *Regularity of a free boundary in potential theory with application to the Pompeiu problem,* Ann. of Math. **151** (2000), no. 1, 269–292.

[CPS] L. Caffarelli, A. Petrosyan, H. Shahgholian, *Regularity of a free boundary in parabolic potential theory,* submitted.

[E] L.C. Evans, *Partial Differential Equations,* Amer. Math. Soc., Providence, R.I., 1998.

[HS] A. Hakobyan, H. Shahgholian, *Generalized mean value property for caloric functions,* (Survey), submitted.

[Sh] H. Shahgholian, *On quadrature domains and the Schwarz potential,* J. of Math. Anal. and Appl. **171** (1992) pp 61–78).

[SU] H. Shahgholian, N. Uraltseva, *Regularity properties of a free boundary near contact points with the fixed boundary.* Duke Math. J. **116** (2003), no. 1, 1–34.

10. Further study of quadrature domains on curved surfaces

SERGUEI SHIMORIN, Mathematics Department, Royal Institute of Technology, S-10044 Stockholm, Sweden.

In the recent work of H. Hedenmalm and S. Shimorin, they study mean value disks on hyperbolic Riemann surfaces. These can be interpreted as domains of the Hele-Shaw flow starting at a point and at the same time they are the simplest examples of quadrature domains on curved surfaces. Even for these domains, some natural questions are still open, for example, it is not known if they are geodesically star-shaped. Another interesting question is how far the assumption on hyperbolicity of the metric can be relaxed so that mean value disks still remain topologically trivial. Extensions of the results to dimensions higher than 2 is also an open problem.

A further possible direction of research is the study of mean value circles on curved surfaces. This corresponds to the following problem in the plane: given a weight function $\omega(z)$ defined in the plane, one searches for closed simple curves γ_t having the property

$$\int_{\gamma_t} h\omega\, ds = th(0)$$

for all bounded harmonic functions h.

Operator Theory:
Advances and Applications, Vol. 156, 1–25

What is a Quadrature Domain?

Björn Gustafsson and Harold S. Shapiro

Abstract. We give an overview of the theory of quadrature domains with indications of some if its ramifications.

Mathematics Subject Classification (2000). Primary 30-02; Secondary 31-02, 32Dxx, 32Sxx, 35A21, 35Jxx, 35R35, 47B20, 76D27.

Keywords. Quadrature domain, quadrature identity, quadrature surface, Schwarz function, Schwarz potential, mean value property, Schottky double, balayage, mother body, free boundary problem, special point, Hele-Shaw flow, exponential transform, hyponormal operator, subnormal operator, Friedrichs operator, Bergman kernel.

1. Introduction

With this introductory paper the authors wish to give a general overview of the theory of quadrature domains, along with indications of some of its ramifications. There will be no proofs, but an extensive bibliography. More details for most of the material known up to 1992 can be found in the book [146]. We shall start by saying a few words of the birth of the theory.

The word 'quadrature' goes back to the latin noun 'quadratura', which means 'making square shaped', 'constructing squares' or, more specifically, 'the division of land into squares' [43]. Accordingly, 'quadrature' in mathematics traditionally refers to constructive or numerical methods for determining areas, and (more recently) for computing integrals in general.

In the theory discussed in this article 'quadrature' has a related meaning. For example, a 'quadrature identity' will typically be an exact formula for the integral of harmonic or analytic functions in terms of simpler functionals, like point evaluations. The domain of integration is then a quadrature domain.

Specifically we shall call a bounded domain Ω in the complex plane a **(classical) quadrature domain** if there exist finitely many points $a_1, \dots, a_m \in \Omega$ and

Paper supported by the Swedish Research Council VR.

coefficients $c_{kj} \in \mathbf{C}$ so that

$$\int_\Omega f\,dA = \sum_{k=1}^{m} \sum_{j=0}^{n_k-1} c_{kj} f^{(j)}(a_k) \tag{1.1}$$

for all integrable analytic functions f in Ω (dA denotes area measure). The identity (1.1) is then called a **quadrature identity** and the integer $n = \sum_{k=1}^{m} n_k$ is the **order** of the quadrature identity (assuming $c_{k,n_k-1} \neq 0$).

Embryonic theories of quadrature domains can be found already in papers of C. Neumann [100], [101] from the beginning of the last century. See [146] for discussions. However, the first person to more systematically study general quadrature identities of the kind (1.1) seems to be P. Davis, in the 1960s, see [26] and references therein. Independently, D. Aharonov and H.S. Shapiro [1], [3] discovered a few years later that solutions of certain extremal problems for univalent functions map the unit disc onto domains which allow quadrature identities for (somewhat restricted) classes of analytic functions. (Not having all analytic functions available caused some technical problems, which were finally resolved in [4], [5].) In [2] the authors started studying such quadrature domains for their own sake.

Around the same time and partially influenced by the Aharonov-Shapiro ideas, Y. Avci, a doctoral student of M. Schiffer, was writing his doctoral thesis [8] about quadrature identities, and in another corner of the world M. Sakai was working with related matters from a more potential theoretic point of view [116], [118], [119], [121]. In both cases, like in the Aharonov-Shapiro case, the motivating problems were extremal problems for analytic functions (see the introduction of [8] and the appendix of [121]).

After this initiation in the 1970s (essentially), the subject was pursued mainly by H.S. Shapiro and his students and by M. Sakai. Part of the subject was developed in parallel with certain topics in fluid mechanics, mainly Hele-Shaw flow, which was linked to quadrature domain theory via a discovery by S. Richardson [112].

After a conference, organized by D. Khavinson, at the University of Arkansas in 1988 and the subsequent book [146] a kind of rebirth took place in the mid 1990s with new influences from operator theory [102], [103], [104], [98]. Also, more links to fluid mechanics were discovered, see [23] for an overview.

2. Classical quadrature domains

We shall find it convenient in the sequel to write the quadrature identity (1.1) simply as

$$\int_\Omega f\,dA = \sum_{k=1}^{n} c_k f(a_k), \tag{2.1}$$

where in the sequence $a_1, \dots, a_n \in \Omega$ repetitions are allowed, a repeated a_k being interpreted as the occurence of derivatives of f at a_k.

In the simplest case, $n = 1$, it is known that discs $\mathbf{D}(a,r)$ are the only quadrature domains [36], [37], [2], [121], [146] and the quadrature identity then reduces to the ordinary mean value property for analytic functions:

$$f(a) = \frac{1}{|\mathbf{D}(a,r)|} \int_{\mathbf{D}(a,r)} f\,dA. \tag{2.2}$$

Thus a quadrature identity can be thought of as a generalized mean value property and a quadrature domain as a generalized disc (at least if all the c_k are positive).

By choosing $f(\zeta) = \frac{1}{z-\zeta}$ for $z \in \mathbf{C} \setminus \Omega$ in (2.1) one realizes that the Cauchy transform

$$\hat{\chi}_\Omega(z) = -\frac{1}{\pi} \int_\Omega \frac{dA(\zeta)}{z-\zeta} \tag{2.3}$$

of the characteristic function of a quadrature domain Ω is a rational function outside Ω. To be exact, there exists a rational function $R(z)$ of the form

$$R(z) = \frac{Q(z)}{P(z)}, \tag{2.4}$$

where $P(z) = \prod_{k=1}^{n}(z - a_k)$ and where $Q(z)$ is a polynomial of degree $n-1$, such that

$$\hat{\chi}_\Omega = R \quad \text{on } \mathbf{C} \setminus \Omega. \tag{2.5}$$

Conversely, by an approximation theorem of L. Bers [14] the linear combinations of the above Cauchy kernels are dense in the integrable analytic functions, so (2.5) holding for some rational function R is actually equivalent to Ω being a quadrature domain.

The main tool for the approximation used in [14] is a system of ingeniously constructed mollifiers, nowadays often named the "Ahlfors-Bers mollifiers". These mollifiers play an important role in the theory of quadrature domains, also in higher dimensions. See, e.g., [121], [48], [82] for their use. Alternative techniques for proving Bers-type approximations are a lemma by V. Havin [64], [143], [146] and "quasi-balayage" [147].

3. The Schwarz function

Assume that Ω is a quadrature domain (2.1). Since $\frac{\partial}{\partial \bar{z}}\hat{\chi}_\Omega = \chi_\Omega$ in the sense of distributions the function $S(z)$ defined on $\overline{\Omega}$ by

$$S(z) = \bar{z} + R(z) - \hat{\chi}_\Omega(z) \tag{3.1}$$

is meromorphic in Ω, and by (2.5),

$$S(z) = \bar{z} \quad \text{for } z \in \partial\Omega. \tag{3.2}$$

Thus $S(z)$ is a one-sided **Schwarz function** of $\partial\Omega$. To be a two-sided Schwarz function one requires that $S(z)$ is defined and analytic in a full neighbourhood of $\partial\Omega$. The use of such a function associated to an analytic curve can be traced back to G. Herglotz [72]. It was named, after H.A. Schwarz' reflection principle, in [27]. Full accounts are given in [26], [146].

A two-sided Schwarz function, defined in some neighbourhood of $\partial\Omega$, exists if and only if $\partial\Omega$ is analytic, while a one-sided Schwarz function allows for certain singular points of $\partial\Omega$. A complete characterization of those boundaries allowing a one-sided Schwarz function was given by M. Sakai in [127].

The previous argument also goes the other way around, so a bounded domain Ω is a quadrature domain if and only if there exists a meromorphic function $S(z)$ in Ω, continuous up to $\partial\Omega$, so that (3.2) holds. This result was obtained in [26] under some smoothness assumptions on $\partial\Omega$ and in full generality in [2].

Indeed, one of the major achievements in [2] is that the authors prove, without any regularity assumptions whatsoever, that $\partial\Omega$ is a subset of an algebraic curve: there exists a nontrivial polynomial $Q(z,w)$ such that

$$\partial\Omega \subset \{z \in \mathbf{C} : Q(z,\bar{z}) = 0\}. \tag{3.3}$$

From this, additional regularity of $\partial\Omega$ follows easily.

Differentiating (3.2) gives

$$s(z)dz = d\bar{z} \quad \text{along } \partial\Omega, \tag{3.4}$$

where $s(z) = S'(z)$. The relation (3.4) holds for some meromorphic function $s(z)$ (not necessarily having a single-valued primitive) if and only if a quadrature identity of the kind

$$\int_\Omega f\,dA = \sum_{k=1}^{m}\sum_{j=0}^{n_k-1} c_{kj} f^{(j)}(a_k) + \sum_{i=1}^{r} b_i \int_{\gamma_i} f dz \tag{3.5}$$

holds, where the γ_i are (smooth) closed or nonclosed curves, compactly contained in Ω, and $b_i \in \mathbf{C}$. See [2], [8], [44], [81].

4. Riemann surfaces (the Schottky double)

One way to understand the algebraicity (3.3) of $\partial\Omega$ is as follows. The relation (3.2), with $S(z)$ meromorphic in Ω, may be interpreted as saying that the pair of functions $(S(z),\bar{z})$ constitutes a meromorphic function on the **Schottky double** [133], [135] $\hat{\Omega}$ of Ω, namely the compact Riemann surface obtained by completing Ω with a backside $\tilde{\Omega}$ having the opposite conformal structure, and glueing the two together along $\partial\Omega$. (We are here assuming *a priori* some mild regularity of $\partial\Omega$.) Then $S(z)$ represents the values of the function on Ω and $\bar{z}$ the values on $\tilde{\Omega}$.

The opposite pair $(z,\overline{S(z)})$ will also represent a meromorphic function on $\hat{\Omega}$, and since any two meromorphic functions on a compact Riemann surface are related by a polynomial equation it follows that there exists a polynomial $Q(z,w)$ such that

$$Q(z,S(z)) = 0 \quad (z \in \Omega).$$

In particular $Q(z,\bar{z}) = 0$ on $\partial\Omega$, i.e., we obtain (3.3) again.

More detailed investigations [44] show that $Q(z, w)$ (if chosen without extraneous factors) has degree exactly n in each of z and w (total degree $2n$) and that the difference set in (3.3) consists of at most finitely many points (called special points, see Section 12). Additional information on the structure of $Q(z, w)$ can be derived from properties of the exponential transform, see Section 13 and also [103], [107], [53], [55].

If Ω is any simply connected bounded domain, then Ω is a quadrature domain if and only if any conformal map $g : \mathbf{D} \to \Omega$ is a rational function. This was proved in [26] under some regularity assumptions on $\partial\Omega$ and in [2] without such assumptions. The idea for the nontrivial direction is to use the fact that $z \mapsto \overline{S(z)}$ ($z \in \Omega$) is the anticonformal reflection in Ω to extend the mapping function g from $\mathbf{D}$ to the entire Riemann sphere.

For multiply connected domains there are analogous results [44]: Let D be a finitely connected domain representing a certain conformal type and let $(g, \overline{h})$ represent a meromorphic function on the Schottky double $\hat{D}$, so that g and h are meromorphic in D and $g = \overline{h}$ on ∂D. If then g is holomorphic and univalent on D it will map D onto a quadrature domain, because the relation $g = \overline{h}$ on ∂D will become a relation of the form (3.2) in the image domain.

Other techniques to construct quadrature domains, based on Riemann surface ideas (automorphic functions, Schottky groups, Poincaré series, etc.) have been developed by Avci [8], Richardson, [113], Crowdy and Marshall [20], [21], [22].

5. Subharmonic quadrature domains

When all the c_k in (2.1) are positive (with the a_k then distinct) it is natural to think of a quadrature domain Ω satisfying (2.1) as something obtained by glueing the discs $\mathbf{D}(a_k, \sqrt{\frac{c_k}{\pi}})$ together in a potential theoretic way, or as the result of some kind of balayage (sweeping) process applied to the measure

$$\mu = \sum_{k=1}^{n} c_k \delta_{a_k}. \tag{5.1}$$

Here δ_a denotes the unit point mass at a. However, it turns out that this picture is fully correct only if one requires (2.1) to hold in the stronger sense that the inequality

$$\int_\Omega f\, dA \geq \sum_{k=1}^{n} c_k f(a_k) \tag{5.2}$$

holds for all integrable subharmonic functions f in Ω. Then (2.1) will automatically hold for all harmonic f (because both f and $-f$ will be subharmonic).

The importance of considering such subharmonic quadrature domains was realized by M. Sakai [119], [121]. The following facts illustrate why subharmonic quadrature domains are natural.

(i) A subharmonic quadrature domain Ω is uniquely determined, up to null-sets, by its measure μ. Such a statement is not true if only (2.1) is required to hold for, e.g., harmonic test functions.

(ii) There are natural ways to construct this unique Ω from μ by a sort of balayage we call **partial balayage** (because the measure is not swept completely, only down to a certain density). The first construction of partial balayage, by M. Sakai, was rather involved [119], [121]. Later more streamlined methods were found, based on techniques such as minimization of energy functionals, variational inequalities or Perron family arguments [122], [123], [45], [48], [57], [51]. See also Section 7 below. Numerical schemes for similar processes of 'equigravitational mass scattering' go back at least to the work of D. Zidarov in the 1960s (see [163] and references therein).

(iii) The geometry of Ω reflects that of μ very well (which need not be the case for analytic or harmonic quadrature domains) [57], [130], [59], [60]. For example, each inward normal ray from $\partial\Omega$ in (5.2) intersects the the convex hull K of the support of μ, and Ω can be written as a union of discs with centers in K. Thus, if μ is concentrated to a small set and the total mass of μ is large, then Ω has to be nearly circular.

6. Harmonic quadrature domains

An intermediate test class for (2.1) is the class of integrable harmonic functions in Ω. In this case we assume that the c_k are real and that the a_k are distinct (no derivatives). Thus μ in (5.1) is a signed measure and (2.1) may be written

$$\int_\Omega h\, dA = \int h\, d\mu, \tag{6.1}$$

to hold for all h harmonic and integrable in Ω.

It becomes natural at this point to allow more general measures μ than in (5.1), for example arbitrary signed measures with compact support in Ω, and also to spell out the various quadrature identities in terms of Newtonian potentials and statements of graviequivalence.

The Newtonian (or logarithmic) potential of any (signed) measure μ is

$$U^\mu(z) = \frac{1}{2\pi}\int \log\frac{1}{|z-\zeta|}\, d\mu(\zeta),$$

where the normalization is chosen so that $-\Delta U^\mu = \mu$. Its gradient ∇U^μ is, apart from a constant factor and a complex conjugation, the same thing as the Cauchy transform $\hat{\mu}$ (defined analogously to (2.3)). Let U^Ω denote the potential of the measure $\chi_\Omega\, dA$.

The linear combinations of the functions $h(\zeta) = \log \frac{1}{|z-\zeta|}$ for $z \in \mathbf{C} \setminus \Omega$ and their first order derivatives are dense in $HL^1(\Omega)$ [121], so (6.1) is equivalent to

$$U^{\Omega} = U^{\mu} \quad \text{on } \mathbf{C} \setminus \Omega, \tag{6.2}$$

$$\nabla U^{\Omega} = \nabla U^{\mu} \quad \text{on } \mathbf{C} \setminus \Omega. \tag{6.3}$$

Here (6.3) is a consequence of (6.2) except possibly at certain singular points on $\partial\Omega$.

In terms of the above notations Ω is a quadrature domain for analytic functions if and only if (6.3) alone holds (because this equation is the same as $\hat{\chi}_{\Omega} = \hat{\mu}$, cf. (2.5)). Also, Ω is a quadrature domain for subharmonic functions if and only if (6.2) holds together with

$$U^{\Omega} \leq U^{\mu} \quad \text{in all } \mathbf{C}. \tag{6.4}$$

In this case (6.3) follows automatically because $U^{\mu} - U^{\Omega}$ attains its minimum on $\mathbf{C} \setminus \Omega$.

There is an interesting difference between allowing signed measures μ in (6.1) or just positive ones: quadrature domains for signed measures with support in a small disc are very flexible, whereas those for positive measures are subject to strong geometric restrictions. For example, the following is true: given any disc $\mathbf{D}(a, r)$ (think of $r > 0$ as being small) and any smoothly bounded domain $D \supset \overline{\mathbf{D}(a, r)}$ one can find domains Ω arbitrarily close to D (with respect to Hausdorff distance for example), which are quadrature domains for harmonic functions with respect to signed measures of the form (5.1) and with support in $\mathbf{D}(a, r)$ [49], [131]. See also [12].

On the other hand, if μ is a positive measure with support in $D(a, r)$ and $R \geq 2r$, where R is defined by $\pi R^2 = \int d\mu$, then any quadrature domain for harmonic functions for Ω is actually a subharmonic quadrature domain and is almost circular, for example $\mathbf{D}(a, R - r) \subset \Omega \subset \mathbf{D}(a, R + r)$. See [130].

Assume that (6.1) holds for harmonic h and a signed measure μ. Let $K \subset \Omega$ be a compact set containing supp μ. Then (6.1) continues to hold if μ is swept, by classical balayage, to ∂K. The question naturally arises whether the swept measure on ∂K will be positive if K is chosen large enough. This question was solved in the affirmative in [62] in the two-dimensional case, but the corresponding question in higher dimensions remains open. See [150] for some recent progress.

7. Free boundary problems and PDE

Keeping the notations of the previous section, the difference

$$u = U^{\mu} - U^{\Omega}$$

is sometimes called the **modified Schwarz potential** of $\partial\Omega$ when (6.2), (6.3) hold. It satisfies

$$\begin{cases} \Delta u = 1 - \mu & \text{in } \Omega, \\ u = |\nabla u| = 0 & \text{on } \partial\Omega. \end{cases} \tag{7.1}$$

The Schwarz potential itself (i.e., nonmodified) is

$$w(z) = \frac{1}{2}|z|^2 - 2u(z).$$

It is harmonic in $\Omega \setminus \operatorname{supp}\mu$, agrees together with its first derivatives with $\frac{1}{2}|z|^2$ on $\partial\Omega$ and it is a real-valued potential of the Schwarz function $S(z)$ in the sense that

$$2\frac{\partial w}{\partial z} = S(z).$$

The system (7.1) can be looked upon in several ways. If Ω is given, but not μ, then considering (7.1) in a small neighbourhood (in Ω) of $\partial\Omega$ we have the elliptic equation $\Delta u = 1$ with the Cauchy data $u = |\nabla u| = 0$ on $\partial\Omega$. This is an ill-posed problem which admits a local solution if and (essentially) only if $\partial\Omega$ is analytic ("if" by the Cauchy-Kovalevskaya theorem, "(essentially) only if" by regularity theory of free boundaries [88], [15], [16], [39], [127], [129]). When this local solution u exists it is natural to try to extend it as far as possible, satisfying $\Delta u = 1$. This is never possible throughout Ω (since, by Green's formula, $\int_\Omega \Delta u dA = \int_{\partial\Omega} \frac{\partial u}{\partial n} ds = 0$), so eventually one has to allow for some singularities, in (7.1) represented by μ. The ideal case is of course that the singularities can be confined to finitely many points, as in (5.1), and in general it is natural to look for μ in (7.1) having as small support as possible. Having achieved this one may think of μ as a potential theoretic skeleton of Ω. See for example [163], [90], [91], [132], [50], [149] for constructions of such a μ using essentially real variable methods (potential theory).

If $\partial\Omega$ is globally analytic in a good enough sense (e.g., is algebraic) another way to try to construct μ, or $\operatorname{supp}\mu$, is to complexify the whole picture, from $\mathbf{R}^2$ (or, more generally, $\mathbf{R}^N$) to $\mathbf{C}^2$ (resp. $\mathbf{C}^N$) and to consider (7.1) near $\partial\Omega$ as a holomorphic Cauchy problem for the complexified Laplacian. The complexification of $\partial\Omega$ always contains points (in the complex) which are characteristic for the Laplacian. These characteristic points create singularities for the Cauchy problem in (7.1) (cf. [92]) and the idea is that these singularities propagate along bicharacteristics of the complex Laplacian, to hit the real at points of $\operatorname{supp}\mu$. In the case of an ellipse the latter points turn out to be the ordinary foci, so in general $\operatorname{supp}\mu$ (or some part of it) could be thought of as a generalized "focal set" of $\partial\Omega$.

Indeed, in the two dimensional case, if the boundary is an algebraic curve, already Herglotz remarked in [72] that the singularities of its analytically continued Schwarz function are the focal points of that curve, as defined earlier by Plücker – this focal concept is purely algebraic, however, and is not yet related to propagation of singularities for the holomorphic Cauchy problem.

A programme along the above lines has been advocated by D. Khavinson and H.S. Shapiro [86] and has been performed in certain cases. For example, in two dimensions it works out well [87], and G. Johnsson [76], [77] has obtained almost complete results for quadric surfaces in any number of dimensions. See further [144], [87], [84], [34], [151] and the books [146], [85]. However, for algebraic boundaries of degree higher than two this programme is fraught with difficulties (especially so in more than two dimensions) since in that setting even the purely algebraic-geometric theory of focal sets is largely undeveloped.

Returning to (7.1) in general, another way to view it is to consider μ as given. Then u is uniquely determined by only the Dirichlet data in (7.1), i.e., without the condition $|\nabla u| = 0$ on $\partial\Omega$. Thus in order to have also this condition satisfied, $\partial\Omega$ must be able to adjust itself. In other words, we have a free boundary problem for $\partial\Omega$.

If we add the requirement $u \geq 0$, which means that Ω is to be a quadrature domain for subharmonic functions, then this free boundary problem is an "obstacle problem". In order to explain the terminology, choose a function ψ (representing the obstacle) satisfying $\Delta\psi = \mu - 1$, e.g.,

$$\psi(z) = -\frac{1}{4}|z|^2 + \frac{1}{2\pi}\int \log|z-\zeta|d\mu(\zeta).$$

The obstacle problem in question is the problem of finding the smallest superharmonic function v satisfying $v \geq \psi$. Equivalently, the function v is to minimize the energy $\int_D |\nabla v|^2 dA$ among all $v \geq \psi$ which agree with ψ outside a sufficiently large disc D.

The above obstacle problem has a unique solution v [89], [39], [114], [122], and in terms of it, $u = v - \psi$ solves (7.1) with Ω given by

$$\Omega = \{v > \psi\} = \{u > 0\},$$

provided this open set covers the support of μ ($\operatorname{supp}\mu \subset \Omega$). The latter is not always the case, but it holds if μ is big enough on its support, e.g., if $\mu \geq 0$ is of the form (5.1).

The free boundary point of view on quadrature domains has been particularly emphasized and developed by H. Shahgholian, and also L. Karp. See for example [137], [83], [17].

8. Quadrature domains for arc length and quadrature surfaces

A variant in the definition of a quadrature domain is to replace Lebesgue measure in Ω by arc length measure ds on $\partial\Omega$:

$$\int_{\partial\Omega} f\, ds = \sum_{k=1}^{n} c_k f(a_k). \tag{8.1}$$

Such identities were studied already in [8].

Let $T(z) = \frac{dz}{ds}$ denote the unit tangent vector on $\partial\Omega$, oriented so that Ω is to the left. Then inserting $ds = \frac{dz}{T(z)}$ in (8.1) one finds that Ω is a quadrature domain for arc length (8.1) if and only if $1/T(z)$ has a meromorphic extension to all of Ω. The definition of $T(z)$ can be written formally as

$$\frac{1}{T(z)}\sqrt{dz} = \sqrt{d\bar{z}} \quad \text{along } \partial\Omega.$$

Here, the notion of a half-order differential (like $\sqrt{dz}$) can be made precise [134] and the above statement can be reformulated: Ω is a quadrature domain for arc length if and only if the half-order differential $\sqrt{dz}$ has a meromorphic extension (as a half-order differential) to the Schottky double $\hat{\Omega}$ [46].

In terms of conformal maps $g : D \to \Omega$ from a standard domain D this gives that Ω is a quadrature domain for arc length if and only if $\sqrt{dg}$ extends meromorphically as a half-order differential to $\hat{D}$. In the simply connected case with $D = \mathbf{D}$, the unit disc, the statement becomes very explicit: $\Omega = g(\mathbf{D})$ is such a quadrature domain if and only if g' is the square of a rational function. This result was first obtained in [148].

The free boundary problem corresponding to quadrature identities for arc length and their counterpart in higher dimensions, **quadrature surfaces** [141], is called the Bernoulli problem (because of an interpretation of it within hydrodynamics). The equations replacing (7.1) are

$$\begin{cases} \Delta u = -\mu \quad \text{in } \Omega, \\ u = 0 \quad \text{on } \partial\Omega, \\ |\nabla u| = 1 \quad \text{on } \partial\Omega. \end{cases}$$

Similar remarks apply to the above system as to (7.1), e.g., given μ, solutions (u, Ω) can be constructed by certain variational methods. Indeed, for $\mu > 0$ of the form (5.1) an open set Ω satisfying (8.1) is obtained as $\Omega = \{u > 0\}$ where $u \geq 0$ is any minimizer of the nonconvex functional

$$F(u) = \int_{\mathbf{C}} (|\nabla u|^2 + \chi_{\{u>0\}})dA - 2\int u d\mu.$$

See [142], [61]. Techniques using methods of sub- and supersolutions can also be used [71].

9. Inverse problems

As a model case, we consider quadrature identities for harmonic functions as in (6.1). Corresponding to the two points of view discussed after (7.1), namely as attempts of maps $\mu \mapsto \Omega$ ("balayage") respectively $\Omega \mapsto \mu$ ("inverse balayage"), there are two natural uniqueness questions:

1) To what extent is Ω uniquely determined by μ?
2) To what extent is μ uniquely determined by Ω?

Let us consider the second question first. If we require μ to be of the form (5.1) then certainly μ is uniquely determined by Ω (and the relation (6.1)). On the other hand, quadrature domains for measures as in (5.1) are exceptionally rare, and it is natural to try to relax (5.1) while retaining some degree of uniqueness. Attempts in this direction were made in [50], leading to concepts of potential theoretic skeletons, or "**mother bodies**", which are reasonably unique when they exist.

The basic requirements are that Ω should be a quadrature domain for subharmonic functions for μ, that $\operatorname{supp}\mu$ should have Lebesgue measure zero and that it should not disconnect any part of Ω from the complement of $\overline{\Omega}$. As an example, convex polyhedra (in any number of dimensions) turn out to have unique mother bodies, while nonconvex polyhedra have at most finitely many of them [58]. (It should be remarked that the definition of a mother body allows the support of μ to reach $\partial\Omega$).

Question 1) above is closely related to general inverse problems in potential theory, e.g., the following (see [162], [75] for more information):

Do there exist two different domains Ω_1 and Ω_2 such that

$$U^{\Omega_1} = U^{\Omega_2} \quad \text{outside } \Omega_1 \cup \Omega_2\,? \tag{9.1}$$

By (6.2), a negative answer to 1) implies a negative answer to (9.1), and conversely, at least under some mild extra assumptions, every counterexample for (9.1) can be made into an example of nonuniqueness for 1) (compare [48], [150]).

It is easy to give counterexamples for (9.1) if one allows multiply connected domains (a disc and a concentric annulus with the same area will do), but as far as we know there are no examples with $\overline{\Omega}_j$ both having connected complements (if Ω_j are just simply connected, which is a little weaker, there are examples [124]). However, for several similar questions [71] (arc length instead of area measure), [99] (bodies of nonconstant density) there do exist counterexamples, so we do not really expect any uniqueness for question 1). Still it would be of interest to have a definite answer to the question whether there can be two different simply connected quadrature domains both satisfying the same identity (2.1). Further results on the uniqueness question can be found in [121], [125], [48], [139], for example.

A weaker version of (9.1) is obtained by asking the potentials to agree only in a neighbourhood of infinity. This is equivalent to asking the complex moments to agree

$$\int_{\Omega_1} z^m dA = \int_{\Omega_2} z^m dA \quad \text{for all } m = 0, 1, 2, \ldots.$$

For this uniqueness question there are explicit counterexamples (with $\mathbf{C} \setminus \overline{\Omega}_j$ connected) in form of circular polygons [117], [162].

If one requires all the real moments $\int_{\Omega_j} x^m y^n dA$ to agree, or equivalently

$$\int_{\Omega_1} z^m \bar{z}^n dA = \int_{\Omega_2} z^m \bar{z}^n dA \quad \text{for all } m, n = 0, 1, 2, \ldots,$$

then certainly $\Omega_1 = \Omega_2$ (up to nullsets). In case Ω is a (classical) quadrature domain (2.1) then Ω can be effectively recovered from the knowledge of these moments by an algorithm based on the exponential transform. See [56].

10. Fluid dynamics (Hele-Shaw flow)

In [23] D. Crowdy discusses connections between quadrature domains and several different problems in fluid dynamics. Here we shall just look at one of these, the Hele-Shaw flow problem, which is the one which so far has been most important.

A Hele-Shaw evolution is, in the simplest case, a family of domains $\Omega(t)$ containing the origin (say) and which develops in time t according to the rule that the normal velocity of the boundary $\partial\Omega(t)$ is proportional to the normal derivative of the Green function (with logarithmic pole at the origin), in other words, proportional to the density of harmonic measure. (Thus Hele-Shaw evolution may be named "motion by harmonic measure".) Physically it models the growth of a blob of a viscous fluid (like oil) confined within a thin gap between two parallel planes and when more fluid is injected (or sucked) at one point (the origin).

One easily finds that the Hele-Shaw evolution is characterized by

$$\frac{d}{dt}\int_{\Omega(t)} h\,dA = qh(0) \tag{10.1}$$

holding for any harmonic function h defined in a neighbourhood of $\overline{\Omega(t)}$. Here $q > 0$ is the strength of the source. Allowing $q < 0$ gives the corresponding problem with a sink instead of a source. In any case it follows that the complex moments of $\Omega(t)$ are preserved quantities:

$$\frac{d}{dt}\int_{\Omega(t)} z^m\,dA = 0$$

for any $m \geq 1$. This property was discovered in the seminal paper [112] by S. Richardson.

The Hele-Shaw problem is that of finding $\{\Omega(t) : 0 \leq t < T\}$ (with T as large as possible) when $\Omega(0)$ is given. This problem turns out to be well-posed when $q > 0$, for example it has a unique global ($T = +\infty$) solution in the weak formulation that

$$\int_{\Omega(t)} h\,dA - \int_{\Omega(0)} h\,dA = qth(0) \tag{10.2}$$

holds for all h harmonic in a neighbourhood of $\overline{\Omega(t)}$. Even better, the inequality $\geq$ holds for all subharmonic h ((10.1) can be amplified in the same way), and this is really the standard requirement on the weak solution.

The above discussion about weak solution only applies for $q > 0$. When $q < 0$ the Hele-Shaw problem is ill-posed and not yet fully understood.

From (10.2) it is clear how Hele-Shaw flow is related to quadrature domains: if $\Omega(0)$ is a quadrature domain (2.1), then the $\Omega(t)$ remain quadrature domains for all t, with the origin as a new quadrature point if it was not already such a point from start.

The literature on Hele-Shaw flow is vast, see [42] for a "complete" bibliography up to 1998. A short selection is [69], [70], [41], [115], [112], [35], [28], [45], [73], [156], [94], [153], [68], [63].

11. Unbounded quadrature domains

Nothing prevents us from allowing also unbounded domains Ω in (2.1). Then there are even simpler quadrature domains than discs. Indeed, there is a whole class of **null quadrature domains**, i.e., domains Ω for which

$$\int_{\Omega} f dA = 0 \tag{11.1}$$

holds for all integrable analytic functions f in Ω. This class consists of half-spaces, exteriors of ellipses, exteriors of parabolas and some degenerate cases. See [120] for a complete classification. In higher dimensions there are similar results [40], but they are less complete.

Null quadrature domains are relevant for Hele-Shaw flow problems in which the fluid occupies a full neighbourhood of infinity, e.g., they exactly comprise those initial fluid domains which can be completely emptied by suction at infinity [74], [29], [156].

Null quadrature domains also come up when investigating regularity of free boundaries, namely when finite boundary points to be investigated are "blown up" under scale changes. See [83], [17] for example.

In general, if a quadrature domain (satisfying (2.1) or (6.1)) is not bounded it must occupy a good portion of a neighbourhood of infinity in the sense that

$$\int_{\Omega} \frac{dA}{1+|z|^2} = +\infty.$$

See [121], and for the corresponding condition in higher dimensions [136]. More specifically, the unbounded quadrature domains which are not dense in the complex plane turn out to be exactly those domains which can be obtained from bounded quadrature domains by inversion $z \mapsto 1/(z-a)$ at a point a on the boundary or in the interior [145], [146], [129].

A substantial theory for general unbounded quadrature domains (in any number of dimensions) has been elaborated by L. Karp and A. Margulis [82]. It is partly based on modifications of the Newtonian kernel at infinity, to make it integrable there [120], [79].

12. Special points

If Ω is a quadrature domain (2.1) and $a \in \Omega$ is a point distinct from the quadrature nodes a_k then $\Omega \setminus \{a\}$ is usually not a quadrature domain because the deletion of the point a allows for the new test function $f(z) = \frac{1}{z-a}$. In case $\Omega \setminus \{a\}$ does remain a quadrature domain then a is called a **special point**.

It is fairly immediate that $a \in \Omega$ is a special point if and only if (2.5) or (3.2) continues to hold at $z = a$. Also, as already mentioned, the special points exactly constitute the difference set in (3.3).

The last statements above referred to quadrature domains for analytic functions. Special points in the case of quadrature domains for harmonic functions are defined similarly, and they constitute the set of points in $\Omega \setminus \operatorname{supp} \mu$ at which (6.2) and (6.3) remain valid.

Special points turn out to be important for questions of uniqueness. For example, if for a quadrature domain Ω for harmonic functions (6.1) there are no special points at all, then this Ω is the unique quadrature domain for the measure μ in question. See [121], section 9 (and also [48]).

The special points also provide additional information for determining the polynomial $Q(z, w)$ in (3.3) from the knowledge of the (generally insufficient) data $\{(a_k, c_k)\}$ in (2.1). See [47], [21], [22].

The number of special points were estimated in [47] and [126] after some initial conjectures in [145], where also the terminology was coined.

13. Operator theory and the exponential transform

An operator T on a Hilbert space is called hyponormal if its self-commutator $[T^*, T] = T^*T - TT^*$ is a positive operator [96]. Thirty years ago R.W. Carey and J.D. Pincus [18] found that if $[T, T^*]$ moreover has rank one then T can be characterized up to unitary equivalence by a function $0 \leq \rho \leq 1$, called the principal function and related to T by

$$\det[(T^* - \overline{w})^{-1}(T - z)(T^* - \overline{w})(T - z)^{-1}] = E_\rho(z, w)$$

for large $z, w \in \mathbf{C}$, where

$$E_\rho(z, w) = \exp[-\frac{1}{\pi} \int \frac{\rho(\zeta)\, dA(\zeta)}{(\zeta - z)(\overline{\zeta} - \overline{w})}] \tag{13.1}$$

is the **exponential transform** of ρ.

In [102], [104] M. Putinar discovered a beautiful connection between operator theory and the theory of quadrature domains (this was actually an independent rediscovery of quadrature domains): $E_\rho(z, w)$ is a rational function of the form

$$E_\rho(z, w) = \frac{Q(z, w)}{P(z)\overline{P(w)}}$$

for large z and w if and only if $\rho = \chi_\Omega$ where Ω is a quadrature domain. Then Q is the same as in (3.3) and P is the same as in (2.4)

Using the exponential transform and results from operator theory more insight into the nature of quadrature domains has been gained. For example, the equation $Q(z, \overline{z}) = 0$ for the boundary can be written in the lemniscate form

$$|P(z)|^2 = \sum_{k=0}^{n-1} |Q_k(z)|^2,$$

where $P(z)$ is the same as above (and in (2.4)) and, for each $0 \leq k \leq n-1$, $Q_k(z)$ is a polynomial of degree k. Here $Q_{n-1}(z)$ is up to a constant factor the same as $Q(z)$ in (2.4) (see [53]).

It follows from the above that P and Q_{n-1} are directly (and bijectively) related to the quadrature data $\{(a_k, c_k)\}$ in (2.1). However, even if Ω is a quadrature domain for subharmonic functions, so that it is uniquely determined by $\{(a_k, c_k)\}$, no general method is known for effectively determining the remaining polynomials $Q_0, \dots, Q_{n-2}$ from $\{(a_k, c_k)\}$.

The exponential transform is also useful for quadrature domains in a wider sense, like (6.1) for analytic h. For example, it can be used to prove analyticity of $\partial\Omega$, see [52], thereby providing an alternative approach to [127], [128], [129]. For further results on hyponormal operators and quadrature domains, see [103], [105], [157], [161], [159], and for connections to moment problems also the survey [108].

Other classes of operators which are linked to quadrature domains in different ways are subnormal operators (those operators which can be extended to be normal operators on a larger Hilbert space), see [97], [157], [98], [160], [19], [38], [152] and the Friedrichs operator (essentially the orthogonal projection in $L^2(\Omega)$ of the analytic functions onto the antianalytic ones) [143], [146], [106], [110], [111].

14. The Bergman kernel

In terms of the Bergman kernel $K(z, w)$ for Ω the quadrature identity (2.1) becomes

$$\sum_{k=1}^{n} c_k K(z, a_k) = 1 \quad (z \in \Omega).$$

Quadrature identities for arc length (8.1) are related in the same way to the Szegö kernel. The methods of Avci [8] are largely based on these relations.

Recently, S. Bell [12], [13] has developed further the connections to the Bergman and other kernel functions. To give an example, the Bergman kernel for any domain extends to the Schottky double of the domain as a meromorphic differential:

$$K(z, a)dz = \overline{L(z, a)dz} \quad \text{along } \partial\Omega$$

($a \in \Omega$ fixed), where $L(z, a)$ is the adjoint kernel. Since, by (3.4), dz itself extends to the double if and only if Ω is a quadrature domain in the sense (3.5) it follows that that $K(z, a)$ extends meromorphically to the double as a *function* if and only if Ω is a quadrature domain in that sense. From this conclusions can be drawn concerning algebraicity properties of $K(z, w)$, for example.

The reproducing property of the Bergman kernel says that

$$f(a) = \int_\Omega f(z)\overline{K(z, a)}dA(z)$$

for all $f \in L^2_a(\Omega)$ (the Bergman space) and for any $a \in \Omega$. Choosing here $f(z) = g(z)K(z, a)$ where g is a bounded analytic function gives

$$g(a) = \frac{1}{K(a, a)} \int_\Omega g(z)|K(z, a)|^2 \, dA(z). \tag{14.1}$$

Thus we have a one-point quadrature identity for bounded analytic functions and with weight $|K(z,a)|^2$ (in place of pure Lebesgue measure). The same arguments work if $K(z,w)$ is the reproducing kernel for any subspace of $L^2_a(\Omega)$ which is invariant under multiplication by bounded analytic functions, e.g., the subspace consisting of those functions which vanish on a given finite subset Ω. In case Ω is the unit disc and $a = 0$ then (14.1) is exactly H. Hedenmalm's definition of an inner divisor in Bergman space, see, e.g., [67], [66]. In this way quadrature identities (in a wide sense) are related to Hedenmalm's theory of contractive zero divisors [65], [30] [31]. (For a general domain Ω, the divisor $K(z,a)$ will possibly not be contractive, however.)

We finally remark that some of the extremal problems which initiated the theory of quadrature domains were problems of giving estimates (e.g., of the Gaussian curvature) for metrics of the kind

$$ds^2 = K(z,z)|dz|^2,$$

on Riemann surfaces, actually with $K(z,w)$ the reproducing kernel for those analytic functions which have a single-valued integral (the reduced Bergman kernel). See [118], [119] and the appendix of [121].

15. Other aspects

a) Quadrature identities of the form

$$\int_\Omega f'' dA = \sum_{k=1}^{n} c_j f(a_j)$$

for f analytic in a neighbourhood of Ω hold for polygons $\Omega \subset \mathbf{C}$, where $a_k \in \partial\Omega$ are the vertices of the polygon [24], [26]. To a limited extent such formulas can be extended to higher dimensions [54].

b) P. Ebenfelt has studied behaviour of solutions of the Dirichlet problems on quadrature domains, e.g., to what extent solutions are analytically continuable outside the domain [32], [33].

c) The ellipse is not a classical quadrature domain, but it is a quadrature domain for a measure on the segment joining the foci, and the complement of it is a null quadrature domain. Similarly in higher dimensions. Many authors have studied quadrature related properties for ellipses (and ellipsoids): [74], [29], [140], [80]. See further [146], [85].

d) For other specific types and more examples of quadrature identities and quadrature domains, see [25], [93], [154], [155], [9], [47], [95], [78], [138], [6], [7], [109], [21], [22], plus the books [26], [121], [146].

e) Studies of quadrature domains in a nonlinear setting (for the p-Laplacian) can be found in [10], [11].

16. Notations

The following notations are in common use in the theory of quadrature domains.

- $AL^p(\Omega) = L^p_a(\Omega)$ = the set of p-integrable analytic functions in Ω ($1 \leq p \leq \infty$).
- $HL^p(\Omega) = L^p_h(\Omega)$ = the set of p-integrable harmonic functions in Ω.
- $SL^p(\Omega)$ = the set of p-integrable subharmonic functions in Ω.
- $Q(\mu, AL^p)$ = the class of quadrature domains for the measure (or distribution) μ and the test class of p-integrable analytic functions.
- Similarly for $Q(\mu, HL^p)$, $Q(\mu, SL^p)$ etc. The exact definitions of these classes differ a little between different authors.

References

[1] D. Aharonov, H.S. Shapiro, *A minimal-area problem in conformal mapping – preliminary report*, Research bulletin TRITA-MAT-1973-7, Royal Institute of Technology, 34 pp.

[2] D. Aharonov, H.S. Shapiro, *Domains in which analytic functions satisfy quadrature identities*, J. Analyse Math. **30** (1976), 39–73.

[3] D. Aharonov, H.S. Shapiro, *A minimal-area problem in conformal mapping – preliminary report: Part II*, Research bulletin TRITA-MAT-1978-5, Royal Institute of Technology, 70 pp.

[4] D. Aharonov, H.S. Shapiro, A. Solynin, *A minimal area problem in conformal mapping*, J.Analyse Math. **78** (1999), 157–176.

[5] D. Aharonov, H.S. Shapiro, A. Solynin, *A minimal area problem in conformal mapping. II*, J.Analyse Math. **83** (2001), 239–259.

[6] D. Armitage, M. Goldstein, *Quadrature and harmonic approximation of subharmonic functions in strips*, J. London Math. Soc. **46** (1992), 171–179.

[7] D. Armitage, S. Gardiner, *Best one-sided L^1-approximation by harmonic and subharmonic functions*, Advances in multivariate approximation (Witten-Bommenholz, 1998), 43–56, Math. Res., 107, Wiley-VCH, Berlin, 1999.

[8] Y. Avci, *Quadrature identities and the Schwarz Function*, Doctoral disseertation, Stanford, 1977.

[9] Y. Avci, *Characterization of shell domains by quadrature identities*, J. London Math. Soc. **23** (1980), 123–128.

[10] F. Bahrami, A. Chademan, *Existence of unbounded quadrature domains for the p-Laplace operator*, Bull. Iranian Math. Soc. **14**, (1998), 623–681.

[11] F. Bahrami, A. Chademan, *Existence of a uniform bound for quadrature domains associated to p-Laplacian*, Analysis (Munich) **19**, (1999), 319–326.

[12] S. Bell, *Quadrature domains and kernel function zipping* preprint 2003.

[13] S. Bell, *The Bergman kernel and quadrature domains in the plane*, this volume.

[14] L. Bers: *An approximation theorem*, J. Analyse Math. **14** (1965), 1–4.

[15] L.A. Caffarelli, *The regularity of free boundaries in higher dimensions*, Acta. Math. **139** (1977) 155–184.

[16] L.A. Caffarelli, *Compactness methods in free boundary problems*, Comm. Partial Diff. Eq. **5** (1980), 427–448.

[17] L.A. Caffarelli, L. Karp, H. Shahgholian, *Regularity of a free boundary problem with application to the Pompeiu problem*, Ann. Math. **151**, (2000), 269–292.

[18] R.W. Carey, J.D. Pincus, *An exponential formula for determining functions,* Indiana Univ. Math.J. **23** (1974), 1031–1042.

[19] J. Conway, L. Yang, *Some open problems in the theory of subnormal operators,* Holomorphic spaces (Berkeley, CA, 1995), 201–209, Math. Sci. Res. Inst. Publ., 33, Cambridge Univ. Press, Cambridge, 1998.

[20] D. Crowdy, *Multipolar vortices and algebraic curves*, R. Soc. Lond. Proc. Ser. A Math. Phys. Eng. Sci. **457** (2001), 2337–2359.

[21] D. Crowdy, *Constructing multiply-connected quadrature domains I: algebraic curves,* preprint 2003.

[22] D. Crowdy, J. Marshall, *Constructing multiply-connected quadrature domains,* SIAM J. Appl. Math. **64** (2004), 1334–1359.

[23] D. Crowdy, *Quadrature domains and fluid dynamics*, in this volume.

[24] P.J. Davis, *Triangle formulas in the complex plane*, Math. of Comp. **18** (1964), 569–577.

[25] P.J. Davis, *Double integrals expressed as single integrals or interpolatory functionals,* J. Approx. Theory **5** (1972), 276–307.

[26] P.J. Davis, *The Schwarz Function and its Applications*, Carus Math. Mongraphs No.17, Math. Assoc. Amer., 1974.

[27] P.J. Davis, H. Pollak, *On the analytic continuation of mapping functions*, Trans. Amer. Math. Soc. **87** (1958), 198–225.

[28] E. DiBenedetto, A. Friedman, *The illposed Hele-Shaw model and the Stefan problem for supercooled water*, Trans. Amer. Math. Soc., **282** (1984) 183–204.

[29] E. DiBenedetto, A. Friedman, *Bubble growth in porous media,* Indiana Univ. Math. J. **35** (1986), 573–606.

[30] P. Duren, D. Khavinson, H.S. Shapiro, C. Sundberg, *Contractive zero-divisors in Bergman spaces*, Pacific J. Math. **157** (1993), 37–56.

[31] P. Duren, A. Schuster, *Bergman spaces*, Math. Surveys and Monographs vol. 100, Amer. Math. Soc., 2004.

[32] P. Ebenfelt, *Singularities encountered by the analytic continuation of solutions to Dirichlet's problem*, Complex Variables **20** (1992), 75–91.

[33] P. Ebenfelt, *Analytic continuation of certain domain functions in quadrature domains and domains bounded by ellipses*, Complex Variables **22** (1993), 69–83.

[34] P. Ebenfelt, *Singularities of the solution to a certain Cauchy problem and applications to the Pompeiu problem*, Duke Math. J. **71** (1993), 119–142.

[35] C.M. Elliott, J.R. Ockendon, *Weak and Variational Methods for Moving Boundary Problems*, Pitman, London (1982).

[36] B. Epstein, *On the mean-value property of harmonic functions*, Proc. Amer. Math. Soc. **13** (1962), 830.

[37] B. Epstein, M. Schiffer, *On the mean-value property of harmonic functions*, J. Analyse Math. **14** (1965), 109–111.

[38] N. Feldman, *Subnormal operators, self-commutators and pseudocontinuations*, Integral Equations Operator Theory **37**, (2000), 402–422.

[39] A. Friedman, *Variational Principles and Free Boundaries*, Wiley and Sons, 1982.

[40] A. Friedman, M. Sakai, *A characterization of null quadrature domains in* $\mathbf{R}^n$, Indiana Univ. Math. J. **35** (1986), 607–610.

[41] L.A. Galin, *Unsteady seepage with a free surface*, Dokl. Akad. Nauk SSSR **47** (1945), 250–253. (Russian)

[42] K.A. Gillow and S.D. Howison, *A bibliography of free and moving boundary problems for Hele-Shaw and Stokes flow*, published electronically at URL http://www.maths.ox.ac.uk/~howison/Hele-Shaw.

[43] P.G.W. Glare (ed.) *Oxford Latin Dictionary*, Clarendon Press, Oxford, 1982.

[44] B. Gustafsson, *Quadrature identities and the Schottky double*, Acta Appl. Math. **1** (1983), 209–240.

[45] B. Gustafsson, *Applications of variational inequalities to a moving boundary problem for Hele-Shaw flows*, SIAM J. Math. Anal. **16** (1985), 279–300.

[46] B. Gustafsson: *Applications of half-order differentials on Riemann surfaces to quadrature identities for arc-length*, J. Analyse Math. **49** (1987), 54–89.

[47] B. Gustafsson, *Singular and special points on quadrature domains from an algebraic geometric point of view*, J. Analyse Math. **51** (1988), 91–117.

[48] B. Gustafsson, *On quadrature domains and an inverse problem in potential theory*, J. Analyse Math. **55** (1990), 172–216.

[49] B. Gustafsson: *A distortion theorem for quadrature domains for harmonic functions*, J. Math. Anal. Appl. **202** (1996), 169–182.

[50] B. Gustafsson, *On mother bodies of convex polyhedra*, SIAM J. Math.Anal. **29:5** (1998), 1106–1117.

[51] B. Gustafsson, *Lectures on Balayage*, preprint 2003.

[52] B. Gustafsson, M. Putinar, *An exponential transform and regularity of free boundaries in two dimensions*, Ann. Scuola Norm. Sup. Pisa Cl. Sci (4) **26** (1998), 507–543.

[53] B. Gustafsson, M. Putinar, *Linear analysis of quadrature domains II*, Israel J.Math. **119** (2000), 187–216.

[54] B. Gustafsson, M. Putinar, *On exact quadrature formulas for harmonic functions on polyhedra*, Proc. Amer. Math. Soc. **128** (1999), 1427–1432.

[55] B. Gustafsson, M. Putinar, *Linear analysis of quadrature domains IV*, in this volume.

[56] B. Gustafsson, C. He, P. Milanfar, M. Putinar, *Reconstructing planar domains from their moments*, Inverse Problems **16** (2000), 1053–1070.

[57] B. Gustafsson, M. Sakai, *Properties of some balayage operators with applications to quadrature domains and moving boundary problems*, Nonlinear Anal. **22** (1994), 1221–1245.

[58] B. Gustafsson, M. Sakai, *On potential theoretic skeletons of polyhedra*, Geom. Didicata **76** (1999), 1–30.

[59] B. Gustafsson, M. Sakai, *Sharp estimates of the curvature of some free boundaries in two dimensions*, Ann. Acad. Sci. Fenn. Math. **28** (2003), 123–142.

[60] B. Gustafsson, M. Sakai, *On the curvature of the free boundary for the obstacle problem in two dimensions*, Monatshefte für Mathematik, **142** (2004), 1–5.

[61] B. Gustafsson, H. Shahgholian, *Existence and geometric properties of solutions of a free boundary problem in potential theory*, J. Reine Angew. Math. **473** (1996), 137–179.

[62] B. Gustafsson, M. Sakai, H.S. Shapiro: *On domains in which harmonic functions satisfy generalized mean-value properties*, Potential Anal. **7** (1997), 467–484.

[63] B. Gustafsson, A. Vasilév, *Complex and Potential Analysis in Hele-Shaw cells*, Lecture notes, in preparation, 2004.

[64] V.P. Havin, *Approximation in the mean by analytic functions*, Dokl. Akad. Nauk. SSSR 178, pp. 1025–1028 (Russion; Engl. transl. in Siviet Math. Dokl. **9** (1968), 245–258.

[65] H. Hedenmalm, *A factorization theorem for square area-integrable analytic functions*, J. Reine Angew. Math. **422** (1991), 45–68.

[66] H. Hedenmalm, *Recent progress and open problems in the Bergman space*, in this volume.

[67] H. Hedenmalm, B. Korenblum, K. Zhu, *Theory of Bergman spaces*, Graduate Texts in Mathematics, **199**, Springer-Verlag, New York, 2000.

[68] H. Hedenmalm, S. Shimorin, *Hele-Shaw flow on hyperbolic surfaces*, J. Math Pures Appl. **81** (2002), 187–222.

[69] H.S. Hele-Shaw, *The flow of water*, Nature **58** (1898), 34–36.

[70] H.S. Hele-Shaw, *On the motion of a viscous fluid between two parallel plates*, Trans. Royal Inst. Nav. Archit., London **40** (1898), 218.

[71] A. Henrot, *Subsolutions and supersolutions in free boundary problems* Ark. Mat. **32** (1994), 79–98.

[72] G. Herglotz, *Über die analytische Fortsetzung des Potentials ins Innere der anziehenden Massen, Gekrönte Preisschr. der Jablonowskischen Gesellsch. zu Leipzig*, 56 pp., Reproduced in 'Gustav Herglotz-Gesammelte Schriften', Vandenhoeck & Ruprecht, Göttingen, 1979, pp. 299–355.

[73] Y. Hohlov, S.D. Howison, C. Huntingford, J.R. Ockendon, A.A. Lacey, *A model for non-smooth free boundaries in Hele-Shaw flows*, Quart. J. Appl. Math. **47** (1994), 107–128.

[74] S.D. Howison, *Bubble growth in porous media and Hele Shaw flow*, Proc. Roy. Soc. Edinburgh **A102** (1985), 141–148.

[75] V. Isakov, *Inverse Source Problems*, AMS Math. Surveys and Monographs 34, Providence Rhode Island, 1990.

[76] G. Johnsson, *The Cauchy problem in* $\mathbf{C}^N$ *for linear second order partial differential equations with data on a quadric surface*, Trans. Amer. Math. Soc. **344** (1994), 1–48.

[77] G. Johnsson, *Global existence results for linear partial differential equations*, J. Differential Equations **115** (1995), 416–440.

[78] L. Karp, *Construction of quadrature domains in* $\mathbf{R}^n$ *from quadrature domains in* $\mathbf{R}^2$, Complex variables **17** (1992), 179–188.

[79] L. Karp, *Generalized Newtonian potential and its applications*, J. Math. Anal. Appl. **174** (1993), 480–497.

[80] L. Karp, *On the Newtonian potential of ellipsoids*, Complex Variables **25** (1995), 367–372.

[81] L. Karp, *Multivalued analytic continuation of the Cauchy transform*, preprint 2003.

[82] L. Karp, A. Margulis, *Newtonian potential theory for unbounded sources and applications to free boundary problems*, J. Analyse Math. **70** (1996), 1–63.

[83] L. Karp, H. Shahgholian, *Regularity of a free boundary problem*, J. Geom. Anal. **9**, (1999), 653–669.

[84] D. Khavinson, *Singularities of harmonic functions in* $\mathbf{C}^n$, Several complex variables and complex geometry, Part 3 (Santa Cruz, CA, 1989), 207–217, Proc. Sympos. Pure Math., 52, Part 3, Amer. Math. Soc., Providence, RI, 1991.

[85] D. Khavinson, *Holomorphic Partial Differential Equations and Classical Potential Theory*, Departamento de Análisis Matemático, Universidad de La Laguna, Tenerife, 1996.

[86] D. Khavinson, H.S. Shapiro, *The Schwarz potential in* $\mathbf{R}^n$ *and Cauchy's problem for the Laplace equation*, Research bulletin TRITA-MAT-1986-36, Royal Institute of Thechnology, 112 pp.

[87] D. Khavinson, H.S. Shapiro, *The Vekua hull of a plane domain*, Complex Variables **14** (1990), 117–128.

[88] D. Kinderlehrer, L. Nirenberg, *Regularity in free boundary problems*, Ann. Scuola Norm. Sup. Pisa Cl. Sci. **4** (1977), 373–391.

[89] D. Kinderlehrer, G. Stampacchia, *Introduction to Variational Inequalities and Their Applications*, Academic Press, New York, 1980.

[90] O.I. Kounchev, *The partial balayage materic bodies and optimization problems in gravimetry*, in Inverse Modeling in Exploration Geophysics, Proceedings of the 6th International Mathematical Geophysics Seminar in Berlin, Feb. 3–6, 1988 (eds. A. Vogel, R. Gorenflo, B. Kummer, C.O. Ofoegbu), Vieweg & Sohn, Braunschweig/Wiesbaden.

[91] O.I. Kounchev, *Obtaining materic bodies through concentration and optimization of a linear functional*, in Geophysical Data Inversion Methods and Applications, Proceedings of the 7th International Mathematical Geophysics Seminar in Berlin, Feb. 8–11, 1989 (eds. A. Vogel, R. Gorenflo, C.O. Ofoegbu, B. Ursin), Vieweg & Sohn, Braunschweig/Wiesbaden.

[92] J. Leray, *Uniformisation de la solution du problème linéaire analyticque de Cauchy, près de la varieté qui porte les données de Cauchy*, Bull. Soc. Math. France **85**, 389–430.

[93] A.A. Levin, *An example of a doubly connected domain which admits a quadrature identity*, Proc. Amer. Math. Soc. **60** (1976), 163–168.

[94] A.S. Margulis, *The moving boundary problem of potential theory*, Adv. Math. Sci. Appl. **5(2)** (1995), 603–629.

[95] R. Marrero, I. Maria, *Analytic characterization of annuli*, Rev. Acad. Canaria Cienc. **1** (1990), 147–153.

[96] M. Martin, M. Putinar, *Lectures on Hyponormal Operators*, Birkhäuser, Basel, 1989.

[97] J. McCarthy, L. Yang, *Cyclic subnormal operators with finite-rank self-commutators*, Proc. Roy. Irish Acad. Sect. A **95** (1995), 173–177.

[98] J. McCarthy, L. Yang, *Subnormal operators and quadrature domains*, Adv. Math. **127** (1997), 52–72.

[99] N.S. Nadirashvili, *Universal classes of uniqueness of domains in the inverse problem of Newtonian potential theory*, Soviet Math. Dokl. **44** (1992), 287–290.

[100] C. Neumann, *Über das logarithmische Potential einer gewissen Ovalfläche*, Abh. der math.-phys. Klasse der Königl. Sächs. Gesellsch. der Wiss. zu Leibzig 59 (1907), pp. 278–312.

[101] C. Neumann, *Über das logarithmische Potential einer gewissen Ovalfläche*, Zweite Mitteilung, ibib. vol. 60 (1908), pp. 53–56. Dritte Mitteilung, ibid. pp. 240–247.

[102] M. Putinar, *On a class of finitely determined planar domains*, Math. Res. Lett. **1** (1994), 389–398.

[103] M. Putinar, *Linear analysis of quadrature domains*, Ark. Mat. **33** (1995), 357–376.

[104] M. Putinar, *Extremal solutions of the two-dimensional L-problem of moments*, J.Funct.An. **136** (1996), 331–364.

[105] M. Putinar, *Extremal solutions of the two-dimensional L-problem of moments II*, J.Approx.Th. **92** (1998), 38–58.

[106] M. Putinar, *Matrix analysis and the Friedrichs operator of a quadrature domain*, Linear Algebra Appl. **270** (1998), 215–229.

[107] M. Putinar, *Linear analysis of quadrature domains III*, Math. Anal. Appl. **239** (1999), 101–117.

[108] M. Putinar, *A renormalized Riesz transform and applications*, Advances in Constructive Approximation, to appear.

[109] G. Putinar, M. Putinar, *Root separation on generalized lemniscates*, Hokkaido Math. J. **30** (2001), 705–716.

[110] M. Putinar, H.S. Shapiro, *The Friedrichs operator of a planar domain*, Complex analysis, operators, and related topics, 303–330, Oper. Theory Adv. Appl. **113**, Birkhäuser, Basel, 2000.

[111] M. Putinar, H.S. Shapiro, *The Friedrichs operator of a planar domain II*, Recent advances in operator theory and related topics (Szeged, 1999), 519–551, Oper. Theory Adv. Appl. **127**, Birkhäuser, Basel, 2001.

[112] S. Richardson, *Hele Shaw flows with a free boundary produced by the injection of fluid into a narrow channel*, J. Fluid Mech. **56** (1972), 609–618.

[113] S. Richardson, *Hele-shaw flows with time-dependent free boundaries involving a multiply connected fluid region*, European J. Appl. Math. **12** (2001), 571–599.

[114] J.F. Rodrigues, *Obstacle problems in Mathematical Physics*, North-Holland, Amsterdam 1987.

[115] P.G. Saffman, F.P. Taylor, *The penetration of a fluid into a porous medium or Hele-Shaw cell containing a more viscous fluid*, Proc. Royal Soc. London Ser. A **245** (1958), 312–329.

[116] M. Sakai, *On basic domains of extremal functions*, Kodai Math. Sem. Rep. **24** (1972), 251–258.

[117] M. Sakai, *A moment problem on Jordan domains*, Proc. Amer. Math. Soc. **70** (1978), 35–38.

[118] M. Sakai, *Analytic functions with finite Dirichlet integrals on Riemann surfaces*, Acta Math. **142** (1979), 199–220.

[119] M. Sakai, *The submeanvalue property of subharmonic functions and its application to the estimation of the Gaussian curvature of the span metric*, Hiroshima Math. J. **9** (1979), 555–593

[120] M. Sakai, *Null quadrature domains*, J. Analyse Math. **40** (1981), 144–154.

[121] M. Sakai, *Quadrature Domains*, Lect. Notes Math. **934**, Springer-Verlag, Berlin-Heidelberg 1982.

[122] M. Sakai, *Application of variational inequalities to the existence theorem on quadrature domains*, Trans. Amer. Math. Soc. **276** (1983), 267–279.

[123] M. Sakai, *Solutions to the obstacle problem as Green potentials*, J. Analyse Math. **44** (1984/85) 97–116.

[124] M. Sakai, *Domains having null complex moments*, Complex Variables **7** (1987), 313–319.

[125] M. Sakai, *Finiteness of the family of simply connected quadrature domains*, pp. 295–305 in "Potential Theory" (editors J. Kral, I. Netuka, J. Vesely), Plenum Publishing Corporation, 1988.

[126] M. Sakai, *An index theorem on singular points and cusps of quadrature doamins*, Holomorphic functions and moduli, Vol. I (Berkeley, CA, 1986), 119–131, Math. Res. Inst. Publ., 10, Springer, New York, 1988.

[127] M. Sakai, *Regularity of boundary having a Schwarz function*, Acta Math. **166** (1991), 263–297.

[128] M. Sakai, *Regularity of free boundaries in two dimensions*, Ann. Scuola Norm Sup. Pisa Cl. Sci. (4) **20** (1993), 323–339.

[129] M. Sakai, *Regularity of boundaries of quadrature domains in two dimensions*, SIAM J. Math. Anal. **24** (1994), 341–364.

[130] M. Sakai, *Sharp estimates of the distance from a fixed point to the frontier of a Hele-Shaw flow*, Potential Anal. **8** (1998), 277–302.

[131] M. Sakai, *Linear combinations of harmonic measures and quadrature domains of signed measures with small supports*, Proc. Edinburgh Math. Soc. (2) **42** (1999), 433–444.

[132] T. Savina, B. Sternin, V. Shatalov, *Notes on "mother body" problem in geographics*, preprint 1995.

[133] F. Schottky *Über die conforme Abbildung mehrfach zusammenhängender ebener Flächen*, Crelles Journal **83** (1877), 300–351.

[134] M. Schiffer, N.S. Hawley, *Half-order differentials on Riemann surfaces*, Acta Math. **115** (1966), 199–236.

[135] M. Schiffer, D.C. Spencer, *Functionals of Finite Riemann Surfaces*, Princeton Univ. Press, Princeton, 1954.

[136] H. Shahgholian, *Unbounded quadrature domains in* $\mathbf{R}^n$ $(n \geq 3)$, J.Analyse Math. **57** (1991), 281–291, 297–298.

[137] H. Shahgholian, *On quadrature domains and the Schwarz potential*, J. Math. Anal. Appl. **171** (1992), 61–78.

[138] H. Shahgholian, *A characterization of the sphere in terms of single-layer potentials*, Proc. Amer. Math. Soc. **115** (1992), 1167–1168.

[139] H. Shahgholian, *Convexity and uniqueness in an inverse problem of potential theory*, Proc. Amer. Math. Soc. **116** (1992), 1097–1100.

[140] H. Shahgholian, *On the Newtonian potential of heterogeneous ellipsoids*, SIAM J. Math. Anal. **24** (1993), 341–364.

[141] H. Shahgholian, *Quadrature surfaces as free boundaries* Ark. Mat. **32** (1994), 475–492.

[142] H. Shahgholian, *Existence of quadrature surfaces for positive measures with finite support* Potential Analysis **3** (1994), 245–255.

[143] H.S. Shapiro, *Domains allowing exact quadrature identities for harmonic functions – an approach based on p.d.e.*, Anniversary volume on approximation theory and functional analysis (Oberwolfach, 1983), 335–354, Internat. Schriftenreihe Numer. Math., 65, Birkhäuser, Basel-Boston, 1984.

[144] H.S. Shapiro, *Global aspects of the Cauchy's problem for the Laplace operator*, Geometrical and algebraical aspects in several complex variables (Cetraro, 1989), 309–324, Sem. Conf., 8, EditEl, Rende, 1991.

[145] H.S. Shapiro, *Unbounded quadrature domains*, Complex analysis, I (College Park, Md., 1985–86), 287–331, Lecture Notes in Math., 1275, Springer-Verlag, Berlin 1987.

[146] H.S. Shapiro, *The Schwarz function and its generalization to higher dimensions*, Uni. of Arkansas Lect. Notes Math. Vol. 9, Wiley, New York, 1992.

[147] H.S. Shapiro, *Quasi-balayage and a priori estimates for the Laplace operator*, Multivariate approximation (Witten-Bommerholz 1996), 203–230, 231–254, Math. Res. **101**, Akademie Verlag, Berlin, 1997.

[148] H.S. Shapiro, C. Ullemar, *Conformal mappings satisfying certain extremal properties, and associated quadrature identities*, Research bulletin TRITA-MAT-1981-6, Royal Institute of Technology, 40 pp.

[149] T. Sjödin, *Mother body for the ellipse*, manuscript.

[150] T. Sjödin, *Quadrature identities and deformation of quadrature domains*, this volume.

[151] B. Sternin, V. Shatalov, *On the problem of balayage in the space* $\mathbf{R}^n$, (Russian) Mat. Zametki **54** (1993), 90–112, 160; translation in Math. Notes **54** (1993), 1246–1260.

[152] S.A. Stewart, D. Xia, *A class of subnormal operators with finite rank self-commutators*, Integral Equations Operator Theory **44** (2002), 370–382.

[153] F.R. Tian, *A Cauchy integral approach to Hele-Shaw flow problems with a free boundary: the case of zero surface tension*, Arch. Rational Mech. Anal. **135** (1996), 175–195.

[154] C. Ullemar, *Symmetric plane domains satisfying two-point quadrature identities for analytic functions*, Research bulletin TRITA-MAT-1977-24, Royal Institute of Technology, 23 pp.

[155] C. Ullemar, *A uniqueness theorem for domains satisfying a quadrature identity for analytic functions*, Research bulletin TRITA-MAT-1980-37, Royal Institute of Technology, 59 pp.

[156] A.N. Varchenko, P.I. Etingof, *Why the Boundary of a Round Drop Becomes a Curve of Order Four*, American Mathematical Society AMS University Lecture Series, Volume 3, Providence, Rhode Island 1992.

[157] D. Xia, *Hyponormal operators with finite rank self-commutators and quadrature domains*, J. Math. Anal. Appl. **203** (1996), 540–559.

[158] D. Xia, *On pure subnormal operators with finite rank self-commutators and related operator-tuples*, Integral Equations Operator Theory **24** (1996), 106–125.

[159] D. Xia, *Trace formulas for some operators related to quadrature domains in Riemann surfaces*, Integral Equations Operator Theory **47** (2003), 123–130.

[160] D.V. Yakubovich, *Subnormal operators of finite type, II Structure theorems*, Rev. Mat. Iberoamericana **14** (1998), 623–681.

[161] D.V. Yakubovich, *A note on hyponormal operators and associated with quadrature domains*, Operator theory, system theory and related topics, Beer-Sheva/Rehovot, 1997), 513–525, Oper. Theory Adv. Appl., **123**, Birkhäuser, Basel, 2001.

[162] L. Zalcman, *Some inverse problems of potential theory*, Contemp. Math. **63** (1987), 337–350.

[163] D. Zidarov, *Inverse Gravimetric Problem in Geoprospecting and Geodesy*, Elsevier, Amsterdam, 1990. (First edition 1968, in Russian.).

Björn Gustafsson and Harold S. Shapiro
Department of Mathematics
The Royal Institute of Technology
S-10044 Stockholm, Sweden
e-mail: gbjorn@math.kth.se
e-mail: shapiro@math.kth.se

Operator Theory:
Advances and Applications, Vol. 156, 27–59

Recent Progress and Open Problems in the Bergman Space

Alexandru Aleman, Håkan Hedenmalm and Stefan Richter

In celebration of Harold Seymour Shapiro's 75-th birthday

Abstract. The aim of this work is to provide a survey of interesting open problems in the theory of the Bergman spaces.

1. Initial remarks

The following text is a modified and updated version of the problem collection [40], which was written in 1993 but became publicly available only in 1995. It was a survey of various open problems; a general survey of the field was provided in [42, 43] in 1998, written in 1995 and 1996, respectively. Since then, a number of new developments have taken place, which in turn have led to new questions. We feel it is time to update the problem collection. Most of the problems we mention make sense in the context of p-th power Bergman spaces $L^p_a(\mathbb{D})$, for $0 < p < +\infty$; the reason why we stick to the Hilbert space case $p = 2$ is the simplicity of the presentation. Much of the background material for this survey can be found in the two recent books on Bergman spaces, [46] and [25].

2. The basic projects

Let $L^2_a(\mathbb{D})$ be the usual Bergman space of square area integrable analytic functions on the open unit disk $\mathbb{D}$, with norm

$$\|f\|_{L^2} = \left\{ \int_{\mathbb{D}} |f(z)|^2 \mathrm{d}S(z) \right\}^{1/2}.$$

The second-named author wishes to thank the Göran Gustafsson Foundation for generous support.

Here, dS denotes area measure in $\mathbb{C}$, normalized by a constant factor:

$$dS(z) = dx dy/\pi, \qquad z = x + i y.$$

A closed subspace J of $L^2_a(\mathbb{D})$ is said to be z-invariant, or simply invariant, provided the product zf belongs to J whenever $f \in J$. Here, we use the standard notation z for the coordinate function:

$$z(\lambda) = \lambda, \qquad \lambda \in \mathbb{D}.$$

A sequence $A = \{a_j\}_j$ of points in $\mathbb{D}$, is said to be an $L^2_a(\mathbb{D})$ zero sequence if there exists a function in $L^2_a(\mathbb{D})$ that vanishes precisely on the sequence A, counting multiplicities.

Three important projects for this space are as follows.

Problem 2.1. Characterize the invariant subspaces of $L^2_a(\mathbb{D})$.

Problem 2.2. Characterize the $L^2_a(\mathbb{D})$ zero sequences.

Problem 2.3. Find an effective factorization of the functions in $L^2_a(\mathbb{D})$.

By an effective factorization we mean one that is in some sense equivalent to that of the Hardy spaces, where Blaschke products, singular inner functions, and outer functions are involved.

Of the above three problems, the second and third ones are more likely to find definite answer than the first one. In fact, from one point of view, Problem 2.1 is as difficult as the famous invariant subspace problem in Hilbert space. Indeed, it is shown [50] how to apply the the dilation theory of Apostol, Bercovici, Foiaş, and Pearcy [12] to obtain the following: If we could show that given two z-invariant subspaces I, J in $L^2_a(\mathbb{D})$, with $I \subset J$, and $\dim(J \ominus I) = +\infty$, there exists another invariant subspace K, other than I and J, but contained in J and containing I, then every bounded linear operator on a separate Hilbert space must have a nontrivial invariant subspace. And it is understood that the invariant subspace problem for Hilbert space is really difficult. However, there are plenty of more reasonable subquestions regarding the invariant subspace lattice for $L^2_a(\mathbb{D})$. For instance, we might be better able to characterize an invariant subspace if we know something about its so-called weak spectrum (see [37, 66] for a definition). As an example of this, we mention that Aharon Atzmon [14, 15] has obtained a complete description of invariant subspaces in $L^2_a(\mathbb{D})$ with one-point spectra, also in wide classes of radially weighted Bergman spaces on $\mathbb{D}$. Another question which is tractable is to ask for a description of the maximal invariant subspaces in $L^2_a(\mathbb{D})$; see Section 10 for details.

There has been some progress on Problem 2.2. Charles Horowitz [54] obtained several interesting results. For instance, he proved that there are $L^2_a(\mathbb{D})$ zero sequences $A = \{a_j\}_j$ of non-Blaschke type, that is, having

$$\sum_j (1 - |a_j|) = +\infty,$$

and that every subsequence of a zero sequence is a zero sequence as well. He also showed that the union of two zero sequences for $L^2_a(\mathbb{D})$ need not be a zero sequence. Another important feature of the zero sequences that was apparently known before Horowitz' work is the angular dependency: inside any given Stolz angle, the zero set must meet the Blaschke condition, although it does not have to be met globally [72]. Boris Korenblum [57] (see also [58]) found a characterization of the zero sequences for the larger topological vector space $A^{-\infty}$ of holomorphic functions f in $\mathbb{D}$ with the growth bound

$$|f(z)| \leq \frac{C(f)}{(1-|z|)^N}, \qquad z \in \mathbb{D},$$

for some positive real number N, and a positive constant $C(f)$ that may depend on the given function f. The description is in terms of Blaschke sums over star domains formed as unions of Stolz angles, as compared with the logarithmic entropy of the collection of the vertices of the Stolz angles on the unit circle. As a step toward the characterization of the zero sequences for $A^{-\infty}$, Korenblum obtains estimates which apply to the Bergman space $L^2_a(\mathbb{D})$, but there is a substantial gap in the constants, which cannot be brought down to be smaller than a factor of 2 with his methods. In another vein, Emile LeBlanc [61] and Gregory Bomash [18] obtained probabilistic conditions on zero sets. Kristian Seip found a way to almost bridge the gap between the Korenblum's necessary and sufficient conditions for a sequence to be the zero set of of a Bergman space function, which allowed him to obtain a complete description of the sampling and interpolating sequences [69]; the main facts which connect these classes of sequences are as follows: interpolation implies "uniform zero sequence under Möbius translations", whereas sampling means "uniform non-zero sequence under Möbius translations". By sharpening his methods further, Seip later obtained a description of the zero sequences for $L^2_a(\mathbb{D})$ of Korenblum type, where a small gap still remained [70]. A complete characterization of the zero sequences for $L^2_a(\mathbb{D})$ remains elusive. A slightly different approach to this theorem of Seip is supplied in the book of Hedenmalm, Korenblum, and Zhu [46, Ch. 4].

As for Problem 2.3, the method of *extremal functions*, or in other words, *inner divisors*, has met with great success. The inner divisors constitute a modification of the classical inner functions from the Hardy space theory.

Definition 2.4. A function $\varphi \in L^2_a(\mathbb{D})$ is said to be an inner divisor for $L^2_a(\mathbb{D})$ if

$$h(0) = \int_{\mathbb{D}} h(z)|\varphi(z)|^2 \mathrm{d}S(z)$$

holds for all bounded harmonic functions h on $\mathbb{D}$.

We note here that if normalized area measure $\mathrm{d}S$ on $\mathbb{D}$ is replaced by normalized arc length measure in the above definition, we have a rather unusual, though equivalent, definition of the concept of an inner function in $H^2(\mathbb{D})$. The Hardy space $H^2(\mathbb{D})$ consists by definition of all analytic functions f in the unit disk $\mathbb{D}$

satisfying

$$\|f\|_{H^2} = \left\{ \sup_{0<r<1} \int_{-\pi}^{\pi} |f(re^{i\theta})|^2 \frac{d\theta}{2\pi} \right\}^{1/2} < +\infty.$$

In analogy with finite Blaschke products, we define the *finite zero divisors* as follows.

Definition 2.5. An inner divisor for $L^2_a(\mathbb{D})$ is said to be a finite zero divisor for $L^2_a(\mathbb{D})$ if it extends continuously to the closed unit disk $\bar{\mathbb{D}}$ and has a finite zero set in $\mathbb{D}$. If A is this finite zero set, counting multiplicities, we shall denote this function by φ_A.

It is not hard to show that if an inner divisor φ extends continuously to the closed unit disk $\bar{\mathbb{D}}$, then it can only have finitely many zeros. Indeed, one of the basic facts (see Hedenmalm's paper [34]) about inner divisors is that $|\varphi(z)| \geq 1$ holds on $\mathbb{T} = \partial\mathbb{D}$, whence the claim is immediate. Also, up to a unimodular constant multiple, any finite zero divisor is uniquely determined by its sequence of zeros. In addition, to every finite sequence A in $\mathbb{D}$, there exists a finite zero divisor φ_A vanishing precisely on A inside $\mathbb{D}$.

Definition 2.6. An inner divisor φ is said to be a zero divisor for $L^2_a(\mathbb{D})$ if it is the limit (as $N \to +\infty$), in the topology of uniform convergence on compact subsets of $\mathbb{D}$, of a sequence of finite zero divisors φ_{A_N}, with $A_1 \subset A_2 \subset A_3 \subset \ldots \mathbb{D}$.

We note in passing that a zero divisor for $L^2_a(\mathbb{D})$ is uniquely determined, up to multiplication by a unimodular constant factor, by its sequence of zeros, counting multiplicities. We shall frequently write G_A for the zero divisor associated with the zero sequence A.

Given an inner divisor φ for $L^2_a(\mathbb{D})$, we denote by Φ_φ the (potential) function

$$\Phi_\varphi(z) = \int_{\mathbb{D}} G(z,\zeta)\left(|\varphi(\zeta)|^2 - 1\right) dS(\zeta), \qquad z \in \mathbb{D};$$

here, $G(z,\zeta)$ stands for the Green function for the Laplacian Δ:

$$G(z,\zeta) = \log\left|\frac{\zeta - z}{1 - \bar{\zeta} z}\right|^2, \qquad (z,\zeta) \in \mathbb{D}^2.$$

Throughout this paper, we use the slightly nonstandard Laplacian

$$\Delta_z = \frac{1}{4}\left(\frac{\partial^2}{\partial x^2} + \frac{\partial^2}{\partial y^2}\right), \qquad z = x + iy,$$

and we regard locally integrable functions u on $\mathbb{D}$ as distributions via the dual action

$$\langle u, f \rangle = \int_{\mathbb{D}} u(z)\, f(z)\, dS(z),$$

for test functions f. The function Φ_φ solves the boundary value problem

$$\begin{cases} \Delta\Phi_\varphi(z) = |\varphi(z)|^2 - 1, & z \in \mathbb{D}, \\ \Phi_\varphi(z) = 0, & z \in \mathbb{T}, \end{cases}$$

and it is interesting to note that in terms of the function Φ_φ, the condition that φ be an inner divisor may be written in a more explicit form: $\nabla\Phi_\varphi = 0$ on $\mathbb{T}$ (in a weak sense if Φ_φ is not continuously differentiable up to the boundary $\mathbb{T}$). Here, ∇ denotes the gradient operator.

The following result was proved in [34, 38]. It proved to be the starting point for further development.

Theorem 2.7. *If φ is an inner divisor for $L^2_a(\mathbb{D})$, then the function Φ_φ meets*

$$0 \le \Phi_\varphi(z) \le 1 - |z|^2, \qquad z \in \mathbb{D},$$

and we have the isometry

$$\|\varphi f\|^2_{L^2} = \|f\|^2_{L^2} + \int_{\mathbb{D}} |f'(z)|^2 \Phi_\varphi(z)\,\mathrm{d}S(z), \tag{2.1}$$

valid for all $f \in H^2(\mathbb{D})$. As a consequence, we have

$$\|f\|_{L^2} \le \|\varphi f\|_{L^2} \le \|f\|_{H^2}, \qquad f \in H^2(\mathbb{D}).$$

Problem 2.8. Does the isometry (2.1) in Theorem 2.7 extend to all $f \in L^2_a(\mathbb{D})$, with the understanding that if one of the sides equals $+\infty$, then so does the other? If not, then for which inner divisors is this so?

It is possible to prove that the isometry (2.1) holds for zero divisors φ. However, it turns out that the first question of Problem 2.8 has a negative answer in general; see Borichev's and Hedenmalm's paper [20] for a (complicated) counterexample.

Theorem 2.7 has the following consequence (see [34, 38]).

Corollary 2.9. *Let A be a zero sequence for the space $L^2_a(\mathbb{D})$. Then the zero divisor φ_A possesses no other zeros in $\mathbb{D}$ than those of the sequence A. Furthermore, it has the property that every function $f \in L^2_a(\mathbb{D})$ that vanishes on A admits a factoring $f = \varphi_A\, g$, with $g \in L^2_a(\mathbb{D})$, and $\|g\|_{L^2} \le \|f\|_{L^2}$.*

This means that although the zero divisors are not isometric divisors as in the Hardy space setting, at least they are *contractive* divisors.

3. Properties of zero divisors

It is a consequence of Theorem 2.7 that if φ is an inner divisor for $L^2_a(\mathbb{D})$, we have

$$\|f\|_{L^2} \le \|\varphi f\|_{L^2}, \qquad f \in H^2(\mathbb{D}),$$

which may be written as $1 \prec \varphi$, in the notation introduced by Korenblum [59]. The precise definition of $\varphi \prec \psi$, for $\varphi, H \in L^2_a(\mathbb{D})$, is

$$\|\varphi f\|_{L^2} \le \|\psi f\|_{L^2}, \qquad f \in H^\infty(\mathbb{D}).$$

The zero divisor for the empty zero sequence is $\varphi_\emptyset = 1$, so we may interpret the statement $1 \prec \varphi_A$ which follows from Corollary 2.9 (or Theorem 2.7) as $\varphi_\emptyset \prec \varphi_A$. Here, as always, φ_A is the zero divisor associated with a zero sequence A. Maybe

$\varphi_\emptyset \prec \varphi_A$ should be thought of as a consequence of the fact that $\emptyset \subset A$ holds for all A? This hints that the following may be true.

Theorem 3.1. (Hedenmalm, Jakobsson, Shimorin) *If A, B are two zero sequences for $L^2_a(\mathbb{D})$ having $A \subset B$, then $\varphi_A \prec \varphi_B$.*

This result was conjectured by Hedenmalm in 1992, and finally proved by Hedenmalm, Jakobsson, and Shimorin in [47, 48] (see also [51]). It turns out that it is related to a certain maximum principle for biharmonic operators on negatively curved surfaces.

We write here Φ_A instead of Φ_{φ_A}, for a given zero sequence A. In view of Theorem 2.7, Theorem 3.1 is a consequence of the following result.

Theorem 3.2. (Hedenmalm, Jakobsson, Shimorin) *If A, B are two finite zero sequences for $L^2_a(\mathbb{D})$ having $A \subset B$, then $\Phi_A \le \Phi_B$ holds on $\mathbb{D}$.*

Remark 3.3. Once Theorem 3.2 has been obtained, a limit process argument asserts that

$$\Phi_A(z) \le \Phi_B(z), \qquad z \in \mathbb{D},$$

holds for general Bergman zero sequences A, B with $A \subset B$.

The following result connects Theorem 3.2 with Problem 2.8.

Proposition 3.4. *The isometry*

$$\|\varphi_A f\|^2_{L^2} = \|f\|^2_{L^2} + \int_{\mathbb{D}} |f'(z)|^2 \Phi_A(z)\, \mathrm{d}S(z)$$

holds for all holomorphic functions f on $\mathbb{D}$, and all zero sequences A in $\mathbb{D}$ for the space $L^2_a(\mathbb{D})$.

Proof. We first treat the case when the sequence A is finite. If f is analytic on $\mathbb{D}$, and r has $0 < r < 1$, consider the dilation f_r of f,

$$f_r(z) = f(rz), \qquad z \in \mathbb{D},$$

which clearly belongs to the space $H^2(\mathbb{D})$. By the isometry of Theorem 1.6, we have

$$\|\varphi_A f_r\|^2_{L^2} = \|f_r\|^2_{L^2} + \int_{\mathbb{D}} |f_r'(z)|^2 \Phi_A(z)\, \mathrm{d}S(z).$$

Since A is finite, we know that φ_A is bounded away from 0 and ∞ in a small neighborhood of the circle $\mathbb{T}$ [34], and by Theorem 2.7, $\Phi_A \ge 0$. Therefore, if we let r tend to 1 in the above identity, with the understanding that if one side takes the value $+\infty$, then so does the other, we obtain in the limit

$$\|\varphi_A f\|^2_{L^2} = \|f\|^2_{L^2} + \int_{\mathbb{D}} |f'(z)|^2 \Phi_A(z)\, \mathrm{d}S(z).$$

We turn to the remaining case of infinite A. We then write $A = \{a_j\}_{j=1}^{\infty}$, and denote by A_N the finite subsequence $\{a_j\}_{j=1}^{N}$. By the above argument, we have the isometry

$$\|\varphi_{A_N} f\|_{L^2}^2 = \|f\|_{L^2}^2 + \int_{\mathbb{D}} |f'(z)|^2 \Phi_{A_N}(z)\, \mathrm{d}S(z), \qquad f \in \mathcal{O}(\mathbb{D}), \tag{3.1}$$

for all positive integers N, where $\mathcal{O}(\mathbb{D})$ denotes the Fréchet space of all holomorphic functions on $\mathbb{D}$. We write $g = \varphi_A\, f$, and apply (3.1) to the function g/φ_{A_N}, to get

$$\|g\|_{L^2}^2 = \|g/\varphi_{A_N}\|_{L^2}^2 + \int_{\mathbb{D}} |\left(g/\varphi_{A_N}\right)'(z)|^2 \Phi_{A_N}(z)\, \mathrm{d}S(z).$$

Letting $N \to +\infty$, an application of Fatou's lemma yields

$$\|g/\varphi_A\|_{L^2}^2 + \int_{\mathbb{D}} |\left(g/\varphi_A\right)'(z)|^2 \Phi_A(z)\, \mathrm{d}S(z) \le \|g\|_{L^2}^2.$$

Remembering that g was the function $\varphi_A\, f$, the inequality

$$\|f\|_{L^2}^2 + \int_{\mathbb{D}} |f'(z)|^2 \Phi_A(z)\, \mathrm{d}S(z) \le \|\varphi_A f\|_{L^2}^2 \tag{3.2}$$

follows.

We now proceed to obtain the claimed isometry. By the above, all we need to do is to obtain the reverse inequality in (3.2). By Theorem 3.2 and the monotone convergence theorem, the right hand side of the identity (3.1) converges to

$$\|f\|_{L^2}^2 + \int_{\mathbb{D}} |f'(z)|^2 \Phi_A(z)\, \mathrm{d}S(z)$$

as $N \to +\infty$, and by Fatou's lemma,

$$\|\varphi_A f\|_{L^2} \le \limsup_{N\to+\infty} \|\varphi_{A_N} f\|_{L^2}, \qquad f \in \mathcal{O}(\mathbb{D}).$$

We conclude that

$$\|\varphi_A f\|_{L^2}^2 \le \|f\|_{L^2}^2 + \int_{\mathbb{D}} |f'(z)|^2 \Phi_A(z)\, \mathrm{d}S(z), \qquad f \in \mathcal{O}(\mathbb{D}).$$

The proof of Proposition 3.4 is complete. □

Remark 3.5. If we denote by $\mathcal{H}(A)$ the Hilbert space of holomorphic functions in $\mathbb{D}$ with norm

$$\|f\|_{\mathcal{H}(A)}^2 = \|f\|_{L^2}^2 + \int_{\mathbb{D}} |f'(z)|^2 \Phi_A(z)\, \mathrm{d}S(z),$$

we may reformulate the assertion of Proposition 3.4 as follows: if f is holomorphic in $\mathbb{D}$, and A is a zero sequence for $L_a^2(\mathbb{D})$, then $\varphi_A f \in L_a^2(\mathbb{D})$ if and only if $f \in \mathcal{H}(A)$, and

$$\|f\|_{\mathcal{H}(A)} = \|\varphi_A f\|_{L^2}.$$

4. Connections with potential theory and partial differential equations

It were Peter Duren, Dmitry Khavinson, Harold Shapiro, and Carl Sundberg [22, 23] who found the connection between the potential function Φ_φ for a given inner divisor and the biharmonic Green function for the disk $\mathbb{D}$. The biharmonic Green function is defined by the expression

$$\Gamma(z,\zeta) = |z-\zeta|^2 G(z,\zeta) + \left(1-|z|^2\right)\left(1-|\zeta|^2\right), \qquad (z,\zeta)\in\mathbb{D}^2,$$

and a calculation shows that it is positive on the bidisk $\mathbb{D}\times\mathbb{D}$. Here, $G(z,\zeta)$ is the usual Green function for Δ:

$$G(z,\zeta) = \log\left|\frac{\zeta - z}{1-\bar\zeta z}\right|^2, \qquad (z,\zeta)\in\mathbb{D}^2.$$

The biharmonic Green function solves the PDE boundary value problem

$$\begin{cases} \Delta_z^2\Gamma(z,\zeta) = \delta_\zeta(z), & z\in\mathbb{D},\\ \Gamma(z,\zeta) = 0, & z\in\mathbb{T},\\ \nabla_z\Gamma(z,\zeta) = 0, & z\in\mathbb{T}, \end{cases}$$

for a given interior point $\zeta\in\mathbb{D}$. The reason is basically that an application of Green's formula yields the representation

$$\Phi_\varphi(z) = \int_{\mathbb{D}} \Gamma(z,\zeta)\,\Delta_\zeta\left(|\varphi(\zeta)|^2-1\right)\mathrm{d}S(\zeta) = \int_{\mathbb{D}} \Gamma(z,\zeta)\,|\varphi'(\zeta)|^2\,\mathrm{d}S(\zeta) \geq 0.$$

The way to obtain Theorem 3.2 (and hence Theorem 3.1 as well) in a similar manner would be to take advantage the positivity of a similar weighted biharmonic operator on $\mathbb{D}$, which corresponds to the bilaplacian on a certain Riemann surface which we map to the unit disk via conformal mapping. The branch points for the Riemann surface correspond to the zeros of the finite zero divisor φ.

Theorem 4.1. (Hedenmalm, Jakobsson, Shimorin) *The Green function for the singular fourth order elliptic operator $\Delta|\varphi|^{-2}\Delta$ is positive on $\mathbb{D}\times\mathbb{D}$ for every finite zero divisor φ for the space $L^2_a(\mathbb{D})$. Here, we mean by the Green function the solution $\Gamma_{|\varphi|^2}(z,\zeta)$ to the problem*

$$\begin{cases} \Delta_z|\varphi(z)|^{-2}\Delta_z\Gamma_{|\varphi|^2}(z,\zeta) = \delta_\zeta(z), & z\in\mathbb{D},\\ \Gamma_{|\varphi|^2}(z,\zeta) = 0, & z\in\mathbb{T},\\ \nabla_z\Gamma_{|\varphi|^2}(z,\zeta) = 0, & z\in\mathbb{T}. \end{cases}$$

We supply the argument connecting Theorems 3.2 and 4.1. Let A and B be two finite sequences of points in the disk, having $A\subset B$. The difference function $\Phi_{B,A} = \Phi_B - \Phi_A$ solves the overdetermined problem

$$\begin{cases} \Delta\Phi_{B,A}(z) = |\varphi_B(z)|^2 - |\varphi_A(z)|^2, & z\in\mathbb{D},\\ \Phi_{B,A}(z) = 0, & z\in\mathbb{T},\\ \nabla\Phi_{B,A}(z) = 0, & z\in\mathbb{T}, \end{cases}$$

and as we divide both sides of the top line by $|\varphi_A(z)|^2$, we get

$$|\varphi_A(z)|^{-2}\Delta\Phi_{B,A}(z) = |\varphi_B(z)/\varphi_A(z)|^2 - 1, \qquad z \in \mathbb{D}.$$

Since we have overdetermined boundary values, we are at liberty to apply another Laplacian, which results in

$$\begin{cases} \Delta\,|\varphi_A(z)|^{-2}\Delta\,\Phi_{B,A}(z) = \left|\left(\dfrac{\varphi_B}{\varphi_A}\right)'(z)\right|^2, & z \in \mathbb{D}, \\ \Phi(z) = 0, & z \in \mathbb{T}, \\ \nabla\Phi(z) = 0, & z \in \mathbb{T}. \end{cases}$$

Note here that we used the fact that the quotient φ_B/φ_A is holomorphic on $\mathbb{D}$. Finally, we see that in terms of the weighted biharmonic Green function $\Gamma_{|\varphi_A|^2}(z,\zeta)$, we may express $\Phi_{B,A}$ as

$$\Phi_{B,A}(z) = \int_D \Gamma_{|\varphi_A|^2}(z,\zeta)\left|\left(\frac{\varphi_B}{\varphi_A}\right)'(\zeta)\right|^2 \mathrm{d}S(\zeta), \qquad z \in \mathbb{D},$$

and the positivity of $\Phi_{B,A}$ is now immediate.

The connection between this weighted situation and general hyperbolic geometry is explained extensively in [48].

The first results suggesting the validity of Theorem 4.1 were calculations made in [39] and [40], dealing with a multiple zero at the origin and a single zero at an arbitrary point of $\mathbb{D}$, respectively. Then a multiple zero arbitrarily located in the disk was considered, based on the computations of Hansbo [31]. This convinced us the claim must be true, but it took a lot of additional effort to carry out the proof, which involves the construction of a Hele-Shaw flow [51], to be able to take advantage of an additional invariance property. We recall that the finite zero divisor φ has the property

$$h(0) = \int_{\mathbb{D}} h(z)\,|\varphi(z)|^2\,\mathrm{d}S(z),$$

for all bounded harmonic functions h on $\mathbb{D}$. Now, let $0 < t < 1$, and suppose $D(t)$ is a subdomain of $\mathbb{D}$ containing the origin for which

$$t\,h(0) = \int_{D(t)} h(z)\,|\varphi(z)|^2\,\mathrm{d}S(z),$$

holds for all bounded harmonic functions h on $D(t)$; note that in the limit, we should get $D(1) = \mathbb{D}$. It turns out that this property uniquely determines the domain $D(t)$, and that these domains increase with the parameter t in a predictable fashion (at least this was proved in a slightly smoother situation in [51]). This offers the opportunity to vary t, which is done quite successfully in [48]. An important question remained.

Conjecture 4.2. For a finite zero divisor φ, the weighted biharmonic Green function has the property

$$\frac{\partial}{\partial n(z)} \Delta_z \Gamma_{|\varphi|^2}(z,\zeta) \geq 0, \qquad (z,\zeta) \in \mathbb{T} \times \mathbb{D},$$

the normal derivative being taken in the exterior direction.

It turns out that if this claim could be verified, then the following quite pleasant property of the functions φ_B/φ_A would follow.

Conjecture 4.3. Let A, B be two finite zero divisors, with $A \subset B$. Then the argument of $\varphi_B(z)/\varphi_A(z)$ increases monotonically as z is moved counterclockwise along the unit circle $\mathbb{T}$. In particular, if $B \setminus A$ is a single point, then φ_B/φ_A is star-shaped univalent function on $\mathbb{D}$.

We turn to a possible generalization of Theorem 4.1 in a different direction. After all, it would be valuable to be able have the entire factorization theory of $L^2_a(\mathbb{D})$ transferred to the weighted Bergman spaces $L^2_a(\mathbb{D}, \omega_\alpha)$, defined as consisting of analytic functions f in $\mathbb{D}$ with

$$\|f\|_{L^2(\omega_\alpha)} = \left(\int_{\mathbb{D}} |f(z)|^2 \, \omega_\alpha(z) \, \mathrm{d}S(z)\right)^{1/2},$$

where the weight is

$$\omega_\alpha(z) = (\alpha+1)\left(1-|z|^2\right)^\alpha,$$

and $-1 < \alpha < +\infty$. It follows from the results in [48] that we have the same factorization theory in these spaces also with respect to Korenblum domination as long as $-1 < \alpha \leq 0$. Moreover, it is known from the counterexample of Hedenmalm and Zhu [52] that this is not so for $1 < \alpha < +\infty$. There is abundant evidence suggesting that on the remaining interval we also have the same factorization theory involving Korenblum domination (but not going so far as Conjecture 4.3, though), but so far this has not been substantiated. The following is what we need to conclude the issue.

Conjecture 4.4. Let ω be a positive C^∞-smooth weight in $\bar{\mathbb{D}}$, with the reproducing property

$$h(0) = \int_{\mathbb{D}} h(z) \, \omega(z) \, \mathrm{d}S(z),$$

for all bounded harmonic functions h in $\mathbb{D}$. Suppose

$$\Delta \log \omega(z) \geq \Delta \log\left(1-|z|^2\right) = -\frac{1}{\left(1-|z|^2\right)^2}, \qquad z \in \mathbb{D}.$$

Then the Green function $\Gamma_\omega(z,\zeta)$ for the weighted biharmonic operator $\Delta\omega^{-1}\Delta$ is positive on $\mathbb{D} \times \mathbb{D}$.

In support of the conjecture, we only mention that it is valid for the weight

$$\omega(z) = \omega_1(z) = 2\,(1-|z|^2), \qquad z \in \mathbb{D}.$$

Proposition 4.5. *The Green function $\Gamma^1(z,\zeta)$ for the weighted biharmonic operator $\Delta(1-|z|^2)^{-1}\Delta$ has the explicit form*

$$\Gamma^1(z,\zeta) = \left(|z-\zeta|^2 - \frac{1}{4}\left|z^2-\zeta^2\right|^2\right) G(z,\zeta) + \frac{1}{8}(1-|z|^2)(1-|\zeta|^2)$$
$$\times\left(7-|z|^2-|\zeta|^2-|\zeta z|^2-4\,\mathrm{Re}\,(\bar\zeta z)-2\,(1-|z|^2)(1-|\zeta|^2)\,\mathrm{Re}\,\frac{1+\bar\zeta z}{1-\bar\zeta z}\right),$$

for $(z,\zeta)\in\mathbb{D}^2$.

It is quickly verified that $\Gamma^1>0$ on $\mathbb{D}^2$. However, for fixed $\zeta\in\mathbb{D}$, the function $\Gamma^1(\cdot,\zeta)$ is very flat at the boundary $\mathbb{T}$; it satisfies

$$\Gamma^1(z,\zeta) = O\big((1-|z|)^3\big) \qquad \text{as} \quad |z|\to 1.$$

The verification of the above explicit formula for $\Gamma^1(z,\zeta)$ obtained by Hedenmalm was based on several rather tedious computations, which were not laid out in [40]. In order to derive the formula for $\Gamma^1(z,\zeta)$, the following identity was used:

$$\Delta_z\Delta_\zeta\,\Gamma^1(z,\zeta) = (1-|z|^2)\,\delta_0(z-\zeta) - 2\,(1-|z|^2)(1-|\zeta|^2)\,Q^1(z,\zeta),$$

where $Q^1(z,\zeta)$ is the reproducing kernel for the Hilbert space $L^2_h(\mathbb{D},\omega_1)$ of harmonic functions in $\mathbb{D}$ with norm

$$\|f\|^2_{L^2(\omega_1)} = \int_{\mathbb{D}} |f(z)|^2\,\omega_1(z)\,\mathrm{d}S(z) = 2\int_{\mathbb{D}} |f(z)|^2\,(1-|z|^2)\,\mathrm{d}S(z) < +\infty.$$

In addition, it was necessary to use that since the weight ω_1 is radial, we have the following nice relation between the analytic and harmonic Bergman kernels:

$$Q^1(z,\zeta) = 2\,\mathrm{Re}\,\frac{1}{(1-z\bar\zeta)^3} - 1.$$

More recently, Rachel Weir [84] found another way to derive the formula for $\Gamma^1(z,\zeta)$, which in addition produces an explicit formula for the Green function for the weighted biharmonic operator $\Delta(1-|z|^2)^{-2}\Delta$.

In the forthcoming paper [49], Hedenmalm and Yolanda Perdomo use Proposition 4.5 as a tool to obtain the following result.

Theorem 4.6. *Let ω be a positive C^∞-smooth weight in $\bar{\mathbb{D}}$, with*

$$\Delta\log\omega(z) \ge \Delta\log\left(1-|z|^2\right) = -\frac{1}{\left(1-|z|^2\right)^2}, \qquad z\in\mathbb{D}.$$

Then the reproducing kernel $K_\omega(z,\zeta)$ for the weighted Bergman space $L^2_a(\mathbb{D},\omega)$ is zero-free throughout

$$(\bar{\mathbb{D}}\times\mathbb{D})\cup(\mathbb{D}\times\bar{\mathbb{D}}).$$

Problem 4.7. How sharp is the above result? By how much may the assumptions be weakened?

5. Generators of zero-based subspaces

As before, we write φ_A for the zero divisor associated with a zero sequence A.

Problem 5.1. Does φ_A generate

$$\mathcal{I}(A) = \{f \in L^2_a(\mathbb{D}) : f = 0 \text{ on } A\}$$

as an invariant subspace? In other words, do the functions φ_A, $z\varphi_A, z^2\varphi_A, \dots$ span a dense subspace of $\mathcal{I}(A)$?

This problem was solved by Aleman, Richter, and Sundberg in [10] in the affirmative. The main tool was a dilation property of the biharmonic Green function Γ, which showed that

$$\varphi_A f_r \to \varphi_A f \qquad \text{as} \quad r \to 1^-,$$

in the norm of $L^2_a(\mathbb{D})$, for any $f \in L^2_a(\mathbb{D})$ with $\varphi_A f \in L^2_a(\mathbb{D})$ as well. Here, $f_r(z) = f(rz)$ is the dilation of f. This way to do it is essentially the approach that is suggested by Problem 5.4 and Remark 5.5 (see below). This was later worked out by Ali Abkar [2] (see also [3]). The Aleman-Richter-Sundberg theorem [10] is actually much more general, and applies to all invariant subspaces rather than just the zero-based ones. An elegant operator-theoretic proof of this abstract Beurling-type theorem was later found by Serguei Shimorin [80, 82].

A further approach to solving Problem 5.1 is to try to apply Theorem 4.1, which leads to the following question.

Problem 5.2. Is the function $\varphi_{A\cup\{0\}}/\varphi_A$ always bounded in the disk $\mathbb{D}$?

This problem was solved in the affirmative by Shimorin [79]; Hedenmalm later modified the approach to obtain a sharp off-diagonal estimate of weighted Bergman kernels, with logarithmically subharmonic weights that reproduce for the origin. Actually, the function

$$\left|\frac{\varphi_{A\cup\{\alpha\}}(z)}{\varphi_A(z)}\right| \le 1 + \frac{1-|\alpha|^2}{|1-\bar{\alpha}z|} \le 3$$

for each $\alpha \in \mathbb{D}$, as was shown by Aleman and Richter [8]. Note that this gives the bound 2 for $\alpha = 0$. A constant bound that depends on α is supplied by

$$\left|\frac{\varphi_{A\cup\{\alpha\}}(z)}{\varphi_A(z)}\right| \le \frac{2+|\alpha|}{\sqrt{2-|\alpha|^2}} \le 3,$$

which also follows from the work in [8]. This gives the optimal bound $\sqrt{2}$ for $\alpha = 0$. See also Section 12 for further details on the methods.

By Theorem 4.1, we have $\varphi_A \prec \varphi_{A\cup\{0\}}$, which entails that

$$|z\varphi_A(z)| \le |\varphi_{A\cup\{0\}}(z)|, \qquad z \in \mathbb{D},$$

holds. To see this, check it for finite sequences A (the argument for this is analogous to what was used in Proposition 1.3 [34]), and we then approximate a general zero sequence with finite subsequences. It follows that $\varphi_{A\cup\{0\}}/\varphi_A$ belongs to the

Nevanlinna class of holomorphic quotients of bounded analytic functions. But, by Shimorin's result, we know it is bounded itself.

Remark 5.3. We should shed some light on the connection between Problems 5.1 and 5.2. To do this, we assume for simplicity that the point 0 does not belong to the given zero sequence A, and denote by Y_n the orthogonal projection onto $\mathcal{I}(A \cup O_n)$ of the function z^n. Here, O_n stands for the the sequence that consists of n copies of the point 0. The assumption that 0 not belong to A prevents Y_n from collapsing to 0. We now claim that the functions Y_n, $n = 0, 1, 2, \ldots$, span a dense subspace of $\mathcal{I}(A)$. To this end, suppose $f \in \mathcal{I}(A)$ is orthogonal to all the functions Y_n. It is convenient here to introduce $X_n = z^n - Y_n$, which for each n is orthogonal to $\mathcal{I}(A \cup O_n)$, by the way we defined the element Y_n. If we knew f belonged to $\mathcal{I}(A \cup O_n)$ for some particular n, we would then also have $\langle f, X_n \rangle_{L^2} = 0$, and since by assumption $\langle f, Y_n \rangle_{L^2} = 0$, we see that

$$\frac{f^{(n)}(0)}{(n+1)!} = \langle f, z^n \rangle_{L^2} = 0.$$

We conclude that f must also belong to $\mathcal{I}(A \cup O_{n+1})$. The initial assumption $f \in \mathcal{I}(A \cup O_n)$ is fulfilled for $n = 0$, so by induction, f belongs to the intersection of all the spaces $\mathcal{I}(A \cup O_n)$, which is $\{0\}$. This shows that $f = 0$, and hence the claim is verified. It is known [34, 23] that the zero divisor for $A \cup O_n$ is

$$\varphi_{A \cup O_n} = Y_n / \|Y_n\|_{L^2},$$

so by the above argument, the functions $\varphi_{A \cup O_n}$, $n = 0, 1, 2, \ldots$, span a dense subspace of $\mathcal{I}(A)$. If we could only demonstrate that every $\varphi_{A \cup O_n}$ belongs to the invariant subspace generated by φ_A, this would provide an alternative route toward obtaining an affirmative solution to Problem 5.1. This is where it is good that Problem 5.2 has an affirmative solution as well. For, we then know that $\varphi_{A \cup O_{n+1}} / \varphi_{A \cup O_n}$ is bounded for each $n = 0, 1, 2, \ldots$, so that the function $\varphi_{A \cup O_n} / \varphi_A$ is bounded, too, for each $n = 0, 1, 2, \ldots$. It is now immediate that $\varphi_{A \cup O_n}$ belongs to the invariant subspace generated by φ_A, and hence φ_A generates all of $\mathcal{I}(A)$ as an invariant subspace.

We really do not understand the process of adding another zero. Nevertheless, for the function $\lambda_A(z) = \bar{\varphi}_A(0)\, \varphi_A(z)$, there is the iterative formula

$$\lambda_{A \cup \{\beta\}}(z) = L_A(z) - \frac{\lambda_A(\beta)}{\lambda_{\varphi_\beta(A)}(0)}\, \varphi_\beta'(z)\, \lambda_{\varphi_\beta(A)}(\varphi_\beta(z)),$$

where φ_β denotes the Möbius mapping

$$\varphi_\beta(z) = \frac{\beta - z}{1 - \bar{\beta} z}, \qquad z \in \mathbb{D}.$$

The starting point for the iterative process is $\lambda_\emptyset = 1$, and the formula connecting φ_A with λ_A may be written

$$\varphi_A(z) = \frac{\lambda_A(z)}{\sqrt{L_A(0)}}, \qquad z \in \mathbb{D}.$$

The next problem is also related to Problem 5.1.

Problem 5.4. Suppose ω is a continuous function on $\mathbb{D}$ which satisfies

$$0 \le \omega(z) \le C\,(1-|z|^2), \qquad z \in \mathbb{D},$$

for some positive constant C, which is super-biharmonic:

$$\Delta^2 \omega(z) \ge 0, \qquad z \in \mathbb{D}.$$

Must then the polynomials be dense in the weighted Bergman space $L^2_a(\mathbb{D}, \omega)$ of all holomorphic functions in $\mathbb{D}$ with

$$\int_{\mathbb{D}} |f(z)|^2 \omega(z)\, \mathrm{d}S(z) < +\infty\,?$$

Remark 5.5. We shall now try to indicate the relationship between Problems 5.1 and 5.4. Note first that in view of Remark 3.5, φ_A generates $\mathcal{I}(A)$ if and only if the closure of polynomials is dense in the space $\mathcal{H}(A)$ (defined in the indicated remark). By the elementary estimates

$$\frac{2}{3}\,\|f\|^2_{L^2} \le \int_{\mathbb{D}} |f'(z)|^2 (1-|z|^2)^2 \mathrm{d}S(z) \le 2\,\|f\|^2_{L^2},$$

valid for $f \in L^2_a(\mathbb{D})$ with $f(0) = 0$, we have that the norm in $\mathcal{H}(A)$ is comparable to

$$\|f\|^2_* = |f(0)|^2 + \int_{\mathbb{D}} |f'(z)|^2 \omega_A(z) \mathrm{d}S(z),$$

where we denote by ω_A the function

$$\omega_A(z) = (1-|z|^2)^2 + \Phi_A(z), \qquad z \in \mathbb{D}.$$

We see that the polynomials are dense in $\mathcal{H}(A)$ if and only if they are dense in $L^2_a(\mathbb{D}, \omega_A)$. It is now clear that φ_A generates the invariant subspace $\mathcal{I}(A)$ if and only if approximation by the polynomials is possible in $L^2_a(\mathbb{D}, \omega_A)$. The constructed function ω_A has

$$0 \le (1-|z|^2)^2 \le \omega_A(z) \le (1-|z|^2)^2 + 1 - |z|^2 \le 2\,(1-|z|^2), \qquad z \in \mathbb{D},$$

by Theorem 2.7. Moreover, since $\Delta^2 \Phi_A \ge 0$, we also have

$$\Delta \omega_A(z) = \Delta^2 \Phi_A(z) + \Delta^2 (1-|z|^2)^2 \ge \Delta^2 (1-|z|^2)^2 = 4 \ge 0, \qquad z \in \mathbb{D}.$$

Polynomial approximation problems are, generally speaking, rather difficult. It turns out that the Problem 5.4 as stated above has an affirmative answer, and the reason is that the biharmonic Green function has certain "good" dilation properties (see [3] and [2]; the problem is actually partially solved already in [10]).

Proposition 5.6 below represents our current level of understanding on the general topic of weighted polynomial approximation. Given a positive continuous weight function ω on the unit disk $\mathbb{D}$, having

$$\int_{\mathbb{D}} \omega(z)\, \mathrm{d}S(z) < +\infty, \tag{5.1}$$

we denote by $L^2_h(\mathbb{D},\omega)$ the Hilbert space of harmonic functions f on $\mathbb{D}$ having

$$\|f\|_{L^2(\omega)} = \left(\int_{\mathbb{D}} |f(z)|^2\omega(z)\,\mathrm{d}S(z)\right)^{1/2} < +\infty.$$

It has been known for a long time [64, p. 131], [56, p. 343], that the analytic polynomials are dense in $L^2_a(\mathbb{D},\omega)$ for radial weights ω. The corresponding statement is also true for the space $L^2_h(\mathbb{D},\omega)$, and moreover, we can get the result for weights that do not deviate too much from radial weights. To obtain such a result, it is useful to consider for a parameter $0<\lambda<1$ and a function $f\in L^2_h(\mathbb{D},\omega)$ the dilation f_λ of f:

$$f_\lambda(z) = f(\lambda z), \qquad z\in\mathbb{D},$$

and observe that every dilation f_λ of f is definitely approximable by harmonic polynomials (or analytic polynomials, if $f\in L^2_a(\mathbb{D},\omega)$), so that if we could show that $f_\lambda\to f$ in the norm of $L^2(\mathbb{D},\omega)$, the desired conclusion would follow. Another condition which is known to assure that we have polynomial approximation is due to Dzhrbashian [64, p.133], and requires that the weight should (almost) fall on every radius emanating from the origin. If we merge these two ideas, we obtain the following result. First, however, we need to recall some terminology: an integrable function $\nu\geq 0$ on the unit circle $\mathbb{T}$ meets the Muckenhoupt (A_2) condition provided that

$$\mathfrak{A}_2(\nu) = \sup_I\left\{|I|^{-2}\int_I \nu\,ds\int_I\frac{1}{\nu}\,ds\right\} < \infty,$$

the supremum being taken over all arcs I on $\mathbb{T}$, where ds denotes arc length measure on $\mathbb{T}$, normalized so that the total length of $\mathbb{T}$ is 1.

Proposition 5.6. *Suppose ω is a positive continuous function on the unit disk $\mathbb{D}$, which meets the integrability condition* (5.1). *Suppose, moreover, that $\{\lambda_j\}_1^\infty$ is a sequence of numbers in the interval $]0,1[$, converging to 1. For r, $0<r<1$, let $\omega_r(z)=\omega(rz)$, and for $0<r,s<1$, introduce the quantity*

$$\mathfrak{Q}[\omega](r,s) = \sup\left\{\frac{\omega(rz)}{\omega(sz)} : z\in\mathbb{T}\right\}.$$

If the weight ω satisfies

$$\limsup_{r\to 1-}\,\sup_j\,\min\left\{\mathfrak{Q}[\omega](r,\lambda_j r),\mathfrak{A}_2(\omega_r)\right\} < \infty,$$

then the dilations f_{λ_j} of f converge to f as $j\to+\infty$ in the norm of $L^2_h(\mathbb{D},\omega)$, for every $f\in L^2_h(\mathbb{D},\omega)$. As a consequence, under this condition on ω, we see that the harmonic polynomials are dense in $L^2_h(\mathbb{D},\omega)$, and the analytic polynomials are dense in $L^2_a(\mathbb{D},\omega)$.

Proof. We follow the general line of argument of [56, pp. 343–344]. Given an ε, $0<\varepsilon$, take ρ, $0<\rho<1$, so close to 1 that

$$\int_{\rho<|z|<1} |f(z)|^2\omega(z)\,\mathrm{d}S(z) < \varepsilon, \tag{5.2}$$

and

$$\int_{\lambda_j \rho<|z|<\lambda_j} |f(z)|^2 \omega(z)\, \mathrm{d}S(z) < \lambda_j^2 \varepsilon. \tag{5.3}$$

By choosing ρ possibly even closer to 1, we may assume that

$$\min\left\{\mathfrak{Q}[\omega](r,\lambda_j r), \mathfrak{A}_2(\omega_r)\right\} \le C, \qquad \rho<r<1,\ j=1,2,3,\dots,$$

for some constant C, $0<C<+\infty$. We now plan to estimate the size of

$$\int_{\rho<|z|<1} \left|f(\lambda_j z)\right|^2 \omega(z)\, \mathrm{d}S(z).$$

We fix an r with $\rho<r<1$, and note that if it is $\mathfrak{Q}[\omega](r,\lambda_j r)$ that is $\le C$, we have

$$\omega(re^{\mathrm{i}\theta}) \le C\,\omega(\lambda_j re^{\mathrm{i}\theta}),$$

and hence

$$\int_{-\pi}^{\pi} \left|f(\lambda_j re^{\mathrm{i}\theta})\right|^2 \omega(re^{\mathrm{i}\theta})\, \mathrm{d}\theta \le C \int_{-\pi}^{\pi} \left|f(\lambda_j re^{\mathrm{i}\theta})\right|^2 \omega(\lambda_j re^{\mathrm{i}\theta})\, \mathrm{d}\theta.$$

If, on the other hand, it is $\mathfrak{A}_2(\omega_r)$ that is $\le C$, then by Muckenhoupt's theorem [65, p. 223] (and the control of the constants involved [65, pp. 215, 224]), we have

$$\int_{-\pi}^{\pi} \left|f(\lambda_j re^{\mathrm{i}\theta})\right|^2 \omega(re^{\mathrm{i}\theta})\, \mathrm{d}\theta \le K(C) \int_{-\pi}^{\pi} \left|f(re^{\mathrm{i}\theta})\right|^2 \omega(re^{\mathrm{i}\theta})\, \mathrm{d}\theta,$$

for some constant $K(C)$ that only depends on C. No matter which is the case, we get

$$\int_{\rho}^{1}\int_{-\pi}^{\pi} \left|f(\lambda_j re^{\mathrm{i}\theta})\right|^2 \omega(re^{\mathrm{i}\theta})\, \mathrm{d}\theta\, r dr \le (C+K(C))\,\pi\varepsilon,$$

in view of (5.2) and (5.3). Since ρ, $0<\rho<1$, was fixed, we have that $f_{\lambda_j} \to f$ as $j\to+\infty$ uniformly on the disk $|z|<\rho$, and in particular, we can arrange so that

$$\int_{|z|<\rho} \left|f(\lambda_j z) - f(z)\right|^2 \omega(z)\, \mathrm{d}S(z) < \varepsilon,$$

for all large j, say $j \ge N(\varepsilon)$. If we combine this with the estimate of the integral on the annulus $\rho<|z|<1$, we see that

$$\int_{\mathbb{D}} |f(\lambda_j z) - f(z)|^2 \omega(z)\, \mathrm{d}S(z) < 8\,(1+C+K(C))\,\varepsilon,$$

for $j \ge N(\varepsilon)$. The assertion of the proposition is now immediate. □

6. A Carathéodory theorem for the Bergman space

Recall the statement of the famous Carathéodory theorem.

Theorem 6.1. (Carathéodory) *Every $f \in H^\infty(\mathbb{D})$ with norm ≤ 1 is the normal limit of finite Blaschke products.*

Here, we mean by a *normal limit* the limit in the topology of uniform convergence on compact subsets. The appropriate analog in a Bergman space setting turns out to be as follows.

Conjecture 6.2. Suppose f is analytic on $\mathbb{D}$ with the property

$$\int_{\mathbb{D}} h(z)\,|f(z)|^2\,\mathrm{d}S(z) \le h(0),$$

for all positive and bounded harmonic functions h in $\mathbb{D}$. Then f is the normal limit of finite zero divisors.

In the previous version of this text [40], the formulation of this conjecture was slightly different, in terms of contractive multiplicativity from $H^2(\mathbb{D})$ into $L^2_a(\mathbb{D})$. That formulation has not yet found an answer, and it is quite likely that the answer is negative. In hindsight, the above version seems much more appropriate, and *it was settled affirmatively by Shimorin in* [79]. The proof can also be found in the book [46, Ch. 3].

7. A Frostman theorem for the Bergman space?

We recall Frostman's classical theorem on approximation of inner functions by Blaschke products.

Theorem 7.1. (Frostman) *Every inner function is approximable in the norm of the space $H^\infty(\mathbb{D})$ by Blaschke products.*

Let $\mathfrak{M}(H^2, L^2_a)$ be the space of multipliers $H^2(\mathbb{D}) \to L^2_a(\mathbb{D})$, normed appropriately:

$$\|\varphi\|_{\mathfrak{M}(H^2,L^2_a)} = \sup\left\{\|\varphi f\|_{L^2} : f \in H^2(\mathbb{D}),\ \|f\|_{H^2} \le 1\right\}.$$

Note that in view of Theorem 2.7, every inner divisor is a contractive multiplier $H^2(\mathbb{D}) \to L^2_a(\mathbb{D})$. This suggests that the above multiplier norm might be the appropriate replacement of the norm in $H^\infty(\mathbb{D})$ in the Bergman space setting. The appropriate analogue of Frostman's theorem might therefore be the result of asking the following question.

Problem 7.2. Is every inner divisor for $L^2_a(\mathbb{D})$ approximable by zero divisors in the norm of $\mathfrak{M}(H^2, L^2_a)$?

So far, there appears to be no progress on this problem.

8. Korenblum's maximum principle

Conjecture 8.1. (Korenblum) There exists an absolute constant ε, $0 < \varepsilon < 2^{-1/2}$, such that if $f, g \in L^2_a(\mathbb{D})$ have $|f(z)| \le |g(z)|$ in the annulus $\varepsilon < |z| < 1$, then $\|f\|_{L^2} \le \|g\|_{L^2}$.

One can rather trivially obtain an estimate like $\|f\|_{L^2} \le C(\varepsilon)\|g\|_{L^2}$, with $C(\varepsilon)$ being a constant larger than 1 tending to 1 as $\varepsilon \to 0$. Korenblum claims that $C(\varepsilon) = 1$ is attained for some nonzero value of the parameter ε. One should view Conjecture 8.1 as a suspected property peculiar to square moduli of analytic functions. If one should try to replace this class by, for instance, the collection of exponentials of subharmonic functions, the analogous assertion that $\varphi(z) \le \psi(z)$ on the annulus $\varepsilon < |z| < 1$ should imply

$$\int_{\mathbb{D}} \exp\big(\varphi(z)\big)\, dS(z) \le \int_{\mathbb{D}} \exp\big(\psi(z)\big)\, dS(z)$$

for subharmonic functions φ, ψ fails, no matter how small the positive number ε is. This is so because one can take as $\varphi(z)$ the function $\log|z|$, and as $\psi(z)$ the function that is the maximum of $\log|z|$ and the constant function $\log\varepsilon$.

The condition of Conjecture 8.1 is invariant under multiplication by a bounded holomorphic function, so the assertion of Conjecture 8.1 may be rephrased as $f \prec g$.

Conjecture 8.1 was settled in 1998 by Hayman [33], with a proof that uses surprisingly simple ingredients. He was helped by Hinkkanen to improve his inner radius to $\varepsilon = \frac{1}{25}$. Hinkkanen [53] later improved this further to $\varepsilon = 0.157$.

The properties of the domination relation $\prec$ deserve to be studied in some depth.

Problem 8.2. Suppose $f, g \in L^2_a(\mathbb{D})$ have $f \prec g$ and $g \prec f$. Must then $f = \gamma g$ for a unimodular constant γ?

It turns out that Stefan Richter [67] solved this problem in the affirmative already back in 1988. We outline the argument briefly. Consider the following transform of $|f|^2 - |g|^2$:

$$B\big[|f|^2 - |g|^2\big](\lambda, \mu) = (1 - \lambda\bar{\mu})^2 \int_{\mathbb{D}} \frac{|f(z)|^2 - |g(z)|^2}{(1 - \lambda\bar{z})^2 (1 - \bar{\mu} z)^2}\, dS(z), \qquad (\lambda, \mu) \in \mathbb{D}^2.$$

We know that this function is holomorphic in λ and anti-holomorphic in μ. Moreover, $f \prec g$ and $g \prec f$ together force the equality

$$B\big[|f|^2 - |g|^2\big](\lambda, \lambda) = 0, \qquad \lambda \in \mathbb{D}.$$

This means that the holomorphic function of two variables

$$B\big[|f|^2 - |g|^2\big](\lambda, \bar{\mu})$$

vanishes along the antidiagonal, which is only possible if the function vanishes throughout $\mathbb{D}^2$. But then $|f|^2 = |g|^2$ follows, and the conclusion $f = \gamma g$, with $|\gamma| = 1$, is immediate. The above function is related to the so-called Berezin transform [46, Ch. 2].

9. Cyclic vectors and Shapiro's problem

The space $A^{-\infty}$ consists of all analytic functions f on the unit disk $\mathbb{D}$ satisfying the growth condition

$$|f(z)| \le C(f,\alpha)\,(1-|z|^2)^{-\alpha}, \qquad z \in \mathbb{D},$$

for some positive constants α and $C(f,\alpha)$. The function theory aspects of this space were illuminated extensively by Korenblum in his Acta paper [57]; one rather trivial but interesting observation is that $A^{-\infty}$ is a topological algebra with respect to pointwise multiplication and the natural injective limit topology. In his second Acta paper [58], Korenblum describes completely the closed ideals in $A^{-\infty}$. The Bergman space $L^2_a(\mathbb{D})$ is clearly a subspace of $A^{-\infty}$, but it is not an algebra. The invariant subspaces are the $L^2_a(\mathbb{D})$ analogs of the closed ideals in $A^{-\infty}$. In order to gain some understanding of invariant subspaces, the concept of a *cyclic vector* is basic.

Definition 9.1. A function $f \in L^2_a(\mathbb{D})$ is cyclic in $L^2_a(\mathbb{D})$ if the functions f,zf, $z^2f,\ldots$ span a dense subspace of $L^2_a(\mathbb{D})$.

Problem 9.2. Describe the cyclic elements of $L^2_a(\mathbb{D})$.

Clearly, a cyclic function cannot have any zeros in $\mathbb{D}$. However, this is not the full story; some additional conditions on the behavior near the unit circle must be involved.

After the efforts of Korenblum [60] and Aleman-Richter-Sundberg [10], we now know that the cyclic functions in $L^2_a(\mathbb{D})$ are precisely the so-called $L^2_a(\mathbb{D})$-outer functions, defined in terms of Korenblum domination. However, this equivalence does not help much in a practical situation, because to decide whether a function is $L^2_a(\mathbb{D})$-outer is no easier than to decide whether it is cyclic.

A natural question when one tries to attack Problem 9.2 is the following.

Problem 9.3. (Korenblum) It is known that every cyclic element of $L^2_a(\mathbb{D})$ generates a dense ideal in $A^{-\infty}$, or in other words, it is cyclic in $A^{-\infty}$. Does the converse hold, that is, if $f \in L^2_a(\mathbb{D})$ is cyclic in $A^{-\infty}$, must then f be cyclic in $L^2_a(\mathbb{D})$?

It is known (see [73]) that the answer to Problem 9.3 is yes, if we add the assumption that the function f belong to the Nevanlinna class of holomorphic quotients of bounded analytic functions. This in its turn follows rather easily from the case when f is assumed bounded. Leon Brown and Boris Korenblum [21] have obtained the considerably stronger result that if the function f belongs to a slightly smaller Bergman space $L^p_a(\mathbb{D})$, $2 < p < +\infty$, then the cyclicity of $f \in L^2_a(\mathbb{D})$ in $A^{-\infty}$ implies its cyclicity in $L^2_a(\mathbb{D})$. If a function $f \in L^2_a(\mathbb{D})$ satisfies

$$|f(z)| \ge \varepsilon\left(1-|z|^2\right)^N, \qquad z \in \mathbb{D}, \tag{9.1}$$

for some positive numbers ε, N, then f is invertible in $A^{-\infty}$, and hence cyclic in $A^{-\infty}$.

Problem 9.4. (Shapiro) Suppose $f \in L^2_a(\mathbb{D})$ satisfies (9.1). Must then f be cyclic in $L^2_a(\mathbb{D})$?

This problem was settled in the negative by Borichev and Hedenmalm [20]. The construction involved first finding harmonic functions that grow at an appropriate rate radially, and then forming the zero-free function obtained by harmonic conjugation plus exponentiation.

10. The index of an invariant subspace

The *index* of an invariant subspace $\mathcal{M}$ is defined to be the codimension of $z\mathcal{M}$ in $\mathcal{M}$, i.e., $\operatorname{ind}\mathcal{M} = \dim(\mathcal{M}/z\mathcal{M})$. It is clear that $\{0\}$ is the only subspace of index 0, and it is easy to see that every nonzero singly generated invariant subspace has index 1. Similarly, nontrivial zero based invariant subspaces have index one. In the converse direction it was shown in [10] that every index 1 invariant subspace of $L^2_a(\mathbb{D})$ is singly generated by its extremal function. It is somewhat difficult to see that there are invariant subspaces of the Bergman space of arbitrary index. This was first established in [12], and it is closely related to the connection between the invariant subspace problem and the apparent difficulty of characterizing all invariant subspaces of $L^2_a(\mathbb{D})$ that we have alluded to at the very beginning of this article.

The first explicit example of an invariant subspace with index 2 was constructed in [36]. The author exhibits two zero based invariant subspaces $\mathcal{I}(A)$ and $\mathcal{I}(B)$, the zero sequences A and B being disjoint, which are at a positive angle from each other. That entails that their sum $\mathcal{M} = \mathcal{I}(A) + \mathcal{I}(B)$ is a closed invariant subspace of $L^2_a(\mathbb{D})$, and this easily implies that $\mathcal{M}$ has index 2. This result was generalized in [50], where a construction was given of an invariant subspace $\mathcal{M}$ of infinite index. The $\mathcal{M}$ is a span of zero based invariant subspaces $\mathcal{I}(A_n)$, $n = 1, 2, 3, \ldots$, where A_n is a certain sequence of regularly spaced points in the unit disc that accumulate nontangentially at every boundary point. An alternate way to construct invariant subspaces of arbitrary index was developed in [19] and [1].

The first example raises the following question:

Problem 10.1. Determine for which pairs of disjoint zero sequences A and B the closed linear span $\mathcal{M}$ of $\mathcal{I}(A)$ and $\mathcal{I}(B)$ has index 1.

We pose two related problems that are perhaps simpler:

Problem 10.2. Determine the zero sequences A such that every invariant subspace $\mathcal{M}$ that contains $\mathcal{I}(A)$ has index 1.

Problem 10.3. Determine which $f \in L^2_a(\mathbb{D})$ have the property that every invariant subspace containing f has index 1.

For the first two questions, one would hope for answers in terms of geometric properties of the sequences A and B, while for the third question it would be nice to obtain an answer in terms of the behavior of f near the unit circle.

The index of an invariant subspace is a natural object of investigation, because it is the existence of invariant subspaces of index larger than 1 that drastically distinguishes the situation in $L_a^2(\mathbb{D})$ from the Hardy space situation. However, we note that Problems 10.1 and 10.2 have equivalent formulations that do not mention the index. In fact, in [36], it was shown that given two disjoint zero sequences A and B, we either have that $\mathcal{I}(A)+\mathcal{I}(B)$ is dense in $L_a^2(\mathbb{D})$, or its closure $\mathcal{M}$ has index 2. Thus, the first problem is equivalent to the question of which disjoint zero sequences A and B have the property that $\mathcal{I}(A)+\mathcal{I}(B)$ is dense in $L_a^2(\mathbb{D})$. That is how this question was phrased in the original article. Furthermore and similarly, it follows from the results of [48] that Problem 10.2 is equivalent to which zero sequences A have the property that each invariant subspace $\mathcal{M}$ that contains $\mathcal{I}(A)$ is again a zero based invariant subspace.

There are some partial results known for these problems. First, it follows from [7] that if A and B are two zero sequences for $L_a^2(\mathbb{D})$ such that there is a point $z_0 \in \mathbb{T}$ and a planar neighborhood U of z_0 for which there are nonzero functions $f \in L_a^2(\mathbb{D}) \cap L^s(U \cap \mathbb{D})$ and $g \in L_a^2(\mathbb{D}) \cap L^t(U \cap \mathbb{D})$, $1/s+1/t = 1/2$, such that f is zero at the points in $A \cap U$ and g is zero at the points in $B \cap U$, then the closed linear span of $\mathcal{I}(A)$ and $\mathcal{I}(B)$ in $L_a^2(\mathbb{D})$ has index 1 (also see [85]). The above-mentioned example in [36] shows that this result is sharp in the sense that if $\varepsilon > 0$, then there are two disjoint zero sequences A and B for $L_a^{4-\varepsilon}(\mathbb{D})$ such that the closed linear span of $\mathcal{I}(A)$ and $\mathcal{I}(B)$ in $L_a^2(\mathbb{D})$ has index 2.

There have been a number of papers studying Problems 10.2 and 10.3; see [66, 86, 7, 83, 85, 9, 11]. Generally, one can state that any nonzero function in any index 2 invariant subspace must be very irregular near every boundary point. Also, using the results of Section 2 of [11] and the results of [12], one can show that if an index 1 invariant subspace $\mathcal{M}$ is contained in an invariant subspace of index > 1, then for any $n = 2, 3, \ldots$ or even $n = +\infty$, $\mathcal{M}$ is contained in an invariant subspace of index n.

The most complete results on Problems 10.2 and 10.3 have been obtained in [11], where the majorization function $\rho_{\mathcal{M}}$ of an invariant subspace $\mathcal{M}$ was introduced to study these types of questions:

$$\rho_{\mathcal{M}}(\lambda) = \frac{\sup\{|f(\lambda)| : f \in \mathcal{M}, \|f\| \le 1\}}{\sup\{|f(\lambda)| : f \in L_a^2(\mathbb{D}), \|f\| \le 1\}}, \qquad \lambda \in \mathbb{D}.$$

It is easy to see that if

$$K_\lambda(z) = K(z, \lambda) = \frac{1}{(1 - z\bar{\lambda})^2}$$

denotes the Bergman kernel, and if $P_{\mathcal{M}}$ is the orthogonal projection onto $\mathcal{M}$, then

$$\rho_{\mathcal{M}}(\lambda) = \frac{\|P_{\mathcal{M}} K_\lambda\|}{\|K_\lambda\|}, \qquad \lambda \in \mathbb{D}.$$

Thus, $\rho_{\mathcal{M}}$ is continuous and its values lie between zero and 1. One checks that the analogous definition for the Hardy space $H^2(\mathbb{D})$ yields that the majorization function of an invariant subspace of $H^2(\mathbb{D})$ equals the modulus of the classical inner function that generates it. Thus, in that case the majorization function of any nonzero subspace has nontangential limit equal to 1 at almost every boundary point. Perhaps it should not be surprising that for $L^2_a(\mathbb{D})$ the situation is different, but this is precisely related to the questions about the index.

Theorem 10.4. (Aleman, Richter, Sundberg) *Let $\mathcal{M}$ be an invariant subspace of $L^2_a(\mathbb{D})$ with ind $\mathcal{M} = 1$. Then the following are equivalent:*

(a) *every invariant subspace $\mathcal{N}$ with $\mathcal{M} \subset \mathcal{N}$ has index 1,*
(b) *there is a set $E \subset \mathbb{T}$ of positive measure such that $\rho_{\mathcal{M}}$ has nontangential limit equal to 1 on E,*
(c) *there are $\varepsilon > 0$, a positive angle σ, and a set $E \subset \mathbb{T}$ of positive measure such that $\rho_{\mathcal{M}}(\lambda) \geq \varepsilon$ for every λ in every Stolz angle with opening σ and vertex point in E,*
(d) *there is a set $E \subset \mathbb{T}$ of positive measure such that the extremal function φ of $\mathcal{M}$ has nontangential limits almost everywhere on E.*

The equivalence of (a) and (d) resolves Problem 10.3 for extremal functions f. For general functions, the answer to Problem 10.3 has to be more subtle, because there are functions f which have nontangential limits almost everywhere on $\mathbb{T}$ and such that f is contained in some index 2 invariant subspace (see [11] for such an example). Nevertheless, for many choices of the sequence A in Problem 10.2 or functions f in Problem 10.3 one can estimate the corresponding majorization function and hence use Theorem 10.4 to decide whether or not $\mathcal{I}(A)$ or f are contained in an invariant subspace of index > 1.

For example, let us show that if there exists an open set U such that $U \cap \mathbb{T} \neq \emptyset$, and if $f \in L^2_a(\mathbb{D})$, $f \neq 0$ is bounded in $U \cap \mathbb{D}$, then every invariant subspace containing f must have index 1. This result is from [7], but here we want to indicate how the majorization function and Theorem 10.4 come in. Let $\mathcal{M}$ be the smallest invariant subspace that contains f. As before, we let $K_\lambda(z)$ be the Bergman kernel; then since K_λ is bounded for each $\lambda \in \mathbb{D}$, the function

$$g(z) = \frac{K_\lambda(z)\, f(z)}{||K_\lambda f||}$$

is a function of unit norm in $\mathcal{M}$. The definition of the majorization function implies that

$$\frac{|f(\lambda)|}{||K_\lambda f||/||K_\lambda||} = \frac{|g(\lambda)|}{||K_\lambda||} \leq \rho_{\mathcal{M}}(\lambda) \leq 1.$$

Now, the hypothesis on f implies that for almost every $z \in E = U \cap \mathbb{T}$, f has a nontangential limit $f(z)$. Similarly, it is not difficult to show that the hypothesis implies that

$$\frac{||K_\lambda f||}{||K_\lambda||}$$

converges nontangantially to $|f(z)|$ for almost every $z \in U \cap \mathbb{T}$. Thus, whenever both limits exist and are nonzero, the nontangential limit of $k_{\mathcal{M}}$ must be 1, and the result follows from the equivalence of (a) and (b) of Theorem 10.4.

We already noted above that this argument cannot be extended to cover all functions in $L^2_a(\mathbb{D})$ that have nontangential limits on a set $E \subset \mathbb{T}$ of positive measure, but in [11] an estimate on the majorization function is used to show the following theorem:

Theorem 10.5. (Aleman, Richter, Sundberg) *Let $E \subset \mathbb{T}$ be closed, have positive measure, and finite entropy, that is, if $\{I_n\}_n$ denotes the complementary arcs of E, then*

$$\sum_n |I_n| \log \frac{1}{|I_n|} < +\infty.$$

Let $\Omega_E \subset \mathbb{D}$ be the union of all Stolz angles with fixed opening angle $\sigma > 0$ and vertices at all points of E.

If a function $f \in L^2_a(\mathbb{D})$, $f \neq 0$, is bounded in Ω_E, then every invariant subspace containing f has index 1.

With regards to Problem 10.2, it had been known for awhile that if the (unrestricted) accumulation points of A omit an arc in $\mathbb{T}$, then any invariant subspace containing $\mathcal{I}(A)$ has index 1 (see [37]). Furthermore, it easily follows from Theorem 10.4 that if a sequence $A \subset \mathbb{D}$ is dominating for $\mathbb{T}$, that is, if almost every $z \in \mathbb{T}$ can be approached nontangentially by a subsequence of A, then $\mathcal{I}(A)$ is contained in an invariant subspace of index > 1. If the sequence A is interpolating for $L^2_a(\mathbb{D})$, then the converse to this last statement is true (see [11]). However, in [11], a zero sequence A for the Bergman space is constructed such that the set of nontangential limit points of A in $\mathbb{T}$ has measure 0, yet the majorization function $\rho_{\mathcal{I}(A)}$ is not bounded below in any Stolz angle, and hence $\mathcal{I}(A)$ is contained in invariant subspaces of high index.

11. Maximal invariant subspaces

Let us agree to say that an invariant subspace I in $L^2_a(\mathbb{D})$ is maximal provided every invariant subspace containing it is either I or the whole space $L^2_a(\mathbb{D})$. If I is maximal, then $L^2_a(\mathbb{D})/I$ is a Hilbert space lacking nontrivial invariant subspaces with respect to the induced operator $z[I] : L^2_a(\mathbb{D})/I \to L^2_a(\mathbb{D})/I$, so that if I has codimension larger than 1 (it must then have codimension $+\infty$), we would have an operator on infinite dimensional Hilbert space with only trivial invariant subspaces. If I is maximal and has codimension 1, it has the form

$$I = \{f \in L^2_a(\mathbb{D}) : f(\lambda) = 0\}$$

for some $\lambda \in \mathbb{D}$.

Problem 11.1. Must every maximal invariant subspace of $L^2_a(\mathbb{D})$ have codimension 1?

This problem was answered in the affirmative in [44]. Much more general results were later obtained by Atzmon [15].

Problem 11.2. Let $\mathcal{M}$ be an invariant subspace in $L^2_a(\mathbb{D})$. Suppose $\mathcal{N}$ is another invariant subspace that is contained $\mathcal{M}$, which is maximal with respect to this property. Does it follow that $\mathcal{N}$ has codimension 1 inside $\mathcal{M}$?

As we were told already in the introduction, this is a disguised version of the invariant subspace problem in separable Hilbert space. The really hard case is when the invariant subspace $\mathcal{M}$ has infinite index.

12. Methods based on reproducing kernels

Recall that reproducing kernel K on a set X is a positive definite function of two variables $K : X \times X \mapsto \mathbb{C}$, that is, K satisfies the inequality

$$\sum_{i,j=0}^{n} c_j \, \bar{c}_k \, K(x_i, x_j) \geq 0, \tag{12.1}$$

whenever $x_j \in X$, $c_j \in \mathbb{C}$, for $j = 1, \dots, n$, and n is a positive integer. Moreover, equality holds in (12.1) if and only if $c_j = 0$ for all $j = 1, \dots, n$. The kernel K determines (uniquely) a Hilbert space $\mathcal{H}(K)$ of functions on X which is the completion of the finite linear combinations of functions of the form $K(x, \cdot)$ with respect to the norm defined in (12.1).

While the classical theory of reproducing kernels is well known, there is a more recent development that has attracted the attention of many researchers in operator theory and complex analysis. More precisely, it appears that certain algebraic properties of such a kernel have a strong impact on a number of important problems in these areas like interpolation, Beurling-type theorems and even factorization theory.

The most investigated class of reproducing kernels consists of kernels K with the properties that the related kernel $1 - 1/K$ is positive definite, and $K(z, 0) \equiv 1$. They are called *complete Nevanlinna-Pick kernels*, and appeared first in some unpublished work by Agler [4] (see also [5]) in connection with Nevannlina-Pick and Carathéodory interpolation as well as commutant-lifting theorems. Arveson [13] studied the complete Nevanlinna-Pick kernel

$$K(z, \lambda) = \frac{1}{1 - \langle z, \lambda \rangle_d}$$

on the unit ball of $\mathbb{C}^d$, $d = 1, 2, 3, \dots$, because of its connection to the dilation theory of certain commuting operator tuples (called d-contractions); here, $\langle \cdot, \cdot \rangle_d$ is the Euclidean inner product of $\mathbb{C}^d$. The Halmos-Lax-Beurling theorem can be extended to all Hilbert spaces of analytic functions on the unit ball of $\mathbb{C}^d$ with a complete Nevanlinna-Pick reproducing kernel. This is a strong result of McCullough and Trent [63], completed by the work of Greene, Richter, and Sundberg [29].

The simplest interesting examples of Hilbert spaces of analytic functions on the unit disk with a complete Nevanlinna-Pick reproducing kernel are the Hardy space and the weighted Dirichlet spaces $D_\alpha, 0 \leq \alpha < 1$. They consist of analytic functions f in the unit disk with the property that

$$\|f\|_\alpha^2 = |f(0)|^2 + \int_{\mathbb{D}} |f'(z)|^2 (1 - |z|^2)^\alpha \mathrm{d}S(z) < +\infty.$$

Weighted Dirichlet spaces are dual to the weighted Bergman spaces $L^2_a(\mathbb{D}, \omega_{-\alpha})$ via Cauchy duality. More precisely, if we define the Cauchy dual of $L^2_a(\mathbb{D}, \omega_{-\alpha})$ as the space of all analytic functions g in $\mathbb{D}$ with the property that

$$f \mapsto \lim_{r \to 1} \int_{-\pi}^{\pi} f(re^{\mathrm{i}\theta}) \bar{g}(re^{\mathrm{i}\theta})\, \mathrm{d}\theta$$

defines a bounded linear functional on $L^2_a(\mathbb{D}, \omega_{-\alpha})$, then this space coincides with D_α. This simple fact continues to hold in much greater generality. For example, if μ is a positive radial measure carried by the unit disk such that

$$\mu\big(\{z : r < |z| < 1\}\big) > 0$$

for all r with $0 < r < 1$, then

$$L^2_a(\mathbb{D}, \mu) = L^2(\mu) \cap \mathcal{O}(\mathbb{D})$$

is a Hilbert space of analytic functions whose Cauchy dual defined as above, is a Hilbert space of analytic functions with a complete Nevannlina-Pick reproducing kernel. This follows by a direct application a classical result (see [5]). *The problem is whether this holds true for nonradial measures as well.* As usual, we will only consider positive measures μ carried by $\mathbb{D}$ with the property that given any $z \in \mathbb{D}$, there exists a positive constant $C(z)$ such that the estimate

$$|p(z)|^2 \leq C(z) \int_{\mathbb{D}} |p(w)|^2 \mathrm{d}\mu(w)$$

holds for all polynomials p. In addition, we assume that μ is *reproducing for the origin*, that is,

$$\int_{\mathbb{D}} p(z)\, \mathrm{d}\mu(z) = p(0)$$

holds for all polynomials p.

Problem 12.1. Let μ be a measure as above and let $P^2(\mu)$ be the closure of polynomials in $L^2(\mu)$. Is it true that the reproducing kernel of the Cauchy dual of $P^2(\mu)$ is a complete Nevanlinna-Pick kernel?

The fact that the measure is reproducing at the origin is equivalent to the fact that the reproducing kernel satisfies $K(0, z) = 1$ for all z. Shimorin has pointed out to us that without this assumption, there are simple examples of measures μ such that for the corresponding reproducing kernel K of the space $P^2(\mu)$, $1 - 1/K$ is not positive definite.

A major breakthrough in the theory of Bergman spaces is the following result that appeared first in [47, 48] (see also [80]).

Theorem 12.2. *Let ω be a logarithmically subharmonic weight on $\mathbb{D}$. Then the reproducing kernel K in $L^2_a(\mathbb{D},\omega)$ can be written in the form*

$$K(z,\lambda) = \frac{K(0,\lambda)\,K(z,0)/K(0,0) - \bar{\lambda}z\,L(z,\lambda)}{(1-\bar{\lambda}z)^2}, \tag{12.2}$$

where L is a positive definite function on $\mathbb{D}\times\mathbb{D}$ [i.e., it satisfies (12.1)].

It turns out (see [48] and [80]) that this particular form of the reproducing kernel is equivalent to the having the inequality

$$\|zf+g\|^2 \le 2\left(\|f\|^2 + \|zg\|^2\right) \tag{12.3}$$

for any two functions f, g in the space. This is an important reformulation of Theorem 12.2, because it shows that the reproducing kernel of any invariant subspace of index one in $L^2_a(\mathbb{D},\omega)$ has the form (12.2) as well. The fact that (12.3) holds in $L^2_a(\mathbb{D},\omega)$ follows by a tricky computation based on Green's formula (see [48]). The converse is true for any Hilbert space $\mathcal{H}(K)$ where the operator M_z of multiplication by z is bounded and bounded below. Indeed, by considering the operators

$$T = (M_z^* M_z)^{-1} \qquad \text{and} \qquad R : \mathcal{H}(K)\oplus\mathcal{H}(K) \to \mathcal{H}(K),$$

where R is defined by

$$R(f,g) = M_z f + Tg,$$

it follows from (12.3) that $\|R\| \le \sqrt{2}$. A direct computation then shows that for the positive operator $2 - RR^*$ we have

$$2 - RR^* = 2 - M_z M_z^* - TT^*,$$

and (12.2) follows by letting

$$L(z,\lambda) = \Big(\big(2 - M_z M_z^* - TT^*\big)K(\cdot,\lambda), K(\cdot,z)\Big),$$

since

$$M_z M_z^* K(z,\lambda) = \bar{\lambda}z\,K(z,\lambda)$$

and

$$TT^*K(z,\lambda) = \frac{K(z,\lambda) - K(0,\lambda)\,K(z,0)/K(0,0)}{\bar{\lambda}z}.$$

To illustrate the power of this representation, let us apply it in order to estimate the reproducing kernel. Suppose that the kernel K on $\mathbb{D}$ satisfies (12.2). Note, first, that for $z=\lambda$, we get

$$0 \le K(\lambda,\lambda) = \frac{|K(0,\lambda)|^2/K(0,0) - |\lambda|^2 L(\lambda,\lambda)}{(1-|\lambda|^2)^2}.$$

In particular, we have

$$|\lambda|^2 L(\lambda,\lambda) \le \frac{|K(0,\lambda)|^2}{K(0,0)},$$

and by the Cauchy-Schwarz inequality, we deduce that

$$\left|\bar{\lambda}z\,L(z,\lambda)\right| \le |\lambda z|\,L(\lambda,\lambda)^{1/2}L(z,z)^{1/2} \le |K(0,\lambda)K(0,z)|/K(0,0).$$

But then, using again (12.2), we see that for $\lambda \in \mathbb{D}$,

$$|K(z,\lambda)| \le 2(1-|\lambda|^2)^{-2}\,|K(0,\lambda)K(0,z)|/K(0,0).$$

Consequently, $K(\cdot,\lambda)/K(\cdot,0)$ is a bounded analytic function for fixed $\lambda \in \mathbb{D}$. Now assume, in addition, that the operator of multiplication by z on $\mathcal{H}(K)$, written $M_z|_{\mathcal{H}(K)}$, is a contraction, and apply the von Neumann inequality to conclude that $K(\cdot,\lambda)$, for $\lambda \in \mathbb{D}$, can be approximated by polynomial multiples of $K(\cdot,0)$. Since the linear span of these functions is dense in the space we deduce that $\mathcal{H}(K)$ is generated by its extremal function $K(0,0)^{-1/2}\,K(\cdot,0)$. The argument can be extended with appropriate modifications to invariant subspaces of $L^2_a(\mathbb{D},\omega)$ that have index greater than one. For this one needs operator-valued reproducing kernels, but the idea is essentially the same. Thus, one can obtain a Beurling-type theorem for such weighted Bergman spaces. In the index > 1 case, the elegant method presented here is essentially due to McCullough and Richter.

A further consequence of Theorem 12.2 and the Beurling-type theorem is as follows. Let $\mathcal{M} \subset L^2_a(\mathbb{D})$ be an invariant subspace with index one, denote by $K_{\mathcal{M}}$ its reproducing kernel, and let

$$K(z,\lambda) = (1-\bar{\lambda}z)^{-2}$$

be the original Bergman kernel. Then $K_{\mathcal{M}}$ can be written in the form

$$K_{\mathcal{M}}(z,\lambda) = \frac{K_{\mathcal{M}}(z,0)K_{\mathcal{M}}(0,\lambda)}{K_{\mathcal{M}}(0,0)}\big(1-U(z,\lambda)\big)\,K(z,\lambda), \tag{12.4}$$

where u is a positive definite function. In other words, the normalized reproducing kernel in $\mathcal{M}$ is obtained from the original kernel K by multiplication by the factor $(1-u)$. The surprising fact about this identity is that it implies the contractive divisor property of the extremal function for $\mathcal{M}$,

$$\varphi_{\mathcal{M}}(z) = K_{\mathcal{M}}(0,0)^{-1/2}\,K_{\mathcal{M}}(z,0),$$

that is, the inequality

$$\big\|f/\varphi_{\mathcal{M}}\big\| \le \|f\|, \quad f \in \mathcal{M}.$$

Indeed, to see that division by $\varphi_{\mathcal{M}}$ is a contractive operator from $\mathcal{M}$ into $L^2_a(\mathbb{D})$, it suffices to note that it is the adjoint T^* of the operator T defined on linear combinations of reproducing kernels in $L^2_a(\mathbb{D})$ by the rule

$$T\Big(\sum_j c_j\,K(\cdot,\lambda_j)\Big) = \sum_j c_j\,\frac{K_{\mathcal{M}}(\cdot,\lambda_j)}{\bar{\varphi}_{\mathcal{M}}(\lambda_j)},$$

where K is the Bergman kernel, and $K_{\mathcal{M}}$ is the reproducing kernel for the invariant subspace $\mathcal{M}$. Now, (12.4) actually states that $I-T^*T$ is a positive operator, making T^* a contraction.

Which other reproducing kernels share this property? This question has been studied by McCullough and Richter [62]. They essentially show that if (12.4) holds for the simplest choices of $\mathcal{M}$, that is, for

$$\mathcal{M} = \mathcal{M}_a = \big\{f \in \mathcal{H}(K):\ f(a) = 0\big\}, \qquad a \in \mathbb{D},$$

then K has the form

$$K(z,\lambda) = \frac{1}{1 - \bar{\psi}(\lambda)\,\psi(z)\,\big(1 - V(z,\lambda)\big)}, \tag{12.5}$$

where ψ is analytic in $\mathbb{D}$, $\psi(0) = 0$, $\psi'(0) \neq 0$, while V is a positive definite kernel with $V(0,0) = 0$. Such reproducing kernels K are called Bergman-type kernels and they can also be characterized by a norm inequality in the space $\mathcal{H}(K)$ that resembles (12.3). McCullough and Richter prove that if K is a Bergman-type kernel, then (12.4) holds for all index one invariant subspaces of $\mathcal{H}(K)$. Moreover, for any zero-based invariant subspace $\mathcal{I}(A)$ of $\mathcal{H}(K)$ the normalized reproducing kernel for $\mathcal{I}(A)$ is a Bergman-type kernel as well. Since (12.4) implies the contractive divisor property, this last fact implies that the analogue of Theorem 4.1 holds for every space $\mathcal{H}(K)$, where K is a Bergman-type kernel.

The above algebraic relations between reproducing kernels and their projections onto invariant subspaces have important consequences related to Problem 5.2. To be more precise, for a set $A \subset \mathbb{D}$, let us denote by

$$\mathcal{I}_K(A) = \big\{ f \in \mathcal{H}(K) : \ f = 0 \text{ on } A \big\}$$

the associated invariant subspace given by zeros, and by φ_A its canonical zero divisor, that is the normalized reproducing kernel at the origin for this invariant subspace. In [8], it is shown that if K is a kernel of Bergman-type, then for any point $a \in \mathbb{D}$ we have that $\varphi_{A\cup\{a\}}/\varphi_A$ is bounded in $\mathbb{D}$ and satisfies the inequality

$$\left| \frac{\varphi_{A\cup\{a\}}(z)}{\varphi_A(z)} \right| \leq \frac{2 + |\alpha|}{\sqrt{2 - |\alpha|^2}} \leq 3.$$

If K satisfies (12.2), then the quotient $\varphi_{A\cup\{a\}}(z)/\,\varphi_A(z)$ is bounded in $\mathbb{D}$ and satisfies

$$\operatorname{Re} \frac{(1 - \bar{a}\,z)\,\varphi_{A\cup\{a\}}(z)}{(z - a)\,\varphi_A(z)} > 1, \qquad z \in \mathbb{D}.$$

We should point out here that together with the results proved in [6], this last inequality provides an alternative proof of Theorem 3.1 (see [8]).

Although there is a large overlap between kernels of the type (12.2) and the Bergman type kernels, the two classes are distinct [62]. The Bergman kernel $K(z,\lambda) = (1 - \bar{\lambda} z)^{-2}$, or more generally, the reproducing kernel for $L^2_a(\mathbb{D}, \omega_\alpha)$, with $-1 < \alpha \leq 0$, are of Bergman type and satisfy condition (12.2) as well, so that, the above estimates provide sharp bounds from above and below for the functions $\varphi_{A\cup\{a\}}/\varphi_A$ in these spaces. The following problem has been suggested to us by Shimorin.

Problem 12.3. Let ω be a logarithmically subharmonic weight on $\mathbb{D}$ that is reproducing at the origin. Is the reproducing kernel K in $L^2_a(\mathbb{D}, \omega)$ a Bergman-type kernel?

Shimorin [81] has shown that the answer is affirmative for radial weights ω. For nonradial weights, however, the problem is still open.

References

[1] E. Abakumov and A. Borichev, *Shift invariant subspaces with arbitrary indices in l^p spaces,* J. Funct. Anal. **188** (2002), 1–26.

[2] A. Abkar, *Norm approximation by polynomials in some weighted Bergman spaces.* J. Funct. Anal. **191** (2002), no. 2, 224–240.

[3] A. Abkar, H. Hedenmalm, *A Riesz representation formula for super-biharmonic functions.* Ann. Acad. Sci. Fenn. Math. **26** (2001), no. 2, 305–324.

[4] J. Agler, *Interpolation* (unpublished manuscript).

[5] J. Agler and J. McCarthy, *Complete Nevanlinna-Pick kernels,* J. Funct. Anal. **175** (2000), 111–124.

[6] A. Aleman, H. Hedenmalm, S. Richter, C. Sundberg, *Curious properties of canonical divisors in weighted Bergman spaces,* Entire functions in modern analysis [Boris Levin Memorial Conference] (Tel-Aviv, 1997), 1–10, Israel Math. Conf. Proc., **15**, Bar-Ilan Univ., Ramat Gan, 2001.

[7] A. Aleman and S. Richter, *Some sufficient conditions for the division property of invariant subspaces in weighted Bergman spaces,* J. Funct. Anal. **144** (1997), 542–556.

[8] A. Aleman and S. Richter, *Single point extremal functions in Bergman-type spaces,* Indiana Univ. Math. J. **51** (2002), 581–605.

[9] A. Aleman, S. Richter, and W. Ross, *Pseudocontinuations and the backward shift.* Indiana Univ. Math. J. **47** (1998), no. 1, 223–276.

[10] A. Aleman, S. Richter, C. Sundberg, *Beurling's theorem for the Bergman space,* Acta Math. **177** (1996), 275–310.

[11] A. Aleman, S. Richter, C. Sundberg, *The majorization function and the index of invariant subspaces in the Bergman spaces.* J. Anal. Math. **86** (2002), 139–182.

[12] C. Apostol, H. Bercovici, C. Foiaş, C. Pearcy, *Invariant subspaces, dilation theory, and the structure of the predual of a dual algebra, I,* J. Funct. Anal. **63** (1985), 369–404.

[13] W. Arveson, *Subalgebras of C^*-algebras, III. Multivariable operator theory.* Acta Math. **181** (1998), 159–228.

[14] A. Atzmon, *Entire functions, invariant subspaces and Fourier transforms.* Proceedings of the Ashkelon Workshop on Complex Function Theory (1996), 37–52, Israel Math. Conf. Proc., **11**, Bar-Ilan Univ., Ramat Gan, 1997.

[15] A. Atzmon, *Maximal, minimal, and primary invariant subspaces.* J. Funct. Anal. **185** (2001), 155–213.

[16] S. Bergman, *The kernel function and conformal mapping.* Second, revised edition. Mathematical Surveys, No. **V**. American Mathematical Society, Providence, R.I., 1970.

[17] A. Beurling, *On two problems concerning linear transformations in Hilbert space,* Acta Math. **81** (1949), 39–255.

[18] G. Bomash, *A Blaschke-type product and random zero sets for the Bergman space,* Ark. Mat. **30** (1992), 45–60.

[19] A. Borichev, *Invariant subspaces of a given index in Banach spaces of analytic functions,* J. Reine Angew. Math. **505** (1998), 23–44.

[20] A. Borichev, H. Hedenmalm, *Harmonic functions of maximal growth: invertibility and cyclicity in Bergman spaces.* J. Amer. Math. Soc. **10** (1997), 761–796.

[21] L. Brown, B. Korenblum, *Cyclic vectors in $A^{-\infty}$,* Proc. Amer. Math. Soc. **102** (1988), 137–138.

[22] P.L. Duren, D. Khavinson, H.S. Shapiro, *Extremal functions in invariant subspaces of Bergman spaces.* Illinois J. Math. **40** (1996), 202–210.

[23] P.L. Duren, D. Khavinson, H.S. Shapiro, C. Sundberg *Contractive zero-divisors in Bergman spaces,* Pacific J. Math. **157** (1993), 37–56.

[24] P.L. Duren, D. Khavinson, H.S. Shapiro, C. Sundberg, *Invariant subspaces and the biharmonic equation,* Michigan Math. J. **41** (1994), 247–259.

[25] P.L. Duren, A. Schuster, *Bergman Spaces,* Mathematical Surveys and Monographs, **100**, American Mathematical Society, 2004.

[26] M. Engliš, *A Loewner-type lemma for weighted biharmonic operators,* Pacific J. Math. **179** (1997), 343–353.

[27] M. Engliš, *Weighted biharmonic Green functions for rational weights,* Glasgow Math. J. **41** (1999), 239–269.

[28] P.R. Garabedian, *A partial differential equation arising in conformal mapping,* Pacific J. Math. **1** (1951), 485–524.

[29] D.C.V. Greene, S. Richter, C. Sundberg, *The structure of inner multipliers in spaces with complete Nevanlinna-Pick kernels,* J. Funct. Anal. **194** (2002), no. 2, 311–331.

[30] J. Hadamard, *Œuvres de Jacques Hadamard, Vols. 1–4,* Editions du Centre National de la Recherche Scientifique, Paris, 1968.

[31] J. Hansbo, *Reproducing kernels and contractive divisors in Bergman spaces.* Zap. Nauchn. Sem. S.-Peterburg. Otdel. Mat. Inst. Steklov. (POMI) 232 (1996), Issled. po Linein. Oper. i Teor. Funktsii. **24**, 174–198, 217; translation in J. Math. Sci. (New York) **92** (1998), 3657–3674 .

[32] V.P. Havin, N.K. Nikolski (editors), *Linear and complex analysis, Problem book 3, Part II,* Lecture Notes in Mathematics **1574**, Springer-Verlag, Berlin, 1994.

[33] W.K. Hayman, *On a conjecture of Korenblum.* Analysis (Munich) **19** (1999), 195–205.

[34] H. Hedenmalm, *A factorization theorem for square area-integrable analytic functions,* J. Reine angew. Math. **422** (1991), 45–68.

[35] H. Hedenmalm, *A factoring theorem for a weighted Bergman space,* St. Petersburg Math. J. **4** (1993), 163–174.

[36] H. Hedenmalm, *An invariant subspace of the Bergman space having the codimension two property,* J. Reine Angew. Math. **443** (1993), 1–9.

[37] H. Hedenmalm, *Spectral properties of invariant subspaces in the Bergman space,* J. Funct. Anal. **116** (1993), 441–448.

[38] H. Hedenmalm, *A factoring theorem for the Bergman space,* Bull. London Math. Soc. **26** (1994), 113–126.

[39] H. Hedenmalm, *A computation of Green functions for the weighted biharmonic operators $\Delta|z|^{-2\alpha}\Delta$, with $\alpha > -1$,* Duke Math. J. **75** (1994), 51–78.

[40] H. Hedenmalm, *Open problems in the function theory of the Bergman space,* Festschrift in honour of Lennart Carleson and Yngve Domar (Uppsala, 1993), 153–169, Acta Univ. Upsaliensis Skr. Uppsala Univ. C Organ. Hist., 58, Uppsala Univ., Uppsala, 1995.

[41] H. Hedenmalm, *Boundary value problems for weighted biharmonic operators,* St. Petersburg Math. J. **8** (1997), 661–674.

[42] H. Hedenmalm, *Recent progress in the function theory of the Bergman space.* Holomorphic spaces (Berkeley, CA, 1995), 35–50, Math. Sci. Res. Inst. Publ., **33**, Cambridge Univ. Press, Cambridge, 1998.

[43] H. Hedenmalm, *Recent developments in the function theory of the Bergman space.* European Congress of Mathematics, Vol. I (Budapest, 1996), 202–217, Progr. Math., **168**, Birkhäuser, Basel, 1998.

[44] H. Hedenmalm, *Maximal invariant subspaces in the Bergman space,* Ark. Mat. **36** (1998), 97–101.

[45] H. Hedenmalm, *An off-diagonal estimate of Bergman kernels,* J. Math. Pures Appl. **79** (2000), 163–172.

[46] H. Hedenmalm, B. Korenblum, K. Zhu, *Theory of Bergman spaces.* Graduate Texts in Mathematics, **199**. Springer-Verlag, New York, 2000.

[47] H. Hedenmalm, S. Jakobsson, S. Shimorin, *A maximum principle à la Hadamard for biharmonic operators with applications to the Bergman spaces,* C. R. Acad. Sci. Paris Sér. I Math. **328** (1999), 973–978.

[48] H. Hedenmalm, S. Jakobsson, S. Shimorin, *A biharmonic maximum principle for hyperbolic surfaces.* J. Reine Angew. Math. **550** (2002), 25–75.

[49] H. Hedenmalm, Y. Perdomo, *Mean value surfaces with prescribed curvature form.* J. Math. Pures Appl., to appear.

[50] H. Hedenmalm, S. Richter, K. Seip, *Interpolating sequences and invariant subspaces of given index in the Bergman spaces.* J. Reine Angew. Math. **477** (1996), 13–30.

[51] H. Hedenmalm, S. Shimorin, *Hele-Shaw flow on hyperbolic surfaces.* J. Math. Pures Appl. **81** (2002), 187–222.

[52] H. Hedenmalm and K. Zhu,*On the failure of optimal factorization for certain weighted Bergman spaces,* Complex Var. Theory Appl. **19** (1992), 165–176.

[53] A. Hinkkanen, *On a maximum principle in Bergman space,* J. Analyse Math. **79** (1999), 335-344.

[54] C. Horowitz, *Zeros of functions in the Bergman spaces,* Duke Math. J. **41** (1974), 693–710.

[55] C. Horowitz, *Factorization theorems for functions in the Bergman spaces,* Duke Math. J. **44** (1977), 201–213.

[56] P. Koosis, *The logarithmic integral. I.* Corrected reprint of the 1988 original. Cambridge Studies in Advanced Mathematics, **12**. Cambridge University Press, Cambridge, 1998.

[57] B. Korenblum, *An extension of the Nevanlinna theory,* Acta Math. **135** (1975), 187–219.

[58] B. Korenblum, *A Beurling-type theorem,* Acta Math. **138** (1977), 265–293.

[59] B. Korenblum, *A maximum principle for the Bergman space,* Publ. Mat. **35** (1991), 479–486.

[60] B. Korenblum, *Outer functions and cyclic elements in Bergman spaces.* J. Funct. Anal. **115** (1993), no. 1, 104–118.

[61] E. LeBlanc, *A probabilistic zero set condition for Bergman spaces,* Michigan Math. J. **37** (1990), 427–438.

[62] S. McCullough, S. Richter, *Bergman-type reproducing kernels, contractive divisors and dilations,* J. Funct. Anal. **190** (2002), 447–480.

[63] S. McCullough, T. Trent, *Invariant subspaces and Nevanlinna-Pick kernels,* J. Funct. Anal. **178** (2000), 226–249.

[64] S.N. Mergelyan, *Weighted approximation by polynomials,* American Mathematical Society Translations (series 2) **10** (1958), 59–106.

[65] B. Muckenhoupt, *Weighted norm inequalities for the Hardy maximal function,* Trans. Amer. Math. Soc. **165** (1972), 207–226.

[66] S. Richter, *Invariant subspaces in Banach spaces of analytic functions,* Trans. Amer. Math. Soc. **304** (1987), 585–616.

[67] S. Richter, *Unitary equivalence of invariant subspaces of Bergman and Dirichlet spaces.* Pacific J. Math. **133** (1988), no. 1, 151–156.

[68] S. Richter, *A representation theorem for cyclic analytic two-isometries,* Trans. Amer. Math. Soc. **328** (1991), 325–349.

[69] K. Seip, *Beurling type density theorems in the unit disk,* Invent. Math. **113** (1993), 21–39.

[70] K. Seip, *On a theorem of Korenblum,* Ark. Mat. **32** (1994), 237–243.

[71] K. Seip, *On Korenblum's density condition for the zero sequences of $A^{-\alpha}$.* J. Anal. Math. **67** (1995), 307–322.

[72] H.S. Shapiro, A.L. Shields, *On the zeros of functions with finite Dirichlet integral and some related function spaces.* Math. Z. **80** (1962), 217–229.

[73] A.L. Shields, *Cyclic vectors in Banach spaces of analytic functions,* Operators and Function Theory (Lancaster, 1984), S. C. Power (editor), NATO Adv. Sci. Inst. Ser. C Math. Phys. Sci., **153**, Reidel, Dordrecht, 1985, 315–349.

[74] S.M. Shimorin, *Factorization of analytic functions in weighted Bergman spaces,* St. Petersburg Math. J. **5** (1994), 1005–1022.

[75] S.M. Shimorin, *On a family of conformally invariant operators,* St. Petersburg Math. J. **7** (1996), 287–306.

[76] S.M. Shimorin, *The Green function for the weighted biharmonic operator $\Delta(1-|z|^2)^{-\alpha}\Delta$ and the factorization of analytic functions* (Russian), Zap. Nauchn. Sem. S.-Peterburg. Otdel. Mat. Inst. Steklov. (POMI) **222** (1995), Issled. po Linein. Oper. i Teor. Funktsii 23, 203–221.

[77] S.M. Shimorin, *Single point extremal functions in weighted Bergman spaces,* Nonlinear boundary-value problems and some questions of function theory, J. Math. Sci. **80** (1996), 2349–2356.

[78] S.M. Shimorin, *The Green functions for weighted biharmonic operators of the form $\Delta w^{-1}\Delta$ in the unit disk,* Some questions of mathematical physics and function theory, J. Math. Sci. (New York) **92** (1998), 4404–4411.

[79] S.M. Shimorin, *Approximate spectral synthesis in the Bergman space.* Duke Math. J. **101** (2000), no. 1, 1–39.

[80] S.M. Shimorin, *Wold-type decompositions and wandering subspaces for operators close to isometries,* J. Reine Angew. Math. **531** (2001), 147-189.

[81] S.M. Shimorin, *An integral formula for weighted Bergman reproducing kernels,* Complex Var. Theory Appl. **47** (2002), no. 11, 1015–1028.

[82] S.M. Shimorin, *On Beurling-type theorems in weighted* l^2 *and Bergman spaces.* Proc. Amer. Math. Soc. **131** (2003), no. 6, 1777–1787.

[83] J. Thomson, and L. Yang, *Invariant subspaces with the codimension one property in* $L^t_a(\mu)$. Indiana Univ. Math. J. **44** (1995), 1163–1173.

[84] R.J. Weir, *Construction of Green functions for weighted biharmonic operators.* J. Math. Anal. Appl. **288** (2003), no. 2, 383–396.

[85] Z. Wu and L. Yang, *The codimension-*1 *property in Bergman spaces over planar regions.* Michigan Math. J. **45** (1998), 369–373.

[86] L. Yang, *Invariant subspaces of the Bergman space and some subnormal operators in* $\mathbb{A}_1 \setminus \mathbb{A}_2$. Michigan Math. J. **42** (1995), 301–310.

Alexandru Aleman
Center for Mathematics
Lund University
S-221 00 Lund, Sweden
e-mail: aleman@maths.lth.se

Håkan Hedenmalm
Department of Mathematics
The Royal Institute of Technology
S-100 44 Stockholm, Sweden
e-mail: haakanh@math.kth.se

Stefan Richter
Department of Mathematics
University of Tennessee
Knoxville, TN
37996-1300, USA
e-mail: richter@math.utk.edu

Operator Theory:
Advances and Applications, Vol. 156, 61–78

The Bergman Kernel and Quadrature Domains in the Plane

Steven R. Bell

In honor of Harold Shapiro's 75th

Abstract. A streamlined proof that the Bergman kernel associated to a quadrature domain in the plane must be algebraic will be given. A byproduct of the proof will be that the Bergman kernel is a rational function of z and one other explicit function known as the Schwarz function. Simplified proofs of several other well known facts about quadrature domains will fall out along the way. Finally, Bergman representative coordinates will be defined that make subtle alterations to a domain to convert it to a quadrature domain. In such coordinates, biholomorphic mappings become algebraic.

Mathematics Subject Classification (2000). Primary 30C40; Secondary 30C20.

Keywords. Szegö kernel.

1. Introduction

In this paper, we will recombine a string of results by Aharonov and Shapiro [1], Gustafsson [14], Davis [12], Shapiro [17], and Avci [2] in light of recent results by the author in [7] and [9] to obtain elementary proofs of a number of results about quadrature domains and the classical functions associated to them. In particular, we present an efficient proof of the fact proved in [9] that the Bergman kernel associated to a quadrature domain in the plane is an algebraic function. In fact, we shall show that it is a rational function of z and the Schwarz function, and consequently it is also a rational combination of z and $Q(z)$ where $Q(z)$ is an explicit algebraic function given by

$$Q(z) = \int_{b\Omega} \frac{\bar{w}}{w - z}\, dw.$$

Research supported by NSF grant DMS-0305958.

These results are all a natural outgrowth of the work of Aharonov and Shapiro [1] and Gustafsson [14] and many of the results obtained in those works will come as corollaries in the approach we take here. For example, Aharonov and Shapiro proved that Ahlfors maps associated to quadrature domains are algebraic, and we shall deduce this via the connection between Ahlfors maps, the double, and the Bergman kernel. The new approach we use shall also allow us to view Bergman representative coordinates in a new and very interesting light.

We shall call an n-connected domain Ω in the plane such that no boundary component is a point a *quadrature domain* if there exist finitely many points $\{w_j\}_{j=1}^N$ in the domain and non-negative integers n_j such that complex numbers c_{jk} exist satisfying

$$\int_\Omega f\, dA = \sum_{j=1}^{N}\sum_{k=0}^{n_j} c_{jk} f^{(k)}(w_j) \tag{1.1}$$

for every function f in the Bergman space of square integrable holomorphic functions on Ω. Here, dA denotes Lebesgue area measure. Our results require the function $h(z) \equiv 1$ to be in the Bergman space, and so we shall also assume that the domain under study has *finite area.*

If Ω is an n-connected quadrature domain of finite area in the plane such that no boundary component is a point, then it is well known that the Bergman kernel function associated to Ω satisfies an identity of the form

$$1 \equiv \sum_{j=1}^{N}\sum_{m=0}^{n_j} c_{jm} K^{(m)}(z, w_j) \tag{1.2}$$

where $K^{(m)}(z,w)$ denotes $(\partial^m/\partial\bar{w}^m)K(z,w)$ (and of course $K^{(0)}(z,w) = K(z,w)$) and where the points w_j are the points that appear in the characterizing formula (1.1) of quadrature domains. It can be seen by noting that the inner product of an analytic function against the function $h(z) \equiv 1$ and against the sum on the right hand side of (1.2) agree for all functions in the Bergman space. Hence the two functions must be equal. Note that we must assume that Ω has finite area here just so that $h(z) \equiv 1$ is in the Bergman space.

We first state a theorem about the Bergman kernel of a multiply connected domain in the plane with smooth boundary. Note that, although formula (1.2) is clearly in the background, we do not assume that the domain is a quadrature domain.

Theorem 1.1. *Suppose that Ω is an n-connected bounded domain in the plane whose boundary is given by n non-intersecting simple closed C^∞ smooth real analytic curves. Let $A(z)$ be a function of the form*

$$\sum_{j=1}^{N}\sum_{m=0}^{n_j} c_{jm} K^{(m)}(z, w_j)$$

where the w_j are points in Ω. Let G_1 and G_2 be any two meromorphic functions on Ω that extend meromorphically to the double of Ω and form a primitive pair *for the field of meromorphic functions on the double. There is a rational function* $R(z_1, z_2, w_1, w_2)$ *of four complex variables such that* $K(z,w)$ *is given by*

$$K(z,w) = A(z)\overline{A(w)}R\left(G_1(z), G_2(z), \overline{G_1(w)}, \overline{G_2(w)}\right).$$

Theorem 1.1 is a corollary of Theorem 2.3 of [7]. We shall give a straightforward and direct proof of Theorem 1.1 in §3.

We remark here that two meromorphic functions on a compact Riemann surface are said to form a primitive pair if they generate the field of meromorphic functions, i.e., if every meromorphic function on the Riemann surface is a rational combination of the two. For basic facts about primitive pairs, see Farkas and Kra [13].

If Ω is an n-connected domain in the plane such that no boundary component is a point, then it is a standard construction in the subject to produce a biholomorphic mapping φ which maps Ω one-to-one onto a bounded domain Ω_s bounded by n smooth real analytic curves. The subscript s stands for "smooth" and we shall write $K(z,w)$ for the Bergman kernel of Ω and $K_s(z,w)$ for the Bergman kernel associated to Ω_s. We shall say that a meromorphic function h on Ω *extends meromorphically to the double* of Ω if $h \circ \varphi^{-1}$ is a meromorphic function on Ω_s which extends meromorphically to the double of Ω_s. (This terminology might be considered to be rather non-standard, but it greatly simplifies the statements of many of our results below.) It is easy to verify that this definition does not depend on the choice of φ and Ω_s. We shall also say that G_1 and G_2 form a primitive pair for Ω if $G_1 \circ \varphi^{-1}$ and $G_2 \circ \varphi^{-1}$ extend meromorphically to the double of Ω_s and form a primitive pair for the double of Ω_s. Using this terminology, we may apply Theorem 1.1 to obtain the following result.

Theorem 1.2. *Suppose that* Ω *is an* n*-connected quadrature domain of finite area in the plane such that no boundary component is a point. Then the Bergman kernel function* $K(z,w)$ *associated to* Ω *is a rational combination of any two functions* G_1 *and* G_2 *that form a primitive pair for* Ω *in the sense that* $K(z,w)$ *is a rational combination of* $G_1(z)$, $G_2(z)$, $\overline{G_1(w)}$, *and* $\overline{G_2(w)}$.

We remark that the same conclusion in Theorem 1.2 can be made about the square $S(z,w)^2$ of the Szegő kernel. Furthermore, the classical functions F_j' (see §2 for definitions) are rational functions of G_1 and G_2 and so is any proper holomorphic mapping of Ω onto the unit disc. The reader may see [7] for the details of the more general statement.

The next theorem of Gustafsson [14] reveals that quadrature domains have nice boundaries. It also allows us to smoothly connect a quadrature domain to another domain with a double in the classical sense.

Theorem 1.3. *Suppose that* Ω *is an* n*-connected quadrature domain of finite area in the plane such that no boundary component is a point. Suppose that* φ *is a holomorphic mapping which maps* Ω *one-to-one onto a bounded domain* Ω_s *bounded by* n *smooth real analytic curves. Then* φ^{-1} *extends holomorphically past the boundary* *of Ω_s. Furthermore, φ^{-1} extends meromorphically to the double of Ω_s. It follows that the boundary of Ω is piecewise real analytic and the possibly finitely many non-smooth boundary points are easily described as cusps which point toward the inside of Ω. Furthermore, it follows that φ extends continuously to the boundary of Ω.*

Avci did not state Theorem 1.3 in [2], however, all the elements of a proof are there. If Avci had done things in a different order, he might have needed this theorem and he might very well have spelled out the proof as well. In fact, much of Avci's work in [2] is headed in a direction that could easily have led to many of the results of this paper.

We shall give a short Bergman kernel proof of Theorem 1.3 in §3.

We remark that at a cusp boundary point b of Ω mentioned in Theorem 1.3, the map φ behaves like a principal branch of the square root of $z - b$ mapping the plane minus a horizontal slit from b to the left to the right half plane. This map sends b to zero, and the inverse map φ^{-1} behaves like $b + z^2$ on the right half plane mapping zero to b. The power two in $b + z^2$ is the only power larger than one that makes such a map one-to-one on the right half plane near zero, and this kind of reasoning can be used to show that the derivative of φ^{-1} can have at most a simple zero at a boundary point of Ω_s.

Aharonov and Shapiro [1] showed that the Schwarz function associated to a quadrature domain extends meromorphically to the domain, i.e., that the function $\bar{z}$ agrees on the boundary with a function $S(z)$, known as the Schwarz function, which is meromorphic on the domain and which extends continuously up to the boundary. We will follow Gustafsson [14] and modify somewhat his observation that z and $S(z)$ form a primitive pair (in the special sense we use here) to obtain a quick proof of the following result.

Theorem 1.4. *Suppose that Ω is an n-connected quadrature domain of finite area in the plane such that no boundary component is a point. The function z extends to the double as a meromorphic function. Consequently, there is a meromorphic function $S(z)$ on Ω (known as the Schwarz function) which extends continuously up to the boundary such that $S(z) = \bar{z}$ on $b\Omega$. The functions z and $S(z)$ form a primitive pair for Ω. It also follows that z and*

$$Q(z) = \int_{b\Omega} \frac{\bar{w}}{w - z}\, dw$$

form a primitive pair for Ω. Consequently, $S(z)$ and $Q(z)$ are algebraic functions. It also follows that the boundary of Ω is a real algebraic curve.

Aharonov and Shapiro [1] first showed that the boundary of a quadrature domain is an algebraic curve and Gustafsson [14] later gave a precise description of what these curves must be. Theorems 1.4 and 1.2 can now be combined to yield the following result.

Theorem 1.5. *Suppose that Ω is an n-connected quadrature domain of finite area in the plane such that no boundary component is a point. Then the Bergman kernel* *function $K(z,w)$ associated to Ω is a rational combination of the two functions z and the Schwarz function. It is also a rational function of z and $Q(z)$. Consequently, $K(z,w)$ is algebraic. Furthermore, the Szegő kernel is algebraic, the classical functions F_j' are algebraic, and every proper holomorphic mapping from Ω onto the unit disc is algebraic.*

Similar statements to the theorems above can be made for the Poisson kernel and first derivative of the Green's function. These results follow from formulas appearing in [6] and we do not spell them out here.

It is interesting to note here that, not only is the Bergman kernel of a quadrature domain Ω a rational combination of z and the Schwarz function, but the Schwarz function associated to Ω is a rational combination of $K(z,a)$ and $K(z,b)$ for two fixed points a and b in the domain by virtue of the fact that $S(z)$ extends to the double and because it is possible to find two such functions $K(z,a)$ and $K(z,b)$ of z that form a primitive pair for Ω.

Since $S(z) = \bar{z}$ on the boundary, the next theorem is an easy consequence of Theorem 1.5.

Theorem 1.6. *Suppose that Ω is an n-connected quadrature domain in the plane of finite area such that no boundary component is a point. The Bergman kernel $K(z,w)$ and the square $S(z,w)^2$ of the Szegő kernel are rational functions of z, $\bar{z}$, w, and $\bar{w}$ on $b\Omega \times b\Omega$ minus the boundary diagonal. The functions $F_j'(z)$ are rational functions of z and $\bar{z}$ when restricted to the boundary. Furthermore, the unit tangent vector function $T(z)$ is such that $T(z)^2$ is a rational function of z and $\bar{z}$ for $z \in b\Omega$.*

The Riemann mapping theorem can be viewed as saying that any simply connected domain in the plane that is not the whole plane is biholomorphic to the grandaddy of all quadrature domains, the unit disc. Gustafsson generalized this theorem to multiply connected domains. He proved that any finitely connected domain in the plane such that no boundary component is a point is biholomorphic to a quadrature domain. The circle of ideas we develop in this paper can be used to reformulate these theorems via a tool I would venture to call "Bergman representative coordinates." We shall show that any domain with C^∞ smooth boundary can be mapped by a biholomorphic mapping which is as C^∞ close to the identity map as desired to a quadrature domain. The biholomorphic map will be given in the form of a quotient of linear combinations of the Bergman kernel. This process begins with the following lemma, which was proved in [3]. Let $A^\infty(\Omega)$ denote the subspace of $C^\infty(\overline{\Omega})$ consisting of functions that are holomorphic on Ω.

Lemma 1.7. *Suppose that Ω is a bounded finitely connected domain bounded by simple closed C^∞ smooth curves. The complex linear span of the set of functions of z of the form $K(z,b)$ where b are points in Ω is dense in $A^\infty(\Omega)$.*

If Ω is a domain as in Lemma 1.7, let $\mathcal{K}_2(z)$ denote a finite linear combination $\sum c_j K(z,b_j)$ which is C^∞ close to the function $h(z) \equiv 1$ and let $\mathcal{K}_1(z)$ denote a finite linear combination $\sum a_j K(z, b_j)$ which is C^∞ close to the function $h(z) \equiv z$. I call the quotient $\mathcal{K}_1(z)/\mathcal{K}_2(z)$ a Bergman representative mapping function if it is one-to-one. By taking the linear combinations to be C^∞ close enough to their target functions, any such mapping can be made one-to-one and as C^∞ close to the indentity as desired. The mappings extend meromorphically to the double of Ω because $T(z)K(z,b) = -\overline{T(z)\Lambda(z,b)}$ for z in $b\Omega$. (See (3.1) and §§3–4 for the details of this argument.) Hence, according to Gustafsson [14], the mapping sends Ω to a quadrature domain. Hence, the following theorem holds

Theorem 1.8. *Suppose that Ω is a bounded finitely connected domain bounded by simple closed C^∞ smooth curves. There is a Bergman representative mapping function which is as C^∞ close to the identity map as desired which maps Ω to a quadrature domain.*

Bergman representative mappings as defined here yield a rather fascinating change of coordinates. Indeed, they were used in several complex variables in [10] to locally linearize biholomorphic mappings. In one variable, if $\Phi : \Omega_1 \to \Omega_2$ is a biholomorphic (or merely proper holomorphic mapping) between C^∞-smooth finitely connected domains in the plane, then Theorem 1.8 allows us to make changes of coordinates that are C^∞ close to the identity on each domain in such a way that the mapping Φ in the new coordinates is an *algebraic function.* Indeed, if $\Phi : \Omega_1 \to \Omega_2$ is a proper holomorphic mapping between quadrature domains in the new coordinates, let f_a denote an Ahlfors mapping of Ω_2 onto the unit disc. Then f_a is a proper holomorphic mapping of the smooth quadrature domain Ω_2 onto the unit disc, and is therefore algebraic by Theorem 1.5. Now $f_a \circ \Phi$ is a proper holomorphic mapping of the smooth quadrature domain Ω_1 onto the unit disc, and is therefore algebraic. It follows that Φ itself must be algebraic. It is rather striking that subtle changes in the boundary can make conformal mappings become defined on the whole complex plane.

Since Bergman representative mappings extend to the double and are one-to-one, they are Gustafsson mappings (as defined in [9]), and they can be used to compress the classical kernel functions into a small data set as in [9]. Björn Gustafsson read a preliminary version of this paper and realized that every Gustafsson map can be expressed as a Bergman representative map. He has granted me permission to include his argument in §4 of the present paper.

The points in the quadrature identity associated to the Bergman representative domain in Theorem 1.8 can be arranged to fall in any small disc in the domain. This can be done using similar constructions to those used in [9] and by Gustafsson in [14]. We do not treat this problem here.

Another interesting way to view Theorem 1.8 is as follows. Shrink a smooth domain Ω by moving in along an inward pointing unit vector a fixed short distance. Now use a Bergman representative mapping which is sufficiently C^∞ close to the identity map so that the shrunken domain gets mapped to a domain inside of Ω. This shows that we may lightly "sand" the edges of our original domain to turn it into a quadrature domain. Similarly, by expanding the domain by first moving along an outward pointing normal and repeating this process, we can see that we can "paint" the edges of our domain with an arbitrarily thin coat of paint of variable thickness to turn it into a quadrature domain.

It is reasonable to allow more general functions of the form

$$\sum_{j=1}^{N}\sum_{m=0}^{n_j} c_{jm}K^{(m)}(z,w_j)$$

to appear as the numerator and denominator of a Bergman representative mapping because the individual elements of these functions satisfy the same kind of basic identity as the Bergman kernel itself (see formula (3.1)) and, consequently, such quotients extend to the double and have the same mapping properties as the Bergman representative mappings as we constructed them above. Under this stipulation, we can also prove a converse to Theorem 1.8.

Theorem 1.9. *If Ω is a finitely connected quadrature domain of finite area, then there is a Bergman representative mapping which is equal to the identity map.*

Another approach to representative coordinates is given in [15] by Jeong and Taniguchi. It might be interesting to see what comes of combining the two approaches.

Finally, we remark that, when the results of this paper are combined with the results in [8], it can be seen that the infinitesimal Carathéodory metric associated to a finitely connected quadrature domain of finite area such that no boundary component is a point is given by $\rho(z)|dz|$ where ρ is a rational combination of z, the Schwarz function, and the complex conjugates of these two functions.

Complete proofs of the theorems will be given in §3.

2. Preliminaries

It is a standard construction in the theory of conformal mappings to show that an n-connected domain Ω in the plane such that no boundary component is a point is conformally equivalent via a map φ to a bounded domain $\widetilde{\Omega}$ whose boundary consists of n simple closed C^∞ smooth real analytic curves. Since such a domain $\widetilde{\Omega}$ is a bordered Riemann surface, the double of $\widetilde{\Omega}$ is an easily realized compact Riemann surface. We shall say that an analytic or meromorphic function h on Ω *extends meromorphically to the double of* Ω if $h\circ\varphi^{-1}$ extends meromorphically to the double of $\widetilde{\Omega}$. Notice that whenever Ω is itself a bordered Riemann surface, this notion is the same as the notion that h extends meromorphically to the double of Ω. We shall say that two functions G_1 and G_2 extend to the double and generate the meromorphic functions on the double of Ω, and that they therefore form a primitive pair for the double of Ω, if $G_1\circ\varphi^{-1}$ and $G_2\circ\varphi^{-1}$ extend to the double of $\widetilde{\Omega}$ and form a primitive pair for the double of $\widetilde{\Omega}$.

It is proved in [6] that if Ω is an n-connected domain in the plane such that no boundary component is a point, then almost any two distinct Ahlfors maps f_a

and f_b generate the meromorphic functions on the double of Ω. It is also proved that any proper holomorphic mapping from Ω to the unit disc extends to the double of Ω.

Suppose that Ω is a bounded n-connected domain whose boundary consists of n non-intersecting C^∞ smooth simple closed curves. The Bergman kernel $K(z,w)$ associated to Ω is related to the Szegő kernel via the identity

$$K(z,w) = 4\pi S(z,w)^2 + \sum_{i,j=1}^{n-1} A_{ij} F_i'(z)\overline{F_j'(w)}, \tag{2.1}$$

where the functions $F_i'(z)$ are classical functions of potential theory described as follows. The harmonic function ω_j which solves the Dirichlet problem on Ω with boundary data equal to one on the boundary curve γ_j and zero on γ_k if $k \neq j$ has a multivalued harmonic conjugate. Let γ_n denote the outer boundary curve. The function $F_j'(z)$ is a single valued holomorphic function on Ω which is locally defined as the derivative of $\omega_j + iv$ where v is a local harmonic conjugate for ω_j. The Cauchy-Riemann equations reveal that $F_j'(z) = 2(\partial\omega_j/\partial z)$.

The Bergman and Szegő kernels are holomorphic in the first variable and antiholomorphic in the second on $\Omega \times \Omega$ and they are hermitian, i.e., $K(w,z) = \overline{K(z,w)}$. Furthermore, the Bergman and Szegő kernels are in $C^\infty((\overline{\Omega}\times\overline{\Omega}) - \{(z,z) : z \in b\Omega\})$ as functions of (z,w) (see [4, p. 100]).

We shall also need to use the Garabedian kernel $L(z,w)$, which is related to the Szegő kernel via the identity

$$\frac{1}{i} L(z,a)T(z) = S(a,z) \qquad \text{for } z \in b\Omega \text{ and } a \in \Omega \tag{2.2}$$

where $T(z)$ represents the complex unit tangent vector at z pointing in the direction of the standard orientation of $b\Omega$. For fixed $a \in \Omega$, the kernel $L(z,a)$ is a holomorphic function of z on $\Omega - \{a\}$ with a simple pole at a with residue $1/(2\pi)$. Furthermore, as a function of z, $L(z,a)$ extends to the boundary and is in the space $C^\infty(\overline{\Omega} - \{a\})$. In fact, $L(z,w)$ is in $C^\infty((\overline{\Omega} \times \overline{\Omega}) - \{(z,z) : z \in \overline{\Omega}\})$ as a function of (z,w) (see [4, p. 102]). Also, $L(z,a)$ is non-zero for all (z,a) in $\overline{\Omega} \times \Omega$ with $z \neq a$ and $L(a,z) = -L(z,a)$ (see [4, p. 49]).

For each point $a \in \Omega$, the function of z given by $S(z,a)$ has exactly $(n-1)$ zeroes in Ω (counting multiplicities) and does not vanish at any points z in the boundary of Ω (see [4, p. 49]).

Given a point $a \in \Omega$, the Ahlfors map f_a associated to the pair (Ω, a) is a proper holomorphic mapping of Ω onto the unit disc. It is an n-to-one mapping (counting multiplicities), it extends to be in $C^\infty(\overline{\Omega})$, and it maps each boundary curve γ_j one-to-one onto the unit circle. Furthermore, $f_a(a) = 0$, and f_a is the unique function mapping Ω into the unit disc maximizing the quantity $|f_a'(a)|$ with $f_a'(a) > 0$. The Ahlfors map is related to the Szegő kernel and Garabedian kernel via (see [4, p. 49])

$$f_a(z) = \frac{S(z,a)}{L(z,a)}. \tag{2.3}$$

Note that $f'_a(a) = 2\pi S(a,a) \neq 0$. Because f_a is n-to-one, f_a has n zeroes. The simple pole of $L(z,a)$ at a accounts for the simple zero of f_a at a. The other $n-1$ zeroes of f_a are given by the $n-1$ zeroes of $S(z,a)$ in $\Omega - \{a\}$.

When Ω does not have smooth boundary, we define the kernels and domain functions above as in [5] via a conformal mapping to a domain with real analytic boundary curves.

3. Proofs of the theorems

In this section, we give complete proofs of Theorems 1.1–1.9.

Proof of Theorem 1.1. The Bergman kernel is related to the classical Green's function via ([11, p. 62], see also [4, p. 131])

$$K(z,w) = -\frac{2}{\pi}\frac{\partial^2 G(z,w)}{\partial z \partial \bar{w}}.$$

Another kernel function on $\Omega \times \Omega$ that we shall need is given by

$$\Lambda(z,w) = -\frac{2}{\pi}\frac{\partial^2 G(z,w)}{\partial z \partial w}.$$

(In the literature, this function is sometimes written as $L(z,w)$ with anywhere between zero and three tildes and/or hats over the top. We have chosen the symbol Λ here to avoid confusion with our notation for the Garabedian kernel above.)

The Bergman kernel and the kernel $\Lambda(z,w)$ satisfy an identity analogous to (2.2):

$$\Lambda(w,z)T(z) = -K(w,z)\overline{T(z)} \qquad \text{for } w \in \Omega \text{ and } z \in b\Omega \tag{3.1}$$

(see [4, p. 135]). We remark that it follows from well known properties of the Green's function that $\Lambda(z,w)$ is holomorphic in z and w and is in $C^\infty(\overline{\Omega} \times \overline{\Omega} - \{(z,z) : z \in \overline{\Omega}\})$. If $a \in \Omega$, then $\Lambda(z,a)$ has a double pole at $z=a$ as a function of z and $\Lambda(z,a) = \Lambda(a,z)$ (see [4, p. 134]). Since Ω has real analytic boundary, the kernels $K(z,w)$, $\Lambda(z,w)$, $S(z,w)$, and $L(z,w)$, extend meromorphically to $\overline{\Omega} \times \overline{\Omega}$ (see [4, p. 103, 132–136]). Let $A(z) = \sum_{j=1}^{N}\sum_{m=0}^{n_j} c_{jm}K^{(m)}(z,w_j)$ where the w_j are points in Ω. Notice that A cannot be the zero function because, if it were, it would be orthogonal to all functions in the Bergman space, and consequently every function in the Bergman space would have to satisfy the identity $0 \equiv \sum_{j=1}^{N}\sum_{m=0}^{n_j} \bar{c}_{jm} g^{(m)}(w_j)$, which is absurd. Notice that (3.1) shows that there is a meromorphic function $M(z)$ on Ω which extends meromorphically to a neighborhood of $\overline{\Omega}$ and which has no poles on $b\Omega$ such that

$$A(z)T(z) = \overline{M(z)T(z)} \qquad \text{for } z \in b\Omega. \tag{3.2}$$

Let $\mathcal{B}$ denote the class of holomorphic functions $B(z)$ on Ω that have the property that they extend holomorphically past the boundary and such that there exists a meromorphic function $m(z)$ on Ω which extends meromorphically to a neighborhood of $\overline{\Omega}$ and which has no poles on $b\Omega$ such that

$$B(z)T(z) = \overline{m(z)T(z)} \qquad \text{for } z \in b\Omega.$$

We have shown that $A(z)$ belongs to $\mathcal{B}$. Notice that, if $B(z)$ is in $\mathcal{B}$, then $B(z)/A(z)$ is equal to the complex conjugate of a meromorphic function $m(z)/M(z)$ for $z \in b\Omega$. This shows that $B(z)/A(z)$ extends meromorphically to the double of Ω. Now the Bergman kernel is given by

$$K(z,w) = 4\pi S(z,w)^2 + \sum_{i,j=1}^{n-1} \lambda_{ij} F_i'(z)\overline{F_j'(w)}$$

where

$$S(z,w) = \frac{1}{1 - f_a(z)\overline{f_a(w)}} \left(c_0 S(z,a)\overline{S(w,a)} + \sum_{i,j=1}^{n-1} c_{ij} S(z,a_i)\,\overline{S(w,a_j)} \right) \qquad (3.3)$$

and where $f_a(z)$ denotes the Ahlfors map associated to a (see [5]). The functions F_j' belong to the class $\mathcal{B}$. Indeed

$$F_j'(z)T(z) = -\overline{F_j'(z)T(z)} \qquad \text{for } z \in b\Omega \qquad (3.4)$$

(see [4, p. 80]). Furthermore, functions of z of the form $S(z,a_i)S(z,a_j)$ also belong to the class $\mathcal{B}$ by virtue of identity (2.2). Hence, $K(z,w)$ is given by a sum of terms of the form

$$\frac{B_1(z)\overline{B_2(w)}}{(1 - f_a(z)\overline{f_a(w)})^2}$$

plus a sum of functions of the form $B_1(z)\overline{B_2(w)}$ where B_1 and B_2 belong to $\mathcal{B}$. Therefore, if we divide $K(z,w)$ by $A(z)\overline{A(w)}$, we obtain a function which is a sum of terms of the form

$$\frac{g_1(z)\overline{g_2(w)}}{1 - f_a(z)\overline{f_a(w)})^2}$$

plus a sum of functions of the form $g_1(z)\overline{g_2(w)}$ where g_1 and g_2 extend meromorphically to the double. But f_a also extends meromorphically to the double. Hence, $K(z,w)$ is equal to $A(z)\overline{A(w)}$ times a rational function of $G_1(z)$, $G_2(z)$, $\overline{G_1(w)}$, and $\overline{G_2(w)}$ where G_1 and G_2 are any two functions that form a primitive pair for the double of Ω. This completes the proof of Theorem 1.1. □

Proofs of Theorems 1.2 *and* 1.3. Suppose that Ω is an n-connected quadrature domain of finite area in the plane such that no boundary component is a point. Suppose that φ is a holomorphic mapping which maps Ω one-to-one onto a bounded domain Ω_s bounded by n smooth real analytic curves. Let Φ denote φ^{-1}. The transformation formula for the Bergman kernels under φ can be written in the form

$$\Phi'(z)K(\Phi(z),w) = K_s(z,\varphi(w))\overline{\varphi'(w)}.$$

As mentioned in §1, since Ω is a quadrature domain of finite area, an identity of the form

$$1 \equiv \sum_{j=1}^{N}\sum_{m=0}^{n_j} c_{jm} K^{(m)}(z, w_j)$$

holds. Let $A(z)$ denote the linear combination on the right hand side of this equation. It now follows that

$$\begin{aligned}\Phi'(z) = \Phi'(z)\cdot(A\circ\Phi)(z) &= \Phi'(z)\sum_{j=1}^{N}\sum_{m=0}^{n_j} c_{jm} K^{(m)}(\Phi(z), w_j)\\ &= \sum_{j=1}^{N}\sum_{m=0}^{n_j} c_{jm}\frac{\partial^m}{\partial\bar{w}^m}\left(K_s(z,\varphi(w))\overline{\varphi'(w)}\right)\bigg|_{w=w_j}.\end{aligned}$$

Notice that the function on the last line of this string of equations is a finite linear combination of functions of the form $K_s^{(m)}(z, W)$ where the W are points in Ω_s. Let $A_s(z)$ denote this linear combination. We have shown two things. We have shown that

$$\Phi' = A_s$$

where A_s is a linear combination of functions of z of the form $K_s^{(m)}(z, W)$. We shall use this first fact momentarily to prove Theorem 1.3. Secondly, we have shown that

$$\Phi'(z)\cdot(A\circ\Phi)(z) = A_s(z),$$

and consequently, that

$$A(z) = \varphi'(z)\cdot(A_s\circ\varphi)(z).$$

This second fact will yield Theorem 1.2. Indeed, we may now combine this fact with Theorem 1.1 and the transformation formula for the Bergman kernels to obtain that

$$\begin{aligned}K(z,w) = \frac{K(z,w)}{A(z)\overline{A(w)}} &= \frac{\varphi'(z)K_s(\varphi(z),\varphi(w))\overline{\varphi'(w)}}{\varphi'(z)\cdot(A_s\circ\varphi)(z)\,\overline{\varphi'(w)\cdot(A_s\circ\varphi)(w)}}\\ &= \frac{K_s(\varphi(z),\varphi(w))}{A_s(\varphi(z))\,\overline{A_s(\varphi(w))}}.\end{aligned}$$

If G_1 and G_2 form a primitive pair for the double of Ω_s, then Theorem 1.1 yields that the last function in the string of equations is a rational combination of $G_1\circ\varphi$ and $G_2\circ\varphi$, and this shows that the Bergman kernel $K(z,w)$ is generated by a primitive pair in the generalized sense that we defined in §1. This completes the proof of Theorem 1.2.

We now turn to the proof of Theorem 1.3. We have shown that Φ' is a linear combination of functions of the form $K_s^{(m)}(z, W_j)$ where $m \geq 0$ and $W_j = \varphi(w_j)$. All functions of this form extend holomorphically past the boundary of Ω_s (see [4, p. 41,133]). Hence Φ' extends holomorphically past the boundary of Ω_s and, consequently, so does Φ. It follows that Ω is a bounded domain and that the boundary of Ω is piecewise real analytic. The singular points, if any, in the boundary of Ω are given as the images of boundary points under Φ where Φ' has a simple zero. (Notice that for Φ to be one-to-one on Ω_s, Φ' cannot have any zeroes of multiplicity greater than one on the boundary.) It is clear that φ extends continuously to the boundary of Ω. We will complete the proof of Theorem 1.3 during the course of the next proof when we show that Φ extends meromorphically to the double of Ω_s. □

Proof of Theorem 1.4. As in the preceding proof, we suppose that Ω is an n-connected quadrature domain of finite area in the plane such that no boundary component is a point and we suppose that φ is a holomorphic mapping which maps Ω one-to-one onto a bounded domain Ω_s bounded by n smooth real analytic curves. Again, let Φ denote φ^{-1}. We now claim that Φ extends to the double of Ω_s as a meromorphic function. To see this, we use the inhomogeneous Cauchy integral formula,

$$u(z) = \frac{1}{2\pi i}\int_{w\in b\Omega_s}\frac{u(w)}{w-z}\,dw + \frac{1}{2\pi i}\iint_{w\in\Omega_s}\frac{\partial u/\partial\bar{w}}{w-z}\,dw\wedge d\bar{w}$$

with $\overline{\Phi}$ in place of u to obtain

$$\overline{\Phi(z)} = \frac{1}{2\pi i}\int_{w\in b\Omega_s}\frac{\overline{\Phi(w)}}{w-z}\,dw + \frac{1}{2\pi i}\iint_{w\in\Omega_s}\frac{\overline{\Phi'(w)}}{w-z}\,dw\wedge d\bar{w}.$$

We now let z approach the boundary of Ω_s from the inside and use a method developed in [4] to determine the boundary values of the two integrals. Since the boundary of Ω_s is real analytic and since Φ extends holomorphically past the boundary of Ω_s, the Cauchy-Kovalevskaya theorem (see [4, p. 39]) yields that there is a function v which is real analytic in a neighborhood of the boundary of Ω_s, which vanishes on the boundary, and such that $\frac{\partial v}{\partial\bar{w}} \equiv \overline{\Phi'(w)}$ on a neighborhood of the boundary. We may extend v to be C^∞ smooth in Ω_s. Now apply the inhomogeneous Cauchy Integral Formula to $\overline{\Phi} - v$ to obtain

$$\overline{\Phi(z)} - v(z) = \frac{1}{2\pi i}\int_{w\in b\Omega_s}\frac{\overline{\Phi(w)}}{w-z}\,dw + \frac{1}{2\pi i}\iint_{w\in\Omega_s}\frac{\overline{\Phi'(w)} - \frac{\partial v}{\partial\bar{w}}}{w-z}\,dw\wedge d\bar{w}.$$

The function given as the integral over the boundary of Ω_s is a holomorphic function $H(z)$ on Ω_s which extends C^∞ smoothly to the boundary (see [4, p. 7]). The function given as the integral over Ω_s is also C^∞ up to the boundary of Ω_s because the integrand has compact support. Furthermore, the boundary values of this function agree with the boundary values of

$$\iint_{w\in\Omega_s}\frac{\overline{\Phi'(w)} - \frac{\partial v}{\partial\bar{w}}}{w-z}\,dw\wedge d\bar{w}.$$

as z approaches the boundary of Ω_s from the *outside* of Ω_s. It was shown in the course of proving Theorem 1.3 that Φ' is equal to a linear combination of functions of z of the form $K_s^{(m)}(z, w_j)$. Note that this integral is a constant times the L^2 inner product of $\Phi' - \frac{\partial\bar{v}}{\partial w}$, and the holomorphic function $1/(w-z)$ of w when z is outside of Ω_s. Functions of the form $\partial\bar{v}/\partial w$ are orthogonal to smooth holomorphic functions when v vanishes on the boundary via integration by parts. Hence, the boundary values of the integral over Ω_s are given as a linear combination of terms of the form $1/(w_j - z)^m$, i.e., a rational function of z. Hence, for z in the boundary of Ω_s, we have shown that

$$\overline{\Phi(z)} = H(z) + \mathcal{R}(z)$$

where H is a holomorphic function on Ω_s which is C^∞ smooth up to the boundary and $\mathcal{R}(z)$ is a rational function of z. (In fact, the poles of $R(z)$ only fall at the points $\varphi(w_j)$ where w_j are the points appearing in the quadrature identity for Ω.) This last identity reveals that Φ extends to the double of Ω_s as a meromorphic function, and this completes the proof of Theorem 1.3, as promised.

Since $z = \Phi(\varphi(z))$ and Φ extends meromorphically to the double of Ω_s, it follows that z extends meromorphically to the double of Ω in the generalized sense that we use in this paper.

Let the symbol Φ also denote the function defined on the double of Ω_s which is the meromorphic extension of Φ. Let R denote the antiholomorphic reflection function on the double of Ω_s which maps Ω_s to reflected copy in the double of Ω_s and let $G = \overline{\Phi \circ R}$. It is now an easy matter to see that Φ and G form a primitive pair for the double of Ω_s. Indeed, G has poles only at finitely many points in the double which fall in the Ω_s side. Choose a complex number w_0 sufficiently close to the point at infinity so that the set $\mathcal{S} = G^{-1}(w_0)$ consists of finitely many points in Ω_s such that G takes the value w_0 with multiplicity one at each point in $\mathcal{S}$. Since Φ is one-to-one on Ω_s, it follows that Φ separates the points of $G^{-1}(w_0)$, and this implies that Φ and G form a primitive pair (see Farkas and Kra [13]). Consequently, there is an irreducible polynomial $P(z,w)$ such that $P(\Phi(z), G(z)) \equiv 0$ on the double of Ω_s.

The Schwarz function for Ω is now given as $S(z) = G(\varphi(z))$ and z and $S(z)$ form a primitive pair for Ω. It follows that $S(z)$ is meromorphic on Ω and continuous up to the boundary with boundary values equal to $\bar{z}$. By composing the polynomial identity $P(\Phi(z), G(z)) \equiv 0$ with φ, we see that $P(z, S(z)) \equiv 0$ on Ω, and this shows that $S(z)$ is an algebraic function. When this identity is restricted to the boundary, we see that $P(z, \bar{z}) = 0$ when z is in the boundary, i.e., that the boundary of Ω is contained in an algebraic curve. (Gustafsson refined this argument to show that the boundary is in fact equal to the algebraic curve minus perhaps finitely many points.)

We now turn to examine the function $Q(z)$. Let $p(z)$ denote the sum of the principal parts of $S(z)$ in Ω. Apply the Cauchy integral formula to $S(z) - p(z)$ and use the fact that $S(z) = \bar{z}$ on the boundary to see that $S(z) - p(z) = Q(z)$ plus a linear combination of integrals of the form $\int_{b\Omega} \frac{1}{(w-w_j)^k (w-z)}\, dw$ where the w_j are points in Ω where S has poles. But all such integrals are zero. Hence $S(z) = p(z) + Q(z)$. Since S is algebraic, so is Q. Since z and S generate the meromorphic functions on the double of Ω, and since $p(z)$ is rational, it follows that z and $Q(z)$ also generate the meromorphic functions on the double of Ω. This completes the proof of Theorem 1.4. □

Proof of Theorem 1.5. Theorem 1.4 together with Theorem 1.2 yield that the Bergman kernel $K(z,w)$ for Ω is a rational combination of z, $S(z)$, $\bar{w}$, and $\overline{S(w)}$. Thus, the Bergman kernel is algebraic. Similarly $K(z,w)$ is a rational combination of z and $Q(z)$. Since any proper holomorphic mapping of Ω onto the unit disc extends meromorphically to the double, it follows that all such maps are rational combinations of z and $S(z)$, and consequently, they are algebraic.

It is proved in [5] that if the Bergman kernel is algebraic, then so is the Szegő kernel, and so are the classical functions F_j'. □

Proof of Theorem 1.6. We have seen that the kernels $K(z,w)$ and $S(z,w)^2$ and the proper holomorphic maps to the unit disc and the functions F_j' are all generated by z and $S(z)$, and since these functions are equal to z and $\bar{z}$, respectively, on the boundary, we may deduce most of the rest of the claims made in Theorem 1.6. To finish the proof, note that identity (2.2) yields that

$$T(z)^2 = -\frac{S(a,z)^2}{L(z,a)^2}$$

where a is an arbitrary point chosen and fixed in Ω. The function $S(z,a)^2$ is a rational function of z and $\bar{z}$ on the boundary. Identity (2.3) yields that $L(z,a)^2 = S(z,a)^2/f_a(z)^2$, and so $L(z,a)^2$ is also a rational function of z and $\bar{z}$ on the boundary. Finally, it follows that $T(z)^2$ is a rational function of z and $\bar{z}$ on the boundary. □

Proof of Theorem 1.9. Suppose Ω is a finitely connected quadrature domain of finite area. We know that $\bar{z}$ is equal to the Schwarz function $S(z)$ on the boundary and that $S(z)$ is meromorphic on Ω with finitely many poles. Let h be a holomorphic function on Ω that extends smoothly to the boundary. Notice that

$$\begin{aligned} \iint_\Omega z\,\overline{h(z)}\,dA &= \frac{i}{2}\iint_\Omega z\,\overline{h(z)}\,dz\wedge d\bar{z} \\ &= \frac{i}{4}\iint_\Omega \frac{\partial}{\partial z}\left(z^2\,\overline{h(z)}\right)\,dz\wedge d\bar{z} \\ &= \frac{i}{4}\int_{b\Omega} z^2\,\overline{h(z)}\,d\bar{z} \\ &= \frac{i}{4}\int_{b\Omega} \overline{S(z)^2}\,\overline{h(z)}\,d\bar{z}, \end{aligned}$$

and the Residue Theorem yields that this last integral is equal to a fixed linear combination of values of h and finitely many of its derivatives at the points in Ω where $S(z)$ has poles. Since such functions h are dense in the Bergman space, this shows that the function z is a linear combination of functions of the form $K^{(m)}(z,w_k)$ where w_k are points where $S(z)$ has poles. Since the constant function 1 is also given by a linear combination as in formula (1.2), we may form a quotient to get a Bergman representative mapping which is equal to the function z. □

4. Gustafsson's Theorem

Björn Gustafsson has granted me permission to include here his proof of his discovery that every Gustafsson map on a smoothly bounded domain is a Bergman representative map. After I give Gustafsson's elegant proof, I will offer an alternative, more pedestrian proof that sheds additional light on this phenomenon.

The main tool in Gustafsson's argument is the following lemma.

Lemma 4.1. *Suppose that Ω is a bounded finitely connected domain bounded by simple closed C^∞ smooth curves. If G is a holomorphic function on Ω which extends meromorphically to the double of Ω and has no poles on $\overline{\Omega}$, then G' must be equal to a complex linear combination of functions of z of the form $K^{(m)}(z, w_k)$ where w_k are points in Ω.*

Proof. Let G denote both the function on Ω and its extension to the double $\widehat{\Omega}$ of Ω. Note that dG is a meromorphic differential on $\widehat{\Omega}$, and that since it is exact, it is free of residues. Identity (3.1) can be used in a standard way to establish the well known fact that differentials of the form $K^{(m)}(z, w_k)\,dz$ extend meromorphically to $\widehat{\Omega}$. Since the function $\Lambda(z, w)$ has a double pole at $z = w$ with no residue term, it is possible to choose a linear combination of the extensions of $K^{(m)}(z, w_k)\,dz$ so that the poles of dG on the back side of $\widehat{\Omega}$ are exactly cancelled by the sum. (We shall say more about this argument below.) Let $\mathcal{K}(z)\,dz$ denote such a linear combination. Now $dG - \mathcal{K}\,dz$ is a holomorphic differential on $\widehat{\Omega}$. All the periods of dG are zero, and it is well known that the β-periods of differentials of the form $\mathcal{K}\,dz$ vanish, i.e., the periods that go across Ω from one boundary curve to another and then return along the backside along the reflected curve. (This fact follows from the relationships between the Bergman kernel and the Λ kernel and the Green's function. It is explained in Schiffer and Spencer [16, p. 101–105]. Also, the arguments can be found in Gustafsson's paper [14] in the proofs of Theorems 1 and 2.)

Now a holomorphic differential vanishes if all its β-periods, or if all its α-periods, vanish. Hence $dG - \mathcal{K}\,dz$ is zero and it follows that $G' = \mathcal{K}$ on Ω. This completes the proof. □

I call a function on a bounded finitely connected domain bounded by smooth curves a *Gustafsson mapping* if it is holomorphic and one-to-one on the domain and extends meromorphically to the double and has no poles on the boundary of the domain. Gustafsson mappings effect a conformal change of variables from the given domain to a quadrature domain.

Theorem 4.2. *Suppose that Ω is a bounded finitely connected domain bounded by simple closed C^∞ smooth curves. If g is a Gustafsson function on Ω, then g is equal to a Bergman representative mapping.*

Proof. If g is a Gustafsson map, apply Lemma 4.1 to g and to $\frac{1}{2}g^2$ to get $g' = \mathcal{K}_2$ and $gg' = \mathcal{K}_1$. Since g is one-to-one, the equality $g' = \mathcal{K}_2$ shows that the linear combination $\mathcal{K}_2$ is non-vanishing on $\overline{\Omega}$. Now g is given by the quotient $\mathcal{K}_1/\mathcal{K}_2$, which is a Bergman representative mapping. □

I shall now give an alternative proof of Lemma 4.1 that sheds further light on the property proved in [9] that, on a smooth quadrature domain, if G is a function on the domain that extends meromorphically to the double, then G' is also a function on the domain that extends meromorphically to the double. Suppose that Ω is a bounded finitely connected domain bounded by simple closed C^∞ smooth curves, and suppose that G is a holomorphic function on Ω that extends meromorphically to the double of Ω and has no poles on $\overline{\Omega}$. In this case, there is a meromorphic function H on Ω which extends smoothly up to the boundary such that $G = \overline{H}$ on $b\Omega$. Let $z(t)$ denote a parameterization of a boundary curve of Ω. Since $G(z(t)) = \overline{H(z(t))}$, we may differentiate with respect to t and divide the result by $|z'(t)|$ to obtain that $G'(z)T(z) = \overline{H'(z)T(z)}$ for $z \in b\Omega$. This last identity is very similar to identity (3.1). Indeed, we may differentiate (3.1) with respect to w and rewrite it to obtain

$$K^{(m)}(z,w)T(z) = -\overline{\Lambda^{(m)}(z,w)T(z)}$$

for $z \in b\Omega$, where $\Lambda^{(m)}(z,w)$ denotes the m-th derivative of $\Lambda(z,w)$ with respect to w (and of course $\Lambda^{(0)}(z,w) = \Lambda(z,w)$). The singular part of $\Lambda(z,w)$ is a constant times $(z-w)^{-2}$. Since H' is the derivative of a meromorphic function, the poles of H' are double or more. Hence, there is a unique linear combination $\mathcal{L}$ of the functions $\Lambda^{(m)}(z,w)$ so that the principal parts at the poles of $\mathcal{L}$ agree with the principal parts of H' at each pole in Ω. If $\mathcal{L}(z) = \sum_{j=1}^{N}\sum_{m=1}^{n_j} c_{jm}\Lambda^{(m)}(z,w_j)$, let $\mathcal{K}(z) = -\sum_{j=1}^{N}\sum_{m=1}^{n_j} \bar{c}_{jm}K^{(m)}(z,w_j)$. Notice that $\mathcal{K}(z)T(z) = \overline{\mathcal{L}(z)T(z)}$ on $b\Omega$. Now $(G'-\mathcal{K})T = \overline{(H'-\mathcal{L})T}$ on $b\Omega$ where both $G'-\mathcal{K}$ and $H'-\mathcal{L}$ are holomorphic on Ω and extend smoothly to the boundary. This implies $(G'-\mathcal{K})T$ is both orthogonal to the Hardy space and conjugates of functions in the Hardy space. Consequently, a theorem of Schiffer yields that $G'-\mathcal{K} = \sum_{j=1}^{n-1} c_jF_j'$ for some constants c_j. (See [4, p. 80] for a proof of this result that proves, rather than assumes, that the zeroes of the Szegö kernel are simple zeroes.) Now $\mathcal{K}(z)$ is $\partial/\partial z$ of a linear combination $\mathcal{G}$ of functions of the form $(\partial^m/\partial\bar{w}^m)G(z,w)$ where $G(z,w)$ is the classical Green's function. All linear combinations of this form vanish on the boundary in the z variable. Also, $F_j' = 2(\partial/\partial z)\omega_j$. Hence $G'-\mathcal{K} = \sum_{j=1}^{n-1} c_jF_j'$ yields that $G - \mathcal{G} - 2\sum_{j=1}^{n-1} c_j\omega_j$ is antimeromorphic on Ω. But $G = \overline{H}$ on the boundary. Hence, the boundary values of $\mathcal{G}+2\sum_{j=1}^{n-1} c_j\omega_j$ agree with the boundary values of a function which is antimeromorphic on Ω and which extends smoothly to the boundary. Since this meromorphic function vanishes on one of the boundary curves, it must be identically zero. This forces us to conclude that all the c_j's are zero, and we have proved Gustafsson's theorem that $G' = \mathcal{K}$.

It should be remarked that the argument just given can be run in reverse to yield a converse to Gustafsson's lemma. Indeed, if $\mathcal{K}$ is a linear combination of the form used above that has vanishing periods along all the boundary curves of Ω, then an analytic antiderivative of $\mathcal{K}$ must extend meromorphically to the double.

Gustafsson's lemma can also be routinely generalized to yield the following result. Suppose that Ω is a bounded finitely connected domain bounded by simple closed C^∞ smooth curves. If G is a *meromorphic* function on Ω which extends meromorphically to the double of Ω, then $G'(z)$ must be equal to a function of the form

$$\sum_{k=1}^{N}\sum_{m=1}^{p_k} a_{km}K^{(m)}(z,x_k)+\sum_{k=1}^{M}\sum_{m=1}^{q_k} b_{km}\Lambda^{(m)}(z,y_k),$$

where the x_k are points in $\overline{\Omega}$ and the y_k are points in Ω. The points x_k are used to cancel the poles of the extension of G' on the back side of the double and the y_k are used to cancel the poles of G' on Ω. These results allow the field of meromorphic functions on the double to be handled like a linear space in many instances.

In case the domain under study is a quadrature domain of finite area with smooth boundary, then we know that the Schwarz function $S(z)$ agrees with $\bar{z}$ on the boundary. We can differentiate the identity $z = \overline{S(z)}$ along the boundary as we did above to see that $T(z) = \overline{S'(z)T(z)}$ for z in $b\Omega$. Notice that this identity reveals that $|S'(z)| = 1$ on the boundary. Now, if G is a meromorphic function on Ω which extends meromorphically to the double, we may write $G'T = \overline{H'T}$ on the boundary as we did above and divide by the identity for the Schwarz function to obtain that $G'(z)$ is equal to the conjugate of $H'(z)/S'(z)$ for z in the boundary. This yields another way to see that, on a smooth quadrature domain, any meromorphic function G on the domain that extends meromorphically to the double has the property that its derivative G' also extends meromorphically to the double. This proof also has the virtue that it gives an explicit formula for the extension of G' to the double.

References

[1] D. Aharonov and H.S. Shapiro, *Domains on which analytic functions satisfy quadrature identities.* Journal D'Analyse Mathématique **30** (1976), 39–73.

[2] Y. Avci, *Quadrature identities and the Schwarz function.* Stanford PhD thesis, 1977.

[3] S. Bell, *Non-vanishing of the Bergman kernel function at boundary points of certain domains in* $\mathbb{C}^n$. Math. Ann. **244** (1979), 69–74.

[4] S. Bell, *The Cauchy transform, potential theory, and conformal mapping.* CRC Press, 1992.

[5] S. Bell, *Finitely generated function fields and complexity in potential theory in the plane.* Duke Math. J. **98** (1999), 187–207.

[6] S. Bell, *Ahlfors maps, the double of a domain, and complexity in potential theory and conformal mapping.* J. d'Analyse Mathématique **78** (1999), 329–344.

[7] S. Bell, *Complexity in complex analysis.* Advances in Math. **172** (2002), 15–52.

[8] S. Bell, *Möbius transformations, the Carathéodory metric, and the objects of complex analysis and potential theory in multiply connected domains.* Michigan Math. J. **51** (2003), 351–362.

[9] S. Bell, *Quadrature domains and kernel function zipping.* Arkiv för matematik, in press.

[10] S. Bell and E. Ligocka, *A Simplification and Extension of Fefferman's Theorem on Biholomorphic Mappings.* Invent. Math. **57** (1980), 283–289.

[11] S. Bergman, *The kernel function and conformal mapping.* Math. Surveys 5, Amer. Math. Soc., Providence, 1950.

[12] P. Davis, *The Schwarz function and its applications.* Carus Mathematical Monographs 17, Math. Assoc. of Amer., 1974.

[13] H.M. Farkas and I. Kra, *Riemann Surfaces.* Springer-Verlag, 1980.

[14] B. Gustafsson, *Quadrature identities and the Schottky double.* Acta Applicandae Math. **1** (1983), 209–240.

[15] M. Jeong and M. Taniguchi, *Bell representations of finitely connected planar domains.* Proc. Amer. Math. Soc. **131** (2003), 2325–2328.

[16] M. Schiffer and D. Spencer, *Functionals of finite Riemann surfaces.* Princeton Univ. Press, 1954.

[17] H.S. Shapiro, *The Schwarz function and its generalization to higher dimensions.* Univ. of Arkansas Lecture Notes in the Mathematical Sciences, Wiley, 1992.

Steven R. Bell
Mathematics Department
Purdue University
West Lafayette, IN 47907, USA
e-mail: `bell@math.purdue.edu`

Operator Theory:
Advances and Applications, Vol. 156, 79–111

The Cauchy Transform

Joseph A. Cima, Alec Matheson and William T. Ross

To Prof. Harold S. Shapiro on the occasion of his seventy-fifth birthday.

1. Motivation

In this expository paper, we wish to survey both past and current work on the space of Cauchy transforms on the unit circle. By this we mean the collection $\mathcal{K}$ of analytic functions on the open unit disk $\mathbb{D} = \{z \in \mathbb{C} : |z| < 1\}$ that take the form

$$(K\mu)(z) := \int \frac{d\mu(\zeta)}{1 - \overline{\zeta} z}, \tag{1.1}$$

where μ is a finite, complex, Borel measure on the unit circle $\mathbb{T} = \partial\mathbb{D}$. Our motivation for writing this paper stems not only from the inherent beauty of the subject, but from its connections with various areas of measure theory, functional analysis, operator theory, and mathematical physics. We will provide a survey of some classical results, some of them dating back to the beginnings of complex analysis, in order to prepare a reader who wishes to study some recent and important work on perturbation theory of unitary and compact operators. We also gather up these results, which are often scattered throughout the mathematical literature, to provide a solid bibliography, both for historical preservation and for further study.

Why is the Cauchy transform important? Besides its obvious use, via Cauchy's formula, in providing an integral representation of analytic functions, the Cauchy transform (with $d\mu = dm$ - normalized Lebesgue measure on the unit circle) is the Riesz projection operator $f \to f_+$ on $L^2 := L^2(\mathbb{T}, m)$, where the Fourier expansions of f and f_+ are given by

$$f \sim \sum_{n=-\infty}^{\infty} a_n \zeta^n \text{ and } f_+ \sim \sum_{n=0}^{\infty} a_n \zeta^n.$$

Questions such as which classes of functions on the circle are preserved (continuously) by the Riesz projection were studied by Riesz, Kolmogorov, Privalov, Stein, and Zygmund. For example, the $L^p (1 < p < \infty)$ and Lipschitz classes are preserved under the Riesz projection operator while L^1, L^∞, and the space of continuous functions are not. With the Riesz projection operator, we can define the Toeplitz operators $f \to T_\phi(f) := (\phi f)_+$, where $\phi \in L^\infty$. Questions about continuity, compactness, etc., depend not only on properties of the symbol ϕ but on how the Cauchy transform acts on the underlying space of functions f. The Cauchy transform is also related to the classical conjugation operator $Q\mu = 2\Im(K\mu)$, at least for real measures μ, and similar preservation and continuity questions arise.

Though this classical material is certainly both elegant and important, our real inspiration for wanting to write this survey is the relatively recent work beginning with a seminal paper of Clark [18] which relates the Cauchy transform to perturbation theory. Due to recent advances of Aleksandrov [4] and Poltoratski [65, 66, 67], this remains an active area of research rife with many interesting problems connecting Cauchy transforms to a variety of ideas in classical and modern analysis.

Let us take a few moments to describe the basics of Clark's results. According to Beurling's theorem [25, p. 114], the subspaces of the classical Hardy space H^2 invariant under the unilateral shift $Sf = zf$ have the form ϑH^2, where ϑ is an inner function. Consequently, the backward shift operator

$$S^* f = \frac{f - f(0)}{z}$$

has invariant subspaces $(\vartheta H^2)^\perp$. A description of $(\vartheta H^2)^\perp$, involving the concept of a "pseudocontinuation", can be found in [16, 24, 75]. Clark studied the compression

$$S_\vartheta = P_\vartheta S|(\vartheta H^2)^\perp$$

of the shift S to the subspace $(\vartheta H^2)^\perp$, where P_ϑ is the orthogonal projection of H^2 onto $(\vartheta H^2)^\perp$ (see [60, p. 18]), and determined that all possible rank-one unitary perturbations of S_ϑ (in the case where $\vartheta(0) = 0$) are given by

$$U_\alpha f := S_\vartheta f + \langle f, \frac{\vartheta}{z} \rangle \alpha, \quad \alpha \in \mathbb{T}.$$

Furthermore, U_α is unitarily equivalent to the operator 'multiplication by z', $g \to zg$, on the space $L^2(\sigma_\alpha)$, where σ_α is a certain singular measure on $\mathbb{T}$ naturally associated with the inner function ϑ. This equivalence is realized by the unitary operator

$$\mathcal{F}_\alpha : (\vartheta H^2)^\perp \to L^2(\sigma_\alpha),$$

which maps the reproducing kernel

$$k_\lambda^\vartheta(z) = \frac{1 - \overline{\vartheta(\lambda)}\vartheta(z)}{1 - \overline{\lambda} z}$$

for $(\vartheta H^2)^\perp$ to the function

$$\zeta \to \frac{1 - \overline{\vartheta(\lambda)}\alpha}{1 - \overline{\lambda}\zeta}$$

in $L^2(\sigma_\alpha)$ and extends by linearity and continuity. This measure σ_α arises as follows: For each $\alpha \in \mathbb{T}$ the analytic function

$$\frac{\alpha + \vartheta(z)}{\alpha - \vartheta(z)} \tag{1.2}$$

on the unit disk has positive real part, which, by Herglotz's theorem [25, p. 3], takes the form

$$\Re\left(\frac{\alpha + \vartheta(z)}{\alpha - \vartheta(z)}\right) = \int_{\mathbb{T}} \frac{1 - |z|^2}{|\zeta - z|^2}\, d\sigma_\alpha(\zeta),$$

where the right-hand side of the above equation is the Poisson integral $(P\sigma_\alpha)(z)$ of a positive measure σ_α. Without too much difficulty, one can show that the measure σ_α is carried by the set $\{\zeta \in \mathbb{T} : \vartheta(\zeta) = \alpha\}$ and hence singular with respect to Lebesgue measure on the circle. Moreover, the measures $(\sigma_\alpha)_{\alpha \in \mathbb{T}}$ are pairwise singular.

This idea extends via eq. (1.2) beyond inner functions ϑ to any ϕ in the unit ball of H^∞ to create a family of positive measures $(\mu_\alpha)_{\alpha \in \mathbb{T}}$ associated with ϕ. On the other hand, for a given positive measure μ, the Poisson integral $P\mu$ of μ is a positive harmonic function on the disk and so

$$P\mu = \Re\left(\frac{1 + \phi}{1 - \phi}\right) \tag{1.3}$$

for some $\phi \in \operatorname{ball}(H^\infty)$. That is to say, every positive measure μ is the Clark measure μ_1 for some ϕ in the unit ball of H^∞. It is worth mentioning that ϕ is an inner function if and only if the boundary function for $(1 + \phi)/(1 - \phi)$ is purely imaginary. Using eq. (1.3) as well as Fatou's theorem (which says that the boundary function for $P\mu_1$ is $d\mu_1/dm$ almost everywhere [25, p. 4]), we conclude that ϕ is inner if and only if $\mu_1 \perp m$.

The family $(\mu_\alpha)_{\alpha \in \mathbb{T}}$ of Clark measures for some function $\phi \in \operatorname{ball}(H^\infty)$ also provide a disintegration of normalized Lebesgue measure m on the circle. A beautiful theorem of Aleksandrov [4] says that

$$\int_{\mathbb{T}} \mu_\alpha \, dm(\alpha) = m,$$

where the integral is interpreted in the weak-$*$ sense, that is,

$$\int_{\mathbb{T}} \left[\int_{\mathbb{T}} f(\zeta)\, d\mu_\alpha(\zeta) \right] dm(\alpha) = \int_{\mathbb{T}} f(\zeta)\, dm(\zeta)$$

for all continuous functions f on $\mathbb{T}$. Moreover, if $\Sigma = \{\, \zeta \in \mathbb{T} \colon |\phi(\zeta)| = 1 \,\}$,

$$\int_{\mathbb{T}} \mu_\alpha^s \, dm(\alpha) = \chi_\Sigma \cdot m \quad \text{and} \quad \int_{\mathbb{T}} \mu_\alpha^{ac} \, dm(\alpha) = (1 - \chi_\Sigma) \cdot m,$$

where μ_α^s is the singular part and μ_α^{ac} is the absolutely continuous part of μ_α (with respect to Lebesgue measure m), and again, the above equation is interpreted in the weak-$*$ sense.

One can show that the Cauchy transform of σ_α (a Clark measure for the inner function ϑ)

$$(K\sigma_\alpha)(z) = \int \frac{1}{1-\overline{\zeta}z} d\sigma_\alpha(\zeta)$$

is equal to the function

$$\frac{1}{1-\overline{\alpha}\vartheta(z)}.$$

Furthermore, one can use this to produce the following formula for $\mathcal{F}_\alpha^* : L^2(\sigma_\alpha) \to (\vartheta H^2)^\perp$ in terms of the "normalized" Cauchy transform

$$\mathcal{F}_\alpha^* f = \frac{K(f\, d\sigma_\alpha)}{K(\sigma_\alpha)}.$$

Poltoratski [65] showed that some striking things happen here. The first is that for σ_α-almost every $\zeta \in \mathbb{T}$, the non-tangential limit of the above normalized Cauchy transform exists and is equal to $f(\zeta)$. On the other hand, for $g \in (\vartheta H^2)^\perp$, the non-tangential limits certainly exist almost everywhere with respect to Lebesgue measure on the circle (since $(\vartheta H^2)^\perp \subset H^2$). But in fact, for σ_α-almost every ζ, the non-tangential limit of g exists and is equal to $(\mathcal{F}_\alpha g)(\zeta)$. Remarkable here is the significance of the role Cauchy transforms play not only in the above perturbation problem involving the rank-one perturbations of the model operator S_ϑ but in other perturbation problems as well (see [67] and the references therein).

After a brief historical introduction in §2, we begin in §3 with a description of some of the classical function theoretic properties of Cauchy transforms, viewed as functions on the unit disk. This will include their boundary and mapping properties, as well as their relationships with the classical Hardy spaces H^p. In §4 we explore the general question: Which analytic functions on the disk can be represented as Cauchy transforms of measures on the circle? Though the answer to this question is still incomplete, there is a related result of Aleksandrov, extending work of Tumarkin, which characterizes the analytic functions on $\widehat{\mathbb{C}}\backslash\mathbb{T}$ which are Cauchy transforms of measures on the circle.

The space of Cauchy transforms can be viewed in a natural way as the dual space of the disk algebra, or equivalently, as the quotient space $M/\overline{H_0^1}$. We will describe this duality in §5, including facts about the weak and weak-$*$ topology on this space. In §6 we will describe multiplication and division in the space of Cauchy transforms and examine such questions as: (i) Which bounded analytic functions ϕ on the disk satisfy $\phi\mathcal{K} \subseteq \mathcal{K}$ (the multiplier question)? (ii) If ϑ is inner and divides $K\mu$, that is, $K\mu/\vartheta \in H^p$ for some $p > 0$, does $K\mu/\vartheta$ belong to $\mathcal{K}$ (the divisor question)? In §7 we explore the classical operators on H^p (forward and backward shifts, composition operators, and the Cesàro operator) in the setting of the space of Cauchy transforms. Questions relating to continuity and invariant subspaces will be explored. In §8 we will touch more briefly on such topics as the distribution of boundary values of Cauchy transforms. This will include the Havin-Vinogradov-Tsereteli theorem, and its recent improvement by Poltoratski, as well as Aleksandrov's weak-type characterization using the A-integral. We will also discuss the maximal properties of Cauchy transforms arising in the recent work of Poltoratski.

Conspicuously missing from this survey are results about the Cauchy transform of a measure compactly supported in the plane. Certainly this is an important object to study and there are many wonderful ideas here. However, broadening the survey to include those Cauchy transforms opens up such a vast array of topics from so many other fields of analysis such as potential theory and partial differential equations, that our original theme and motivation for writing this survey - a brief overview of the subject with a solid bibliography for further study - would be lost. We feel that focusing on the Cauchy transform of measures on the circle links the classical function theory with more modern applications to perturbation theory. If one is interested in exploring the Cauchy transform of a measure on the plane we suggest that the books [9, 28, 61] are a good place to start.

2. Some early history

In a series of papers from the mid 1800's [86], Cauchy developed what is known today as the "Cauchy integral formula": If f is analytic in $\{|z| < 1 + \varepsilon\}$ for some $\varepsilon > 0$, then

$$f(z) = \frac{1}{2\pi i} \int_{\mathbb{T}} \frac{f(\zeta)}{\zeta - z} d\zeta, \quad |z| < 1.$$

After Cauchy, others, such as Sokhotski, Plemelj, and Privalov [58], examined the "Cauchy integral"

$$\widehat{\phi}(z) = \frac{1}{2\pi i} \int_{\mathbb{T}} \frac{\phi(\zeta)}{\zeta - z} d\zeta,$$

where the boundary function $f(\zeta)$ in Cauchy's formula is replaced by a "suitable" function ϕ defined only on the unit circle $\mathbb{T}$. Amongst other things, they explored the relationship between the density function ϕ and the limiting values of $\widehat{\phi}$, as $|z| \to 1$, as well as the Cauchy principal-value integral

$$\frac{1}{2\pi i} P.V. \int_{\mathbb{T}} \frac{\phi(\zeta)}{\zeta - e^{i\theta}} d\zeta.$$

In particular, Sokhotski [87] in his 1873 thesis (see also [54, Vol. I, p. 316]) proved the following.

Theorem 2.1 (Sokhotski, 1873). *Suppose ϕ is continuous on $\mathbb{T}$ and, for a particular $\zeta_0 \in \mathbb{T}$, satisfies the condition*

$$|\phi(\zeta) - \phi(\zeta_0)| \leqslant C|\zeta - \zeta_0|^{\alpha}, \quad \zeta \in \mathbb{T}$$

for some positive constants C and α. Then the limits

$$\widehat{\phi}^{-}(\zeta_0) := \lim_{r \to 1^-} \widehat{\phi}(r\zeta_0) \quad \textit{and} \quad \widehat{\phi}^{+}(\zeta_0) := \lim_{r \to 1^-} \widehat{\phi}(\zeta_0 / r)$$

exist and moreover, $\widehat{\phi}^-(\zeta_0) - \widehat{\phi}^+(\zeta_0) = \phi(\zeta_0)$*. Furthermore, the Cauchy principal-value integral*

$$P.V. \int_{\mathbb{T}} \frac{\phi(\zeta)}{\zeta - \zeta_0} d\zeta := \lim_{\varepsilon \to 0} \int_{|\zeta - \zeta_0| > \varepsilon} \frac{\phi(\zeta)}{\zeta - \zeta_0} d\zeta$$

exists and

$$\widehat{\phi}^+(\zeta_0) + \widehat{\phi}^-(\zeta_0) = \frac{1}{\pi i} P.V. \int_{\mathbb{T}} \frac{\phi(\zeta)}{\zeta - \zeta_0} d\zeta.$$

Though Sokhotski first proved these results in 1873, the above formulas are often called the "Plemelj formulas" due to reformulations and refinements of them by J. Plemelj [64] in 1908.

I.I. Privalov, in a series of papers and books[1], beginning with his 1919 Saratov doctoral dissertation [70], began to examine the Plemelj formulas for integrals of Cauchy-Stieltjes type

$$\widehat{F}(z) := \frac{1}{2\pi i} \int_{[0,2\pi]} \frac{1}{1 - e^{-i\theta} z} dF(\theta),$$

where F is a function of bounded variation on $[0, 2\pi]$. Privalov, knowing the recently discovered integration theory of Lebesgue and following the lead of his teacher Golubev [30], developed the Sokhotski-Plemelj formulas for these Cauchy-Stieltjes integrals.

Theorem 2.2 (Privalov, 1919). *Suppose* F *is a function of bounded variation on* $[0, 2\pi]$*. Then for almost every* t*,* $\widehat{F}^-(e^{it})$*, the non-tangential limit of* $\widehat{F}(z)$ *as* $z \to e^{it}$ $(|z| < 1)$ *and* $\widehat{F}^+(e^{it})$*, the non-tangential limit at* $z \to e^{it}$ $(|z| > 1)$*, exist and moreover,* $\widehat{F}^-(e^{it}) - \widehat{F}^+(e^{it}) = F'(t)$*. Furthermore, for almost every* t*, the Cauchy principal-value integral*

$$P.V. \int_{[0,2\pi]} \frac{1}{1 - e^{-i\theta} e^{it}} dF(\theta) := \lim_{\varepsilon \to 0} \int_{|t-\theta| > \varepsilon} \frac{1}{1 - e^{-i\theta} e^{it}} dF(\theta)$$

exists and

$$\widehat{F}^+(e^{it}) + \widehat{F}^-(e^{it}) = \frac{1}{\pi i} P.V. \int_{[0,2\pi]} \frac{1}{1 - e^{-i\theta} e^{it}} dF(\theta).$$

A well-known theorem from Fatou's 1906 thesis [26] (see also [25, p. 39]) states that $\widehat{F}(re^{it}) - \widehat{F}(e^{it}/r) \to F'(t)$ as $r \to 1^-$, whenever $F'(t)$ exists (which is almost everywhere). Privalov's contribution was to prove the existence of the principal-value integral as well as to generalize to the case when the unit circle is replaced with a general rectifiable curve (see [31, 58] for more).

A question explored for some time, and for which there is still no completely satisfactory answer is: Which analytic functions f on $\mathbb{D}$ can be represented as a Cauchy-Stieltjes integral? Perhaps a first step in answering this question would be to determine which functions can be represented as the Cauchy integral of their

---

[1]There is Privalov's famous book [71] as well as a nice survey of his work in [53].

boundary values. The following theorem of the brothers Riesz [72] (see also [25, p. 41]) gives the answer [2].

Theorem 2.3 (F. and M. Riesz). *Let f be analytic on $\mathbb{D}$. Then f has an almost everywhere defined, integrable – with respect to Lebesgue measure on $\mathbb{T}$ – non-tangential limit function f^* and*

$$f(z) = \frac{1}{2\pi i}\int_{\mathbb{T}} \frac{f^*(\zeta)}{\zeta - z} d\zeta, \quad z \in \mathbb{D}$$

if and only if f belongs to H^1.

Though the question as to which analytic functions on $\mathbb{D}$ can be represented as Cauchy integrals of their *boundary functions* has been answered, there is no complete description of these functions, though there are some nice partial results (see Theorem 4.9 below).

To give the reader some historical perspective, we stated these classical theorems of Sokhotski, Plemelj, Riesz, and Privalov, in terms of Cauchy-Stieltjes integrals of functions of bounded variation. However, for the rest of this survey, we will use the more modern, but equivalent, notation of Cauchy integrals of finite, complex, Borel measures on the circle as in eq. (1.1). This is done by equating a function of bounded variation with a corresponding measure and vice-versa (see [39, p. 331] for further details).

3. Some basics about the Cauchy transform

Before moving on, let us set some notation. Throughout this survey, $\mathbb{C}$ will denote the complex numbers, $\widehat{\mathbb{C}} := \mathbb{C} \cup \{\infty\}$ the Riemann sphere, $\mathbb{D} := \{|z| < 1\}$ the open unit disk, $\mathbb{D}^- := \{|z| \leqslant 1\}$ its closure, $\mathbb{T} := \partial\mathbb{D} = \{|z| = 1\}$ its boundary, and $\mathbb{D}_e := \widehat{\mathbb{C}} \setminus \mathbb{D}^-$ will denote the (open) extended exterior disk. $M := M(\mathbb{T})$ will denote the complex, finite, Borel measures on $\mathbb{T}$, M_+ the positive measures in M, $M_a := \{\mu \in M : \mu \ll m\}$ ($dm = |d\zeta|/2\pi$ is normalized Lebesgue measure on the circle), and $M_s := \{\mu \in M : \mu \perp m\}$. The norm on M, the total variation norm, will be denoted by $\|\mu\|$. The Lebesgue decomposition theorem says that $M = M_a \oplus M_s$ and that if $\mu = \mu_a + \mu_s$ ($\mu_a \in M_a, \mu_s \in M_s$), then $\|\mu\| = \|\mu_a\| + \|\mu_s\|$.

For $0 < p \leqslant \infty$ we will let $L^p := L^p(m)$ denote the usual Lebesgue spaces, H^p the standard Hardy spaces, and $\|\cdot\|_p$ the norm on these spaces. When $p = \infty$, H^∞ will denote the bounded analytic functions on $\mathbb{D}$. Recall that every function in H^p has finite non-tangential limits almost everywhere and this boundary function belongs to L^p. In fact, the L^p norm of this boundary function is the H^p norm. We will use N to denote the Nevanlinna class of the disk and N^+ to denote the Smirnov class. We refer the reader to some standard texts about the Hardy spaces [25, 29, 40, 51, 77].

---

[2] A nice survey of these results (and others) can be found in [31, Ch. IX].

The following theorem of F. and M. Riesz [25, p. 41] plays an important role in the theory of Cauchy transforms and will be mentioned many times in this survey.

Theorem 3.1 (F. and M. Riesz theorem). *Suppose* $\mu \in M$ *satisfies*

$$\int \zeta^n d\mu(\zeta) = 0 \quad \textit{whenever } n = 0, 1, 2, \ldots.$$

Then $d\mu = \phi dm$, *where* $\phi \in H^1_0 = \{f \in H^1 : f(0) = 0\}$.

For $\mu \in M$, define the analytic function on $\widehat{\mathbb{C}} \setminus \mathbb{T}$

$$\widehat{\mu}(z) := \int \frac{d\mu(\zeta)}{1 - \overline{\zeta} z}. \tag{3.2}$$

The function $\widehat{\mu}$ is called the *Cauchy transform* of μ. We let

$$K\mu := \widehat{\mu}|\mathbb{D}$$

and set

$$\mathcal{K} := \{K\mu : \mu \in M\}$$

to be the *space of Cauchy transforms.* A few facts are immediate from the definition: (i) $\widehat{\mu}$ has an analytic continuation to $\widehat{\mathbb{C}} \setminus \operatorname{supp}(\mu)$, (ii) $\widehat{\mu}(\infty) = 0$, (iii) $\widehat{\mu}$ has the following series expansions:

$$(K\mu)(z) = \sum_{n=0}^{\infty} \mu_n z^n, \; z \in \mathbb{D}, \qquad \widehat{\mu}(z) = -\sum_{n=1}^{\infty} \frac{\mu_{-n}}{z^n}, \; z \in \mathbb{D}_e, \tag{3.3}$$

where

$$\mu_n := \int e^{-in\theta} d\mu(e^{i\theta}), \quad n \in \mathbb{Z}$$

are the Fourier coefficients of μ, (iv) $\widehat{\mu}$ satisfies the growth condition

$$|\widehat{\mu}(z)| \leqslant \frac{\|\mu\|}{|1 - |z||}, \quad |z| \neq 1. \tag{3.4}$$

For a given $f \in \mathcal{K}$, there are a variety of measures $\mu \in M$ such that $f = K\mu$. For example, by eq. (3.3), $K(\overline{\phi} dm) = 0$ whenever $\phi \in H^1_0$. By the F. and M. Riesz theorem however, these are the only measures for which $K\mu = 0$. For $f \in \mathcal{K}$, let

$$M_f := \{\mu \in M : f = K\mu\}$$

be the set of "representing measures" for f. For $\mu \in M$ decomposed (uniquely) as $\mu = \mu_a + \mu_s$ ($\mu_a \in M_a$ and $\mu_s \in M_s$) we can apply the F. and M. Riesz theorem to show that all measures in M_f have the same singular part μ_s. For $\mu_1, \mu_2 \in M_f$, $d\mu_1 - d\mu_2 = \overline{\phi}\, dm$ for some $\phi \in H^1_0$. Thus M_f can be identified with a coset in the quotient space $M/\overline{H^1_0}$. We will explore M_f further in §5.

A routine argument using Lebesgue's dominated convergence theorem shows that

$$\lim_{r \to 1^-} (1 - r)(K\mu)(r\zeta) = \mu(\{\zeta\})$$

and so $K\mu$ is poorly behaved at the points ζ where $\mu(\{\zeta\}) \neq 0$. This can indeed be a dense subset of $\mathbb{T}$. Despite this seemingly poor behavior of $K\mu$ near $\mathbb{T}$, there is some regularity in the boundary behavior of a Cauchy transform. For $0 < p \leqslant \infty$, let $H^p(\widehat{\mathbb{C}} \setminus \mathbb{T})$ denote the class of analytic functions f on $\widehat{\mathbb{C}} \setminus \mathbb{T}$ for which

$$\|f\|^p_{H^p(\widehat{\mathbb{C}}\setminus\mathbb{T})} := \sup_{r \neq 1} \int_{\mathbb{T}} |f(r\zeta)|^p dm(\zeta) < \infty.$$

A well-known theorem of Smirnov [85] (see also [25, p. 39]) is the following.

Theorem 3.5 (Smirnov). *If $\mu \in M$, then*

$$\widehat{\mu} \in \bigcap_{0<p<1} H^p(\widehat{\mathbb{C}} \setminus \mathbb{T})$$

and moreover, $\|\widehat{\mu}\|_{H^p(\widehat{\mathbb{C}}\setminus\mathbb{T})} \leqslant c_p\|\mu\|$, *where* $c_p = O((1-p)^{-1})$.

The above theorem says that for fixed $0 < p < 1$, the operator $\mu \to K\mu$ is a continuous linear operator from M to H^p. What is the norm of this operator? Equivalently, what is the best constant A_p in the estimate $\|K\mu\|_p \leqslant A_p\|\mu\|$? We do not know the answer to this. However, we can say that

$$\sup\{\|K\mu\|_p : \mu \in M_+, \|\mu\| = 1\} = \|\frac{1}{1-z}\|_p.$$

To see this note that $K\mu$ is subordinate to $\phi(z) = (1-z)^{-1}$. Now use Littlewood's subordination theorem [25, p. 10]. For any complex measure $\mu = (\mu_1 - \mu_2) + i(\mu_3 - \mu_4), \mu_j \geqslant 0$, one can use a slight variation of the above argument four times to prove Smirnov's theorem: $\|K\mu\|_p \leqslant A_p\|\mu\|$. However, the best constant A_p is unknown for general complex measures.

The containment

$$\mathcal{K} \subsetneq \bigcap_{0<p<1} H^p$$

is strict since $f(z) = (1-z)^{-1}\log(1-z) \in H^p$ for all $0 < p < 1$ but does not satisfy the growth condition in eq. (3.4). Also worth pointing out here again is Theorem 2.3 which says that every $f \in H^p$ $(p \geqslant 1)$ can be written as the Cauchy integral of its boundary function. Thus

$$\bigcup_{p\geqslant 1} H^p \subsetneq \mathcal{K} \subsetneq \bigcap_{0<p<1} H^p.$$

The first containment above is strict since $(1-z)^{-1} = K\delta_1$ but does not belong to H^1. For $0 < p < 1$, the boundary functions for H^p functions certainly belong to L^p. However, they may not be integrable on the circle and so their Cauchy integral does not always make sense. There is a remedy for this in the theory of "A-integrals" that will be described in §4, which says that some, but not all, Cauchy transforms can be written as the Cauchy A-integral of their boundary functions.

Smirnov's theorem has been refined in a variety of ways. The first refinement due to M. Riesz [73] (see also [25, p. 54]) deals with the case when the measure μ takes the form $d\mu = fdm$ for $f \in L^p$ $(1 < p < \infty)$. Since the notation $K(fdm)$ can be a bit cumbersome, we use f_+ to denote $K(fdm)$ for $f \in L^1$.

Theorem 3.6 (M. Riesz). *If $1 < p < \infty$ then $f_+ \in H^p$ whenever $f \in L^p$. Moreover, the map $f \to f_+$ from L^p to H^p is continuous and onto.*

For $f \in L^p$ $(1 < p < \infty)$ with Fourier series

$$f \sim \sum_{n=-\infty}^{\infty} f_n \zeta^n, \quad f_n := \int_0^{2\pi} e^{-in\theta} f(e^{i\theta}) \frac{d\theta}{2\pi},$$

it is not too difficult to see that the L^p boundary function for f_+ has Fourier series

$$f_+ \sim \sum_{n=0}^{\infty} f_n \zeta^n.$$

Thus we can think of the Cauchy transform as projection operator $P : L^p \to H^p$, $Pf = f_+$. Hollenbeck and Verbitsky [41] compute the norm of the Riesz projection as

$$\sup\{\|f_+\|_p : \|f\|_{L_p} = 1\} = \frac{1}{\sin(\pi/p)}, \quad 1 < p < \infty.$$

The endpoint cases $p = 1$ and $p = \infty$ are more complicated. For example [25, pp. 63–64], the $f \in L^1$ whose Fourier series is

$$\sum_{n=2}^{\infty} \frac{\cos n\theta}{\log n}$$

has Cauchy transform equal to

$$f_+ = \sum_{n=2}^{\infty} \frac{z^n}{\log n}$$

which does not belong to H^1 [25, p. 48][3]. It is worth remarking here that not only does the Riesz projection $f \to f_+$ fail to be continuous from L^1 onto H^1, but there is no other continuous projection of L^1 onto H^1 [59]. If the function is slightly better than L^1, there is the following theorem of Zygmund [100] (see also [25, p. 58]): If $|f| \log^+ |f| \in L^1$, then $f_+ \in H^1$.

Despite these pathologies with $p = 1$, there is a well-known, and often revisited, theorem about the Cauchy transforms of L^1 functions due to Kolmogorov [50] (see also [51, p. 92]).

Theorem 3.7 (Kolmogorov).
For $f \in L^1$,

$$m(|f_+| > \lambda) \leqslant A \frac{\|f\|_1}{\lambda}, \quad \lambda > 0.$$

[3] If $f = \sum_n a_n z^n \in H^1$, then $\sum_n |a_n|/(n+1) \leqslant \pi \|f\|_1$.

We use the notation $m(|f_+| > \lambda)$ as a shorthand for $m(\{e^{i\theta} : |f_+(e^{i\theta})| > \lambda\})$. Kolmogorov's theorem can be use to give an alternate proof of Theorem 3.5 (see [51, p. 98]). It has also been generalized in several directions. First, the estimate

$$m(|K\mu| > \lambda) = O(\frac{1}{\lambda})$$

holds for any $\mu \in M$, not just the absolutely continuous ones. This result is somewhat folklore by now and a proof can be obtained by making minor changes to the proof of Kolmogorov's theorem in [51, p. 92]. Secondly, for $\mu \ll m$, Kolmogorov's estimate can be improved to

$$m(|K\mu| > \lambda) = o(\frac{1}{\lambda}). \tag{3.8}$$

Indeed, this inequality is true when $d\mu = p\,dm$, where p is a trigonometric polynomial. Now approximate any L^1 function with trigonometric polynomials and use Theorem 3.7. More surprising [93] is that the converse of this is true, namely

$$m(|K\mu| > \lambda) = o(\frac{1}{\lambda}) \Leftrightarrow \mu \ll m.$$

We will see a much stronger version of this in §8.

The $f \in L^\infty$ whose Fourier series is

$$\sum_{n=1}^{\infty} \frac{\sin n\theta}{n}$$

has Cauchy transform

$$f_+ = \sum_{n=1}^{\infty} \frac{z^n}{n} = \log(\frac{1}{1-z})$$

which is not bounded. As was true in the L^1 case, not only is the Riesz projection $f \to f_+$ not a continuous projection of L^∞ onto H^∞, there is no other continuous projection of L^∞ onto H^∞ [40, p. 155]. There is, however, a theorem of Spanne [88] and Stein [90] which characterizes the Cauchy transforms of L^∞ functions.

Theorem 3.9 (Spanne, Stein). *For $f \in L^\infty$, the Cauchy transform f_+ belongs to $BMOA$, the analytic functions of bounded mean oscillation. Moreover, the map $f \to f_+$ is continuous from L^∞ onto $BMOA$.*

See [29, Chapter 6] for a definition of BMO and $BMOA$ and their basic properties. For now, note that $H^\infty \subsetneq BMOA$.

For spaces of smooth functions, there are results about the action of the Cauchy transform. Some examples: (i) $C_+ = VMOA$ (where C denotes the continuous functions on $\mathbb{T}$, $C_+ = \{f_+ : f \in C\}$, and $VMOA$ are the analytic functions of vanishing mean oscillation) [80] (ii) If Λ^n_α $(n = 0, 1, 2, \ldots, 0 < \alpha < 1)$ denotes the Lipschitz classes and Λ^n_* $(n = 0, 1, 2, \ldots)$ denotes the Zygmund classes on $\mathbb{T}$, then $(\Lambda^n_\alpha)_+ \subseteq \Lambda^n_\alpha$ and $(\Lambda^n_*)_+ \subseteq \Lambda^n_*$ [69] (see also [51, p. 110] and [101]) (iii) One can check just by looking at Fourier and power series coefficients that if $f \in C^\infty(\mathbb{T})$, then every derivative of f_+ has a continuous extension to the closed disk $\mathbb{D}^-$.

What about Cauchy transforms of functions from weighted L^p spaces? The question here is the following: Given $1 < p < \infty$, what are the conditions on a measure $\mu \in M_+$ (the non-negative measures in M) such that

$$\int |g_+|^p d\mu \leqslant C \int |g|^p d\mu$$

for all trigonometric polynomials g? Though there has been earlier work on this problem [32, 38] (for example $d\mu$ must take the form $w dm$), the definitive result here is a celebrated theorem of Hunt, Muckenhoupt, and Wheeden [45] (a precise condition on the weight w called the "A_p condition").

In looking at the references for the above results, one notices that they all deal with "conjugate functions" (*fonctions conjuguées*). The relationship between this conjugation operator and the Cauchy transform is as follows. A computation reveals that

$$\frac{1}{1-\overline{\zeta}z} = \frac{1}{2}\left[\,1 + P_z(\zeta) + iQ_z(\zeta)\,\right],$$

where

$$P_z(\zeta) = \frac{1-|z|^2}{|\zeta - z|^2} \qquad Q_z(\zeta) = \frac{2\Im(\overline{\zeta}z)}{|\zeta - z|^2}, \quad \zeta \in \mathbb{T}, z \in \mathbb{D}, \tag{3.10}$$

are the Poisson and conjugate Poisson kernels. For $f \in L^1$,

$$f_+(z) = \frac{1}{2}\left[\int_{\mathbb{T}} f(\zeta)dm(\zeta) + \int_{\mathbb{T}} P_z(\zeta)f(\zeta)dm(\zeta) + i\int_{\mathbb{T}} Q_z(\zeta)f(\zeta)dm(\zeta)\,\right].$$

By Fatou's theorem [25, p. 5],

$$\lim_{\measuredangle z \to e^{i\theta}} \int_{\mathbb{T}} P_z(\zeta)f(\zeta)dm(\zeta) = f(e^{i\theta}) \text{ a.e.}$$

and [101, Vol. I, p. 131] the following limit

$$\widetilde{f}(e^{i\theta}) := \lim_{\measuredangle z \to e^{i\theta}} \int_{\mathbb{T}} Q_z(\zeta)f(\zeta)dm(\zeta)$$

exists for almost all $e^{i\theta}$. Here $\measuredangle$ denotes the non-tangential limit. Moreover,

$$\widetilde{f}(e^{i\theta}) = \lim_{\varepsilon \to 0^+} \frac{1}{2\pi}\int_{|\theta - t| > \varepsilon} \cot\left(\frac{\theta - t}{2}\right) f(e^{it})\, dt \text{ a.e.}$$

One can also define, for $\mu \in M$,

$$(Q\mu)(z) = \int Q_z(\zeta)d\mu(\zeta)$$

and note that the above function has non-tangential limits almost everywhere denoted by $\widetilde{\mu}(e^{i\theta})$. The operator $f \to \widetilde{f}$ (or $\mu \to \widetilde{\mu}$) is called the *conjugation operator*. Thus, for example,

$$f_+(e^{i\theta}) = \frac{1}{2}\left[\int_{\mathbb{T}} f(\zeta)dm(\zeta) + f(e^{i\theta}) + i\widetilde{f}(e^{i\theta})\,\right] \text{ a.e.}$$

Thus questions about the continuity of the operator $f \to f_+$ (on spaces of functions on $\mathbb{T}$) are equivalent to questions about the continuity of the conjugation operator $f \to \widetilde{f}$. The norm of the operator $f \to f_+$ is known when $1 < p < \infty$ (see the remarks following Theorem 3.6) and not quite understood when $0 < p < 1$. For the conjugation operator $f \to \widetilde{f}$ (as an operator from L^p to L^p when $p > 1$ and an operator from L^1 to L^p for $0 < p < 1$), the norm has been computed by Pichorides [63] as: $\tan(\pi/2p)$ if $1 < p \leqslant 2$; $\cot(\pi/2p)$ if $p > 2$; $(\cos(p\pi/2))^{-1/p}\|f\|_1 \leqslant \|\widetilde{f}\|_p \leqslant 2^{1/p-1}(\cos(p\pi/2))^{-1/p}\|f\|_1$ for $0 < p < 1$.

A variant of Kolmogorov's theorem (Theorem 3.7) says that for fixed $0 < p < 1$ and $\mu \in M$, $\widetilde{\mu}$ belongs to L^p and

$$\|\widetilde{\mu}\|_{L^p} \leqslant c_p\|\mu\|.$$

B. Davis [19, 20, 21] computes the best constant c_p, at least for real measures $M_{\mathbb{R}}$ as

$$\sup\{\|\widetilde{\mu}\|_{L^p} : \mu \in M_{\mathbb{R}}, \|\mu\| = 1\} = \|\widetilde{\nu}\|_{L^p},$$

where ν is a measure given by $\nu(\{1\})=1/2, \nu(\{-1\})=-1/2$ and $|\nu|(\mathbb{T}\setminus\{-1,1\})=0$.

Kolmogorov's theorem says that

$$m(|K\mu| > \lambda) \leqslant C\frac{\|\mu\|}{\lambda}, \quad \mu \in M.$$

The best constant C is unknown. However, there is information about the best constant in the related inequality

$$m(|\widetilde{\mu}| > \lambda) \leqslant C\frac{\|\mu\|}{\lambda}.$$

When $\mu \ll dm$, the best constant C is Θ^{-1}, where $\Theta = (1 - 3^{-2} + 5^{-2} - \cdots)/(1 + 3^{-2} + 5^{-2} + \cdots)$. For $\mu \in M_+$, the best constant C is one.

Here is a good place to mention that the conjugation operator Q is closely related to the *Hilbert transform*

$$(Hf)(x) := P.V.\frac{1}{\pi}\int_{-\infty}^{\infty}\frac{f(t)}{x-t}dt,$$

defined almost everywhere for $f \in L^1(\mathbb{R}, dx)$. Results for the conjugation operator frequently have direct analogs for the Hilbert transform in that they are both singular integral operators [29, 91].

4. Which analytic functions are Cauchy transforms?

For an analytic f on $\mathbb{D}$, when is $f = K\mu$? Gathering up the observations from previous sections of this survey, there are the following necessary conditions.

Proposition 4.1. *Suppose $f = K\mu$ for some $\mu \in M$. Then*

1. *f satisfies the growth condition $|f(z)| \leqslant C_\mu(1 - |z|)^{-1}$.*

2. *f has finite non-tangential limits m-almost everywhere on $\mathbb{T}$ and*

$$m(|f| > \lambda) \leqslant C_f/\lambda, \quad \lambda > 0.$$

3. *$f \in H^p$ for all $0 < p < 1$ and $\|f\|_p = O((1-p)^{-1})$.*
4. *If $f = \sum_{n \geqslant 0} a_n z^n$, then $(a_n)_{n \geqslant 1}$ is a bounded sequence of complex numbers.*

Known necessary and sufficient conditions for an analytic function on the disk to be a Cauchy transform are difficult to apply and in a way, the very question is unfair. For example, suppose that f is analytic on $\mathbb{D}$ with power series

$$f(z) = a_0 + a_1 z + a_2 z^2 + \cdots$$

and we want to know if $f = K\mu$ on $\mathbb{D}$ for some $\mu \in M$. Since, from eq. (3.3), $K\mu$ can be written as

$$(K\mu)(z) = \mu_0 + \mu_1 z + \mu_2 z^2 + \cdots,$$

we would be trying to determine (comparing a_n with μ_n) the measure μ from only "half" its Fourier coefficients (the non-negative ones).

If one can settle for a functional analysis condition, there is characterization of $\mathcal{K}$ [33], albeit difficult to apply. The proof follows directly by a duality argument (see §5 below).

Theorem 4.2 (Havin). *Suppose $f = \sum_{k=0}^{\infty} a_k z^k$ is analytic on $\mathbb{D}$. Then the following statements are equivalent.*

1. *There is some constant $C > 0$, depending only on f, such that*

$$\Big| \sum_{k=0}^{p} \lambda_k a_k \Big| \leqslant C \max \Big\{ \Big| \sum_{k=0}^{p} \frac{\lambda_k}{z^{k+1}} \Big| : z \in \mathbb{T} \Big\}$$

for any complex numbers $\lambda_0, \ldots, \lambda_p$.

2. *$f = K\mu$ for some $\mu \in M$.*

Instead of asking whether or not an analytic function defined only on $\mathbb{D}$ is a Cauchy transform, suppose we were to ask if an analytic function f on $\widehat{\mathbb{C}} \setminus \mathbb{T}$ is a Cauchy transform $\widehat{\mu}$ on $\widehat{\mathbb{C}} \setminus \mathbb{T}$ (recall the definition of $\widehat{\mu}$ from eq. (3.2)). This is a more tractable question since we would be comparing the Laurent series of these two functions which would involve knowing *all* of the Fourier series coefficients of μ and not just the non-negative ones as before (see eq. (3.3)). An early result which answers this question is one of Tumarkin [95] (see [55] for a generalization).

Theorem 4.3 (Tumarkin). *Let f be analytic on $\widehat{\mathbb{C}} \setminus \mathbb{T}$ with $f(\infty) = 0$ and set $f_1 = f|\mathbb{D}$ and $f_2 = f|\mathbb{D}_e$. Then $f = \widehat{\mu}$ for some $\mu \in M$ if and only if*

$$\sup_{0<r<1} \int_{\mathbb{T}} |f_1(r\zeta) - f_2(\zeta/r)| dm(\zeta) < \infty. \tag{4.4}$$

The Havin and Tumarkin results are easily seen to be equivalent using the technique of dual extremal problems as pioneered by S. Ja. Khavinson [36, 37] and W. Rogosinski and H. S. Shapiro [74].

There is a refinement of Tumarkin's due to Aleksandrov [2, Thm. 5.3] which even identifies the type of representing measure μ for f. From the previous section, a Cauchy transform $f = \widehat{\mu}$ on $\widehat{\mathbb{C}}\backslash\mathbb{T}$ satisfies the four conditions

$$f(\infty) = 0, \tag{4.5}$$

$$f \in \bigcap_{0<p<1} H^p(\widehat{\mathbb{C}} \setminus \mathbb{T}), \tag{4.6}$$

$$\|f\|_{H^p(\widehat{\mathbb{C}}\setminus\mathbb{T})} = O(\frac{1}{1-p}), \tag{4.7}$$

$$Jf \in L^1, \quad (Jf)(\zeta) := \lim_{r\to 1^-} \left[f(r\zeta) - f(\zeta/r) \right]. \tag{4.8}$$

Theorem 4.9 (Aleksandrov). *Let f be an analytic function on $\widehat{\mathbb{C}} \setminus \mathbb{T}$ satisfying the conditions in eq. (4.5) through eq. (4.8) above. Then $f = \widehat{\mu}$ for some $\mu \in M$. Moreover, if the conditions in eq. (4.5) and eq. (4.6) are satisfied, then*

1. *$f = \widehat{\mu}$ for some $\mu \ll dm$ if and only if*

$$\varliminf_{p\to 1^-} \|f\|_{H^p(\widehat{\mathbb{C}}\setminus\mathbb{T})}(1-p) = 0$$

and $Jf \in L^1$.

2. *$f = \widehat{\mu}$ for some $\mu \perp dm$ if and only if*

$$\varliminf_{p\to 1^-} \|f\|_{H^p(\widehat{\mathbb{C}}\setminus\mathbb{T})}(1-p) < \infty$$

and $Jf = 0$ m-almost everywhere.

There is even a further refinement. Let X be a class of analytic functions on $\mathbb{D}$ and E be a closed subset of $\mathbb{T}$. Let $\mathcal{F}(X, E)$ denote the functions $f \in X$ such that $f = K\mu$, where $\mu \in M$ and has support in E. Under what conditions on E is $\mathcal{F}(X, E) \neq (0)$? When $0 < p < 1$, notice that $\mathcal{F}(H^p, E) \neq (0)$ for every non-empty set E (Theorem 3.5). When $p \geqslant 1$, Havin proves that $\mathcal{F}(H^p, E) \neq (0)$ if and only $m(E) > 0$ [34]. Though the analysis is more complicated, Hruščev [42] answers this question for the disk algebra and various other spaces of functions that are smooth up to the boundary.

We close this section by mentioning the following generalization of the Cauchy integral formula involving the theory of A-integrals as studied by Denjoy, Titchmarsh, Kolmogorov, Ul'yanov, and Aleksandrov. A measurable function $g : \mathbb{T} \to \widehat{\mathbb{C}}$ is *A-integrable* if

$$m(|g| > t) = o(1/t) \tag{4.10}$$

and

$$(A) \int g(\zeta)dm(\zeta) := \lim_{t\to\infty} \int_{|g|<t} g(\zeta)dm(\zeta)$$

exists. For $\mu \in M$ with $\mu \ll m$, $f = K\mu$ has non-tangential boundary values m-almost everywhere and, by eq. (3.8), f satisfies the condition in eq. (4.10). However, this Cauchy transform may not belong to H^1 and so cannot be recovered from its boundary function via the Cauchy integral formula. A theorem of Ul'yanov [96] is the substitute "Cauchy A-integral formula".

Theorem 4.11 (Ul'yanov). *For $\mu \in M$ with $\mu \ll m$, the function $f = K\mu$ is A-integrable and*

$$f(z) = \frac{1}{2\pi i}(A)\int_{\mathbb{T}} \frac{f(\zeta)}{\zeta - z} d\zeta, \quad z \in \mathbb{D}. \tag{4.12}$$

A theorem of Aleksandrov [1] says that if $f \in N^{+}$ (the Smirnov class) satisfies $m(|f| > t) = o(1/t)$ as $t \to \infty$ then eq. (4.12) holds. The condition in eq. (4.10) cannot be weakened since the function $i(1+z)/(1-z)$ cannot be written as the Cauchy A-integral of its boundary values. More about this can be found in the references in [79]. There is even a version of the conjugation operator using A-integrals in [8] and moreover, this book also contains a nice historical treatment of A-integrals.

5. Topology on the space of Cauchy transforms

In this section we consider the space $\mathcal{K}$ of Cauchy transforms as a Banach space. We will show how it arises naturally as the dual space to the disk algebra A and indicate its relationship to the space of measures M.

Let $C = C(\mathbb{T})$ denote the Banach space of complex-valued continuous functions on the unit circle endowed with the supremum norm $\|f\|_{\infty}$. Every $\mu \in M$ determines a bounded linear functional ℓ_{μ} on C by

$$\ell_{\mu}(f) := \int_{\mathbb{T}} f \, \overline{d\mu}$$

with norm equal to the total variation norm $\|\mu\|$ of μ. Conversely, the Riesz representation theorem [77] guarantees that every bounded linear functional on C has such a representation, and in fact the map $\mu \to \ell_{\mu}$ is an isometric isomorphism from M onto C^{*} (the dual space of C). Thus, by tradition, we identify C^{*} with M.

The *disk algebra* A is the closure of the analytic polynomials in C or equivalently, the space of functions analytic on $\mathbb{D}$ that have continuous extensions to $\mathbb{D}^{-}$. Its annihilator $A^{\perp}$ is a closed subspace of $C^{*} \simeq M$ which we identify with the set of measures $\mu \in M$ for which

$$\int g \overline{d\mu} = 0 \text{ for all } g \in A.$$

By the F. and M. Riesz theorem (Theorem 3.1), such annihilating measures μ take the form $d\mu = \overline{f} dm$, where $f \in H^{1}_{0}$ and so we identify $A^{\perp}$ and $\overline{H^{1}_{0}}$. We can identify $C^{*}/A^{\perp}$ with A^{*} via the mapping $\ell_{\mu} + A^{\perp} \to \ell_{\mu}|A$ and furthermore, endowing $C^{*}/A^{\perp}$ with the usual quotient space norm

$$\|\ell_{\mu} + A^{\perp}\| = \operatorname{dist}(\ell_{\mu}, A^{\perp}),$$

the above mapping is an isometric isomorphism [78, pp. 96–97]. Putting this all together, we have

$$A^* \simeq M/\overline{H_0^1},$$

where, as before, $\simeq$ denotes an isometric isomorphism.

Since $K\mu = 0$ if and only if $\mu \in A^\perp \simeq \overline{H_0^1}$, the map $\mu + \overline{H_0^1} \to K\mu$ from $M/\overline{H_0^1}$ to $\mathcal{K}$ is bijective. From here it makes sense to endow $\mathcal{K}$ with the norm of $M/\overline{H_0^1}$, that is

$$\|K\mu\| := \operatorname{dist}(\mu, \overline{H_0^1}) = \inf\{\|d\mu + f dm\| : f \in \overline{H_0^1}\}.$$

Hence $\mathcal{K} \simeq M/\overline{H_0^1}$.

Finally, recall that for $f \in \mathcal{K}$ we let $M_f = \{\mu \in M : f = K\mu\}$ be the set of representing measures for f. Since $K\mu = K\nu$ if and only if $d\mu - d\nu = \overline{f} dm$ ($f \in H_0^1$), it follows that

$$\|f\| = \inf\{\|\nu\| : \nu \in M_f\}.$$

Gathering up these facts, we have the following summary result.

Theorem 5.1. *The norm dual of A can be identified in an isometric and isomorphic way with $\mathcal{K}$ via the sesquilinear pairing*

$$\langle f, K\mu\rangle = \int_{\mathbb{T}} f\, \overline{d\mu}, \tag{5.2}$$

or equivalently, by a power series computation,

$$\langle f, K\mu\rangle = \lim_{r\to 1^-} \sum_{n=0}^{\infty} f_n \overline{\mu_n} r^n, \tag{5.3}$$

where $(f_n)_{n\geqslant 0}$ are the Taylor coefficients of f and $(\mu_n)_{n\geqslant 0}$ are the Fourier coefficients of μ.

From the Lebesgue decomposition theorem, the space of measures M admits a direct sum decomposition

$$M = L^1 \oplus M_s.$$

where L^1 is identified with the absolutely continuous measures M_a. Since $\overline{H_0^1} \subset L^1$, A^* satisfies

$$A^* \simeq L^1/\overline{H_0^1} \oplus M_s.$$

Similarly, the space of Cauchy transforms can be decomposed as

$$\mathcal{K} = \mathcal{K}_a \oplus \mathcal{K}_s,$$

where $\mathcal{K}_a = \{K\mu : \mu \ll m\}$ and $\mathcal{K}_s = \{K\mu : \mu \perp dm\}$. Furthermore, if $\mu = \mu_a + \mu_s$ ($\mu_a \ll m$ and $\mu_s \perp m$), then

$$\|K\mu\| = \|K\mu_a\| + \|K\mu_s\|.$$

We point out a few more items of interest. First,

$$\mu \perp dm \Rightarrow \|K\mu\| = \|\mu\|.$$

Secondly, $\mathcal{K}_a \simeq L^1/\overline{H_0^1}$ and is the norm closure of the analytic polynomials in $\mathcal{K}$. Thirdly, $\mathcal{K}_s \simeq M_s$ and as such is non-separable since M_s is non-separable. Fourth, since, for $f \in \mathcal{K}$,

$$\|f\| = \inf\{\|\nu\| : \nu \in M_f\},$$

there is a sequence of measures $(\nu_n)_{n\geqslant 1}$ in M_f such that $\|\nu_n\| \leqslant \|f\| + 1/n$ for each n. If ν is a weak-$*$ limit point of $(\nu_n)_{n\geqslant 1}$ in M, it follows that $\nu \in M_f$ and $\|\mathcal{K}\mu\| = \|\nu\|$. It is known that the measure ν is *unique* [35, 46]. We denote this unique representing measure ν by ν_f. Since all measures in M_f have the same singular part, a similar theorem holds for $\mathcal{K}_a$. Indeed, if $f \in \mathcal{K}_a$, then there is a unique $h \in L^1$ such that $h\,dm \in M_f$ and $\|f\| = \|h\|_1$.

There is another topology on $\mathcal{K}$ namely the weak-$*$ topology it inherits from being the norm dual of A. The Banach-Alaoglu theorem says that the closed unit ball of $\mathcal{K}$ is compact. In this case, the weak-$*$ topology on bounded sets can be characterized by its action on sequences. To be specific, we say that a sequence $(f_n)_{n\geqslant 1} \subset \mathcal{K}$ converges weak-$*$ to zero if $\langle g, f_n\rangle \to 0$ for every $g \in A$. From the pointwise estimate

$$|f(z)| \leqslant \frac{1}{1-|z|}\|f\| \text{ for all } z \in \mathbb{D}, f \in \mathcal{K}^{\,4} \tag{5.4}$$

one can prove that a sequence $(f_n)_{n\geqslant 1}$ in $\mathcal{K}$ converges to f weak-$*$ if and only if $(f_n)_{n\geqslant 1}$ is uniformly bounded in the norm of $\mathcal{K}$ and $f_n(z) \to f(z)$ for each $z \in \mathbb{D}$ [13, Prop. 2]. Although $\mathcal{K}$ is not separable in the norm topology, it is separable in the weak-$*$ topology. This follows from the same fact about M [5] and the weak-$*$ continuity of the canonical quotient map $\pi : M \to M/\overline{H_0^1} \simeq \mathcal{K}$.

Since $\mathcal{K}$, with the norm topology, is a Banach space, it can be endowed with a weak topology, though the dual space of $\mathcal{K}$ is not a readily identifiable space of analytic functions. There are two interesting theorems regarding this weak topology on $\mathcal{K}$. The first deals with weak completeness. A Banach space is weakly complete (weak Cauchy nets convergence in the space) if and only if it is reflexive (a consequence of Goldstine's theorem [23, p. 13]) and so $\mathcal{K}$ is not weakly complete. However, $\mathcal{K}$ is weakly sequentially complete. In general, a sequence $(x_n)_{n\geqslant 1}$ in a Banach space X is weak Cauchy if the numerical sequence $(\ell(x_n))_{n\geqslant 1}$ converges for each $\ell \in X^*$. Thus X is weakly sequentially complete if every weak Cauchy sequence in X converges weakly to some element of X. Mooney's Theorem [57] asserts that $\mathcal{K}$ is weakly sequentially complete and is a consequence of the following result: Let $(\phi_n)_{n\geqslant 1}$ be a sequence in L^1 such that the limit

$$\lim_{n\to\infty} \int_{\mathbb{T}} \phi_n f\,dm := L(f)$$

[4] $\|f\|$ is the norm in $\mathcal{K}$. We know from eq. (3.4) that $|f(z)| \leqslant \|\mu\|(1-|z|)^{-1}$, where $\mu \in M_f$. When $\mu = \mu_f$, then $\|\mu\| = \|f\|$.

[5] In M, the linear span of the unit point masses $\{\delta_\zeta : \zeta = e^{i\theta}, \theta \in \mathbb{Q}\}$ form a weak-$*$ dense set.

exists for every $f \in H^\infty$. Then there is a function $\phi \in L^1$ such that

$$L(f) = \int_{\mathbb{T}} \phi f \, dm \quad \text{for all } f \in H^\infty.$$

A proof of this can be found in [29, pp. 206–209]. How does this imply Mooney's theorem? Since $\overline{H_0^1}$ is the annihilator of H^∞ in L^1, it follows that H^∞ can be regarded as the dual space of $L^1/\overline{H_0^1}$. The above theorem is then just the statement that $L^1/\overline{H_0^1}$ is weakly sequentially complete. Now notice that $\mathcal{K} \simeq L^1/\overline{H_0^1} \oplus M_s$ and so $\mathcal{K}^* \simeq H^\infty \oplus M_s^*$. Suppose $(\phi_n)_{n\geqslant 1}$ is a weak Cauchy sequence in $\mathcal{K}$. Each ϕ_n has a unique decomposition as $\phi_n = \psi_n + \nu_n$, where $\psi_n \in \mathcal{K}_a$ and $\nu_n \in \mathcal{K}_s$. Because of the direct sum decomposition of $\mathcal{K}^*$, it is evident that each of the sequences $(\psi_n)_{n\geqslant 1}$ and $(\nu_n)_{n\geqslant 1}$ is a weak Cauchy sequence. By the result quoted above, the first sequence converges weakly to some $\psi \in \mathcal{K}_a$. On the other hand, a theorem of Kakutani says that M_s is isometrically isomorphic to $L^1(\Omega, \Sigma, \mu)$ for some abstract measure space (Ω, Σ, μ) [47]. It follows that M_s is weakly sequentially complete since every such space $L^1(\Omega, \Sigma, \mu)$ is weakly sequentially complete [23].

The second theorem we will present on the weak topology in $\mathcal{K}$ is a deep result due independently to Delbaen [22] and Kisliakov [49]. For each $f \in \mathcal{K}$, recall that μ_f is the unique measure such that $f = K\mu_f$ and $\|f\| = \|\mu_f\|$.

Theorem 5.5. *Let W be a weakly compact set in $\mathcal{K}$, and let $\widehat{W} = \{\, \mu_f : f \in W \,\}$. Then $\widehat{W}$ is relatively weakly compact.*

A thorough discussion of this theorem can be found in [62, Ch. 7] or [99]. This theorem derives its significance from the Dunford-Pettis Theorem characterizing weakly compact subsets in M. This characterization says that a set W in M is weakly compact if and only if it is bounded and uniformly absolutely continuous (cf. [23]).

We now talk about a basis for $\mathcal{K}$. A sequence $(x_n)_{n\geqslant 1}$ in a Banach space X is a *Schauder basis* for X if every $x \in X$ can be written uniquely as

$$x = \sum_{n=1}^{\infty} c_n x_n,$$

where c_n are complex numbers and the $=$ sign means convergence in the norm of X. A sequence $(\ell_n)_{n\geqslant 1} \subset X^*$ is called a *weak-$*$ Schauder basis* if every $\ell \in X^*$ can be written uniquely as

$$\ell = \sum_{n=1}^{\infty} d_n \ell_n,$$

where d_n are complex numbers and $=$ means weak-$*$ convergence.

For a Schauder basis $(x_n)_{n\geqslant 1}$, there is a natural sequence $(x_n^*)_{n\geqslant 1}$ of continuous linear functionals defined by

$$x_n^*(f) = c_n, \quad \text{where} \quad f = \sum_{n=1}^{\infty} c_n x_n$$

(remember that the expansion of f is unique and so x_n^* is well-defined). One can prove the following: If $(x_n)_{n\geqslant 1}$ is a Schauder basis for X, then $(x_n^*)_{n\geqslant 1}$ is a weak-* Schauder basis for X^*. If $(x_n)_{n\geqslant 1}$ is a Schauder basis for X, then $(x_n^*)_{n\geqslant 1}$ is a Schauder basis for its closed linear span [23, p. 52].

Let us apply the above to $X = A$ and $X^* \simeq \mathcal{K}$ to identify a weak-* Schauder basis for $\mathcal{K}$ and a Schauder basis for $\mathcal{K}_a \simeq L^1/\overline{H_0^1}$. The result which drives this is one of Bočkarev [10] which produces a Schauder basis $(b_n)_{n\geqslant 1}$ for the disk algebra A. Moreover, by its actual construction, $\{b_n\}_{n\geqslant 1}$ is an orthonormal set in L^2. Let $B_n := (b_n)_+$. We do this to not confuse b_n, the element of the disk algebra A, with B_n, the element of $\mathcal{K}$. The result here is the following.

Proposition 5.6. *$(B_n)_{n\geqslant 1}$ is a weak-$*$ Schauder basis for $\mathcal{K}$ and a Schauder basis for $\mathcal{K}_a$.*

6. Multipliers of the space of Cauchy transforms

For a linear space of analytic functions X on the unit disk, an important collection of functions are those ϕ, analytic on $\mathbb{D}$, for which $\phi f \in X$ whenever $f \in X$. Such ϕ are called *multipliers* of X and we denote this class by $M(X)$. For the Hardy spaces H^p $(0 < p \leqslant \infty)$ it is routine to check that $M(H^p) = H^\infty$. For other classes of functions such as the classical Dirichlet space or the Besov spaces, characterizing the multipliers is more difficult [56, 89]. For the space of Cauchy transforms, the multipliers are not completely understood but they do have several interesting properties.

For example, there is a nice relationship between multipliers and Toeplitz operators. If $\phi \in H^\infty$ and $T_{\overline{\phi}}(f) = P(\overline{\phi} f)$ is the co-analytic Toeplitz operator, then the Riesz theorem (Theorem 3.6) says that P, and hence $T_{\overline{\phi}}$, is a bounded operator on H^p for $1 < p < \infty$. When $p = \infty$, the Riesz projection P is unbounded. However, this following theorem of Vinogradov [97] determines when $T_{\overline{\phi}}$ is bounded on H^∞.

Proposition 6.1. *For $\phi \in H^\infty$, the following are equivalent.*

1. *$\phi \in M(\mathcal{K})$.*
2. *$\phi \in M(\mathcal{K}_a)$.*
3. *$T_{\overline{\phi}} : A \to A$ is bounded.*
4. *$T_{\overline{\phi}} : H^\infty \to H^\infty$ is bounded.*

Moreover, $\|T_{\overline{\phi}} : A \to A\|$ is equal to the multiplier norm, see below, of ϕ.

If ϕ is a multiplier of $\mathcal{K}$, the closed graph theorem, says that the multiplication operator $f \to \phi f$ on $\mathcal{K}$ is continuous. The norm of this operator is denoted by $\|\phi\|$. The weak-$*$ density of the convex hull of $\{\delta_\zeta : \zeta \in \mathbb{T}\}$ in the unit ball of M, together with the identity $(K\delta_\zeta)(z) = (1 - \overline{\zeta}z)^{-1}$ yields [98]

$$\|\phi\| = \sup \{\| \frac{\phi}{1 - \overline{\zeta}z} \|_{\mathcal{K}}: \zeta \in \mathbb{T} \} .$$

Since the constant functions belong to $\mathcal{K}$, any multiplier belongs to $\mathcal{K}$. In fact, one can show that any multiplier ϕ is a bounded function on $\mathbb{D}$ and satisfies $\|\phi\|_\infty \leqslant \|\phi\|$. This standard fact is true for the multipliers for just about any Banach space of analytic functions. Although a usable characterization of the multipliers of $\mathcal{K}$ is unknown, there are some sufficient conditions one can quickly check to see if a given $\phi \in H^\infty$ is a multiplier of $\mathcal{K}$. See [98] for details.

Theorem 6.2 (Goluzina-Havin-Vinogradov). *Let ϕ be an analytic function on $\mathbb{D}$. Then any one of the following conditions imply that ϕ is a multiplier of $\mathcal{K}$.*

1. $\phi \in H^\infty$ *and*

$$\sup \{ \int_{\mathbb{T}} | \frac{\phi(\zeta) - \phi(\xi)}{\zeta - \xi} | \, dm(\xi) : \zeta \in \mathbb{T} \} < \infty.$$

2. ϕ *extends to be continuous on* $\mathbb{D}^-$ *and*

$$\int_0 \frac{\omega(t)}{t} dt < \infty,$$

where ω is the modulus of continuity of ϕ.

3. *The quantity*

$$\sum_{n=0}^{\infty} \frac{|\phi^{(n)}(0)|}{n!} \log(n+2)$$

is finite.

This next theorem says that the multipliers of $\mathcal{K}$ are reasonably regular near the boundary of $\mathbb{D}$. Again, see [98] for details.

Theorem 6.3. *If ϕ is a multiplier of $\mathcal{K}$, then*

1. *ϕ has a finite non-tangential limit for every $\zeta \in \mathbb{T}$.*
2. *The Taylor polynomials are uniformly bounded in H^∞ norm, that is to say,*

$$\sup \{ \Big\| \sum_{k=0}^{n} \frac{\phi^{(k)}(0)}{k!} z^k \Big\|_\infty : n = 0, 1, 2, \dots \} < \infty.$$

3. $(1-z)^2 \phi' \in H^1(\{|z - 1/2| < 1/2\})$.

Statement (1) says that not all Blaschke products are multipliers of $\mathcal{K}$ since there are Blaschke products which do not have radial limits at certain points of the circle. In fact, the Blaschke product with zeros $(a_n)_{n\geqslant 1}$ has a finite non-tangential limit at $\zeta \in \mathbb{T}$ if and only if

$$\sum_{n=1}^{\infty} \frac{1-|a_n|^2}{|\zeta - a_n|} < \infty.$$

This is an old result of Frostman [27]. Statement (3) says that the singular inner function $\exp((z+1)/(z-1))$ is not a multiplier. If not all inner functions are multipliers, which ones are? This question has a complete answer due to a deep and quite difficult theorem of Hruščev and Vinogradov [44].

Theorem 6.4 (Hruščev-Vinogradov). *An inner function ϑ is a multiplier for $\mathcal{K}$ if and only if ϑ is a Blaschke product whose zeros $(a_n)_{n\geqslant 1}$ – repeated according to multiplicity – satisfy the uniform Frostman condition*

$$\sup \Big\{ \sum_{n=1}^{\infty} \frac{1-|a_n|^2}{|\zeta - a_n|} : \zeta \in \mathbb{T} \Big\} < \infty.$$

We can also talk about divisors. Suppose that X is a class of analytic functions contained in the Smirnov class N^+. We say that X has the *F-property* if whenever $f \in X$ and ϑ is an inner function which "divides f", i.e., $f/\vartheta \in N^+$, then $f/\vartheta \in X$. From the factorization theorem for H^p functions [25, p. 24], H^p certainly has the F-property, as do other well-known classes of analytic functions such as $BMOA$, the disk algebra, and the analytic Lipschitz and Besov classes [84]. A theorem of Hruščev and Vinogradov [44, 97] says that the space of Cauchy transforms, as well as their multipliers, also enjoy the F-property.

Theorem 6.5 (Hruščev-Vinogradov).

1. *The space $\mathcal{K}$ has the F-property.*
2. *The space of multipliers of $\mathcal{K}$ has the F-property.*

7. Operators on the space of Cauchy transforms

On the Hardy spaces H^p $(0 < p < \infty)$, there are a variety of operators worthy of study. Several that immediately come to mind are

$$(Sf)(z) = zf(z) \quad \text{(the forward shift)}$$

$$(Bf)(z) = \frac{f - f(0)}{z} \quad \text{(the backward shift)}$$

$$(C_\phi f)(z) = f(\phi(z)) \quad \text{(composition by } \phi : \mathbb{D} \to \mathbb{D})$$

$$(Cf)(z) = \frac{1}{z}\int_0^z \frac{f(t)}{1-t} dt \quad \text{(the Cesàro operator).}$$

We wish to make a few remarks about these operators on the space of Cauchy transforms.

The forward shift. It is easy to see that the forward shift $(Sf)(z) = zf(z)$ is a well-defined operator from $\mathcal{K}$ to itself and, by our dual pairing in eq. (5.3), $B^* = S$, where B is the backward shift operator on A (the disk algebra). One quickly sees that for all $f \in A$, $\|Bf\|_\infty = \|f - f(0)\|_\infty \leqslant 2\|f\|_\infty$. Furthermore, if

$$f_r(z) = \frac{z+r}{1+rz},$$

then $\|f_r - f_r(0)\|_\infty = 1+r$. It follows that the operator norm of $B : A \to A$ is equal to 2 and so the operator norm of $S : \mathcal{K} \to \mathcal{K}$ is also equal to 2. It also follows from duality that S is continuous on $\mathcal{K}$ when $\mathcal{K}$ is endowed with the weak-$*$ topology.

If ϑ is an inner function, then ϑH^p is a (norm) closed S-invariant subspace of H^p for each $0 < p < \infty$ and a theorem of Beurling [25, p. 114] says that all (non-trivial) norm closed S-invariant subspaces of H^p take this form. What is the analog of Beurling's theorem for $\mathcal{K}$? Endowed with the norm topology, $\mathcal{K}$ is non-separable and so characterizing its norm closed S-invariant subspaces is troublesome. However, $\mathcal{K}$, endowed with the weak-$*$ topology, is separable and so characterizing its weak-$*$ closed S-invariant subspaces is a more tractable problem. Since not all inner functions are multipliers of $\mathcal{K}$ (see Theorem 6.4), then $\vartheta\mathcal{K}$ is not always a subset of $\mathcal{K}$ (unlike $\vartheta H^p \subseteq H^p$). However, the subspace

$$\vartheta(\mathcal{K}) := \{f \in \mathcal{K} : f/\vartheta \in \mathcal{K}\}$$

does make sense and is clearly S-invariant, although it is not immediately clear that $\vartheta(\mathcal{K})$ is weak-$*$ closed. A theorem of Aleksandrov [3] is our desired "Beurling's theorem" for $\mathcal{K}$.

Theorem 7.1 (Aleksandrov). *For each inner function ϑ, $\vartheta(\mathcal{K})$ is a weak-$*$ closed S-invariant subspace of $\mathcal{K}$. Furthermore, if $\mathcal{M}$ is a non-zero weak-$*$ closed S-invariant subspace of $\mathcal{K}$, then there is an inner function ϑ, such that $\mathcal{M} = \vartheta(\mathcal{K})$.*

The backward shift. The backward shift

$$(Bf)(z) = \frac{f - f(0)}{z}$$

is a well-defined operator from $\mathcal{K}$ to itself and is the adjoint (under the pairing eq. (5.3)) of the forward shift S on A. Since the operator norm of $S : A \to A$ is equal to one, then the operator norm of $B : \mathcal{K} \to \mathcal{K}$ is also equal to one.

The space $\mathcal{K}_a = \{f_+ : f \in L^1\}$ is the norm closure of the polynomials and hence is separable. A theorem of Aleksandrov [3] (see also [16, p. 99]) characterizes the B-invariant subspaces of $\mathcal{K}_a$.

Theorem 7.2 (Aleksandrov). *If $\mathcal{M}$ is a norm closed B-invariant subspace of $\mathcal{K}_a$, then there is an inner function ϑ such that $f \in \mathcal{M}$ if and only if there is a $G \in N^+(\mathbb{D}_e)$ with $G(\infty) = 0$ and such that*

$$\lim_{r \to 1^-} \frac{f}{\vartheta}(r\zeta) = \lim_{r \to 1^-} G(\zeta/r)$$

for m-almost every $\zeta \in \mathbb{T}$.

The function G is known in the literature as a "pseudocontinuation" of the function f/ϑ (see [75] for more on pseudocontinuations). We compare Aleksandrov's result to the characterization of the backward shift invariant subspaces of H^2 which, as mentioned in the introduction, are all of the form $(\vartheta H^2)^{\perp}$. A theorem of Douglas-Shapiro-Shields [24] says that $f \in (\vartheta H^2)^{\perp}$ if and only if there is a function $G \in H^2(\mathbb{D}_e)$, with $G(\infty) = 0$, such that G is a pseudocontinuation of f/ϑ, i.e., G (from the outside of the disk) has the same radial limits as f/ϑ (from the inside of the disk) almost everywhere.

For a weak-$*$ closed B-invariant subspace $\mathcal{N} \subseteq \mathcal{K}$, the dual pairing eq. (5.3) tells us that $\mathcal{N}_{\perp}$ (the pre-annihilator of $\mathcal{N}$) is an S-invariant subspace of A. Since A is a Banach algebra and polynomials are dense in A, then $\mathcal{N}_{\perp}$ is a closed ideal of A. A result of Rudin [76] (see also [40, p. 82]) characterizes these ideals by their inner factors and their zero sets on the circle. A result in [15] uses the Rudin characterization to describe the corresponding $\mathcal{N}$ using analytic continuation across certain portions of the circle. This result also has connections to an analytic continuation result of Korenblum [52]. Some partial results on the B-invariant subspaces of $\mathcal{K}$, when endowed with the norm topology, can be found in [15].

Composition operators. For an analytic map $\phi : \mathbb{D} \to \mathbb{D}$, define, for a Cauchy transform f, the function

$$(C_{\phi} f)(z) = f(\phi(z)).$$

Clearly $C_{\phi} f$ is an analytic function on $\mathbb{D}$. What is not immediately clear is that $C_{\phi} f$ is a Cauchy transform. The proof of this follows from the Herglotz theorem [25, p. 2] and the decomposition of any measure as the complex linear sum of four positive measures. Furthermore, we note that if G is simply connected and the Riemann map $g : \mathbb{D} \to G$ belongs to $\mathcal{K}$, then any analytic map f of $\mathbb{D}$ into G belongs to $\mathcal{K}$. Indeed, set $\phi = g^{-1} \circ f$ and notice that $f = C_{\phi} g \in \mathcal{K}$. Finally, if f is analytic on $\mathbb{D}$ and $\mathbb{C} \setminus f(\mathbb{D})$ contains at least two oppositely oriented half-lines, then $f \in \mathcal{K}$ [12].

Without too much difficulty, one can show that C_{ϕ} has closed graph and so C_{ϕ} is bounded on $\mathcal{K}$. Bourdon and Cima [12] proved that

$$\|C_{\phi}\| \leqslant \frac{2 + 2\sqrt{2}}{1 - |\phi(0)|}$$

which was improved to

$$\|C_{\phi}\| \leqslant \frac{1 + 2|\phi(0)|}{1 - |\phi(0)|}$$

by Cima and Matheson [14]. Moreover, equality is attained for certain linear fractional maps ϕ.

The theory of Clark measures, as discussed in §1, leads us to a characterization of the compact composition operators on $\mathcal{K}$. Following Sarason, for each self map $\phi : \mathbb{D} \to \mathbb{D}$, we define an operator on M as follows: For μ in M_+ the function $(P\mu)(z)$, the Poisson integral of the measure μ, is positive and harmonic in $\mathbb{D}$.

Hence, the function $v(z) = (P\mu)(\phi(z))$ is also positive and harmonic. By the Herglotz theorem there is $\nu \in M_+$ with $v(z) = (P\nu)(z)$. Let $S_\phi(\mu) = \nu$, where ν is the (positive) measure for which $(P\nu)(z) = (P\mu)(\phi(z))$. Extended S_ϕ linearly to all of M in the obvious way (writing every measure as a complex linear combination of four positive measures).

Theorem 7.3 (Sarason [81]**).** *The operator S_ϕ is compact if and only if each Clark measure μ_α associated with ϕ satisfies $\mu_\alpha \ll m$ for all $\alpha \in \mathbb{T}$. Moreover, S_ϕ maps L^1 continuously to L^1.*

We comment that Sarason's result is equivalent to J. Shapiro's condition for compactness on H^2 [14, 83]. Also worth noting is that if the map ϕ is inner, it can not induce a compact operator in this setting, since the Clark measure σ_α will be carried by $\{\phi = \alpha\}$, and hence is singular. Using these results, one can show the following [14].

Theorem 7.4. *Let ϕ be an analytic self map of the disk. Then:*

1. *C_ϕ is compact on $\mathcal{K}$ if and only if S_ϕ is compact on M.*
2. *If C_ϕ is weakly compact on $\mathcal{K}$, then C_ϕ is compact on $\mathcal{K}$.*

The Cesàro operator. The Cesàro operator, as it originally appeared in the operator setting, was simply the map defined on the sequence space $(\ell^2)_+$ by

$$(a_n)_{n\geqslant 0} \to (b_N)_{N\geqslant 0},$$

where

$$b_N := \frac{1}{N+1}\big(a_0 + a_1 + \cdots + a_N\big), \quad N = 0, 1, 2, \ldots.$$

is the N-th Cesàro mean of the sequence $(a_n)_{n\geqslant 0}$. It is easy to see, equating the ℓ^2 sequence $(a_n)_{n\geqslant 0}$ with the H^2 function $f = \sum a_n z^n$, that this operator, denoted by C, can be viewed on H^2 as the integral operator

$$(Cf)(z) := \frac{1}{z}\int_0^z \frac{f(t)}{(1-t)}dt.$$

Cima and Siskakis [17] prove that the Cesàro operator is continuous from $\mathcal{K}$ to $\mathcal{K}$. An interesting problem to consider is whether or not any of the generalized Cesàro operators on H^p considered in [6, 7] are continuous on $\mathcal{K}$.

8. Distribution of boundary values

For a measurable (with respect to normalized Lebesgue measure m) function f on $\mathbb{T}$, the distribution function

$$\lambda \to m(|f| > \lambda)$$

is seen throughout analysis. For example, there is Chebyshev's inequality

$$m(|f| > \lambda) \leqslant \frac{1}{\lambda^p}\|f\|_p$$

and the "layer cake representation" formula

$$\|f\|_p^p = p\int_0^\infty \lambda^{p-1} m(|f| > \lambda)\,d\lambda.$$

In this brief section, we mention some known results about the distribution functions for $K\mu$ and $Q\mu$ and how they relate in quite surprising ways to μ.

Perhaps the earliest result on the distribution function for $Q\mu$ is one of G. Boole [11] who in 1857 found, in the special case where μ is a finite positive linear combination of point masses, the exact formula

$$m(Q\mu > \lambda) = m(Q\mu < -\lambda) = \frac{1}{\pi}\arctan\frac{\|\mu\|}{\lambda}.$$

Although, for positive measures with an absolutely continuous component (i.e., $d\mu/dm \not\equiv 0$), Boole's formula is not true, it is "asymptotically true" as more recent results of Tsereteli [94] and Poltoratski [66] show.

Theorem 8.1 (Poltoratski). *Let $\mu \in M_+$ and $E_\mu = \{\zeta \in \mathbb{T} : \frac{d\mu}{dm}(\zeta) > 0\}$. Then*

$$\frac{1}{\pi}\arctan\frac{\|\mu\|}{\lambda} - m(E_\mu) \leqslant m(\{\, Q\mu > \lambda \,\} \setminus E_\mu) \leqslant \frac{1}{\pi}\arctan\frac{\|\mu\|}{\lambda},$$

$$\frac{1}{\pi}\arctan\frac{\|\mu\|}{\lambda} - m(E_\mu) \leqslant m(\{\, Q\mu < -\lambda \,\} \setminus E_\mu) \leqslant \frac{1}{\pi}\arctan\frac{\|\mu\|}{\lambda}.$$

The above theorem is an improvement of [94]. Another fascinating aspect of $Q\mu$ (and hence $K\mu$) is a result of Stein and Weiss [92] (see also [48, p. 71]) which says that if χ_U is the characteristic function of some set $U \subset \mathbb{T}$, then the distribution function for $Q(\chi_U dm)$ depends only on $m(U)$ and not on the particular geometric structure of U.

The next series of results deal with reproducing the singular part of a measure from knowledge of the distribution function for $Q\mu$ (or $K\mu$). Probably one of the earliest of such theorems is this one of Hruščev and Vinogradov [43].

Theorem 8.2 (Hruščev and Vinogradov). *For any $\mu \in M$,*

$$\lim_{\lambda\to\infty} \pi\lambda\, m(|K\mu| > \lambda) = \|\mu_s\|.$$

Note that if $\mu \ll m$, then we get the well-known estimate (see eq. (3.8))

$$m(|K\mu| > \lambda) = o(1/\lambda).$$

Recent work of Poltoratski [66] says that not only can one recover the total variation norm of μ_s from the distribution function for $K\mu$, but one can actually recover the measure μ_s. For any measure $\mu \in M$ decomposed as $\mu = (\mu_1 - \mu_2) + i(\mu_3 - \mu_4)$, where $\mu_j \geqslant 0$, let $|\mu|$ be the measure $|\mu| := \sum_{j=1}^4 \mu_j$.

Theorem 8.3 (Poltoratski). *For $\mu \in M$, the measure $\pi\lambda\chi_{\{|K\mu|>\lambda\}} \cdot m$ converges weak-$*$ to $|\mu_s|$ as $\lambda \to \infty$.*

Though we do not have time to go into the details, we do mention that at the heart of these Poltoratski theorems is the Aleksandrov conditional expectation operator and the relationship between the distribution functions for $Q\mu$ and $K\mu$ and Clark measures (as described in §1).

9. The normalized Cauchy transform

If σ_α is a Clark measure for the inner function ϑ, the adjoint $\mathcal{F}_\alpha^*$ of the unitary operator $\mathcal{F}_\alpha \colon (\vartheta H^2)^\perp \to L^2(\sigma_\alpha)$ (as described in §1) is

$$\mathcal{F}_\alpha^* f = \frac{K(f\, d\sigma_\alpha)}{K(\sigma_\alpha)}$$

In general, for $\mu \in M_+$ we can consider the *normalized Cauchy transform*

$$V_\mu f = \frac{K(f\, d\mu)}{K(\mu)}, \quad f \in L^1(\mu).$$

This transform has been studied by several authors, beginning with D. Clark [18], with significant contributions by Aleksandrov [4] and Poltoratski [65, 66, 67].

In this section we wish to describe some recent work of Poltoratski on maximal properties of this normalized transform. Poltoratski's motivation for this study is the famous theorem of Hunt, Muckenhoupt, and Wheeden [29, p. 255] [45], which states that the Cauchy transform is bounded on $L^p(w)$, $1 < p < \infty$, if and only if w is an "A_p-weight". Moreover, if MKf denotes the standard nontangential maximal function of the Cauchy transform Kf, we have similarly $MKf \in L^p(w)$ for all $f \in L^p(w)$ if and only if w is an A_p-weight.

We can ask about the behavior of the normalized Cauchy transform V_μ and the associated maximal operator MV_μ on $L^p(\mu)$. We first note that V_μ is always bounded on $L^2(\mu)$. In fact if $\mu = m$, then $V_\mu(L^2(\mu))$ is just the Hardy space H^2, while if μ is singular $V_\mu(L^2(\mu)) = (\vartheta H^2)^\perp$, where ϑ is the inner function related to μ by the formula

$$K\mu = \frac{1}{1-\vartheta}.$$

If μ is an arbitrary positive measure, $V_\mu(L^2(\mu))$ is the de Branges-Rovnyak space $\mathcal{M}_\vartheta$, where ϑ comes from the above formula, but is no longer inner [82]. In particular, V_μ is a bounded operator from $L^2(\mu)$ to H^2 for any $\mu \in M_+$. When $p \neq 2$ the situation is described by the following theorem of Aleksandrov [5].

Theorem 9.1. *For any $\mu \in M_+$, V_μ is a bounded operator from $L^p(\mu)$ to H^p for $1 < p \leqslant 2$. In general, it is unbounded for $p > 2$. If μ is singular and V_μ is bounded from $L^p(\mu)$ to H^p, then μ is a discrete measure.*

The following theorem of Poltoratski [65] describes the boundary behavior of $V_\mu f$.

Theorem 9.2. *Let $\mu \in M_+$ and $f \in L^1(\mu)$. Then the function $V_\mu f$ has finite nontangential boundary values μ-a.e. These values coincide with f μ_s-a.e.*

The above result shows in particular that if μ is singular, then V_μ is in fact the identity operator on $L^1(\mu)$, and so is certainly a bounded operator from $L^p(\mu)$ to itself. The situation for arbitrary μ and for the maximal operator MV_μ is more involved and was examined by the following theorem of Poltoratski [68].

Theorem 9.3. *For any $\mu \in M_+$, V_μ is a bounded operator from $L^p(\mu)$ to itself for $1 < p \leqslant 2$. In addition, the maximal operator MV_μ is bounded on $L^p(\mu)$ for $1 < p < 2$, and of weak type $(2,2)$ on $L^2(\mu)$.*

Even for singular μ, the maximal operator MV_μ may be unbounded. Indeed, Poltoratski provides an example of a singular measure μ and an $f \in L^\infty(\mu)$ such that $MV_\mu f \notin L^p(\mu)$ for any $p > 2$. Finally, although MV_μ is always of weak type $(2,2)$, it is not known whether or not it is bounded on $L^2(\mu)$ in general. It is also not known under what conditions it is of weak type $(1,1)$.

References

[1] A.B. Aleksandrov, *A-integrability of boundary values of harmonic functions*, Mat. Zametki **30** (1981), no. 1, 59–72, 154. MR **83j**:30039

[2] ———, *Essays on nonlocally convex Hardy classes*, Complex analysis and spectral theory (Leningrad, 1979/1980), Springer, Berlin, 1981, pp. 1–89. MR **84h**:46066

[3] ———, *Invariant subspaces of shift operators. An axiomatic approach*, Zap. Nauchn. Sem. Leningrad. Otdel. Mat. Inst. Steklov. (LOMI) **113** (1981), 7–26, 264, Investigations on linear operators and the theory of functions, XI. MR **83g**:47031

[4] ———, *Multiplicity of boundary values of inner functions*, Izv. Akad. Nauk Armyan. SSR Ser. Mat. **22** (1987), no. 5, 490–503, 515. MR **89e**:30058

[5] ———, *Inner functions and related spaces of pseudocontinuable functions*, Zap. Nauchn. Sem. Leningrad. Otdel. Mat. Inst. Steklov. (LOMI) **170** (1989), no. Issled. Linein. Oper. Teorii Funktsii. 17, 7–33, 321. MR **91c**:30063

[6] A. Aleman and J. Cima, *An integral operator on H^p and Hardy's inequality*, J. Anal. Math. **85** (2001), 157–176. MR 1 869 606

[7] A. Aleman and A. Siskakis, *An integral operator on H^p*, Complex Variables Theory Appl. **28** (1995), no. 2, 149–158. MR **2000d**:47050

[8] N.K. Bary, *A treatise on trigonometric series. Vols. I, II*, Authorized translation by Margaret F. Mullins. A Pergamon Press Book, The Macmillan Co., New York, 1964. MR 30 #1347

[9] S. Bell, *The Cauchy transform, potential theory, and conformal mapping*, CRC Press, Boca Raton, FL, 1992. MR **94k**:30013

[10] S.V. Bočkarev, *Existence of a basis in the space of functions analytic in the disc, and some properties of Franklin's system*, Mat. Sb. (N.S.) **95(137)** (1974), 3–18, 159. MR 50 #8036

[11] G. Boole, *On the comparison of transcendents with certain applications to the theory of definite integrals*, Phil. Trans. Royal Soc. **147** (1857), 745–803.

[12] P. Bourdon and J.A. Cima, *On integrals of Cauchy-Stieltjes type*, Houston J. Math. **14** (1988), no. 4, 465–474. MR **90h**:30095

[13] L. Brown and A.L. Shields, *Cyclic vectors in the Dirichlet space*, Trans. Amer. Math. Soc. **285** (1984), no. 1, 269–303. MR **86d:**30079

[14] J.A. Cima and A. Matheson, *Cauchy transforms and composition operators*, Illinois J. Math. **42** (1998), no. 1, 58–69. MR **98k:**42028

[15] J.A. Cima, A. Matheson, and W.T. Ross, *The backward shift on the space of Cauchy transforms*, to appear, Proc. Amer. Math. Soc.

[16] J.A. Cima and W.T. Ross, *The backward shift on the Hardy space*, American Mathematical Society, Providence, RI, 2000. MR **2002f:**47068

[17] J.A. Cima and A. Siskakis, *Cauchy transforms and Cesàro averaging operators*, Acta Sci. Math. (Szeged) **65** (1999), no. 3-4, 505–513. MR **2000m:**47043

[18] D. Clark, *One dimensional perturbations of restricted shifts*, J. Analyse Math. **25** (1972), 169–191. MR 46 #692

[19] B. Davis, *On the distributions of conjugate functions of nonnegative measures*, Duke Math. J. **40** (1973), 695–700. MR 48 #2649

[20] ———, *On the weak type* (1, 1) *inequality for conjugate functions*, Proc. Amer. Math. Soc. **44** (1974), 307–311. MR 50 #879

[21] ———, *On Kolmogorov's inequalities* $\tilde{f}_p \le C_p \, f_1$, $0 < p < 1$, Trans. Amer. Math. Soc. **222** (1976), 179–192. MR 54 #10967

[22] F. Delbaen, *Weakly compact operators on the disc algebra*, J. Algebra **45** (1977), no. 2, 284–294. MR 58 #2304

[23] J. Diestel, *Sequences and series in Banach spaces*, Springer-Verlag, New York, 1984. MR **85i:**46020

[24] R.G. Douglas, H.S. Shapiro, and A.L. Shields, *Cyclic vectors and invariant subspaces for the backward shift operator.*, Ann. Inst. Fourier (Grenoble) **20** (1970), no. fasc. 1, 37–76. MR 42 #5088

[25] P.L. Duren, *Theory of* H^p *spaces*, Academic Press, New York, 1970. MR 42 #3552

[26] P. Fatou, *Séries trigonométriques et séries de Taylor*, Acta Math. **30** (1906), 335–400.

[27] O. Frostman, *Sur les produits de Blaschke*, Kungl. Fysiografiska Sällskapets i Lund Förhandlingar [Proc. Roy. Physiog. Soc. Lund] **12** (1942), no. 15, 169–182. MR 6,262e

[28] J.B. Garnett, *Analytic capacity and measure*, Springer-Verlag, Berlin, 1972, Lecture Notes in Mathematics, Vol. 297. MR 56 #12257

[29] ———, *Bounded analytic functions*, Academic Press Inc., New York, 1981. MR **83g:**30037

[30] V.V. Golubev, *Single valued analytic functions with perfect sets of singular points*, Ph.D. thesis, Moscow, 1917.

[31] G.M. Goluzin, *Geometric theory of functions of a complex variable*, American Mathematical Society, Providence, R.I., 1969. MR 40 #308

[32] G.H. Hardy and J.E. Littlewood, *Some more theorems concerning Fourier series and Fourier power series*, Duke Math. J. **2** (1936), 354–382.

[33] V.P. Havin, *On analytic functions representable by an integral of Cauchy-Stieltjes type*, Vestnik Leningrad. Univ. Ser. Mat. Meh. Astr. **13** (1958), no. 1, 66–79. MR 20 #1762

[34] ______, *Analytic representation of linear functionals in spaces of harmonic and analytic functions which are continuous in a closed region*, Dokl. Akad. Nauk SSSR **151** (1963), 505–508. MR 27 #2636

[35] ______, *The spaces H^∞ and L^1/H^1_0*, Zap. Naučn. Sem. Leningrad. Otdel. Mat. Inst. Steklov. (LOMI) **39** (1974), 120–148, Investigations on linear operators and the theory of functions, IV. MR 50 #965

[36] S.Ya. Havinson, *On an extremal problem of the theory of analytic functions*, Uspehi Matem. Nauk (N.S.) **4** (1949), no. 4(32), 158–159. MR 11,508e

[37] ______, *On some extremal problems of the theory of analytic functions*, Moskov. Gos. Univ. Učenye Zapiski Matematika **148(4)** (1951), 133–143. MR 14,155f

[38] H. Helson and G. Szegö, *A problem in prediction theory*, Ann. Mat. Pura Appl. (4) **51** (1960), 107–138. MR 22 #12343

[39] E. Hewitt and K. Stromberg, *Real and abstract analysis. A modern treatment of the theory of functions of a real variable*, Springer-Verlag, New York, 1965. MR 32 #5826

[40] K. Hoffman, *Banach spaces of analytic functions*, Dover Publications Inc., New York, 1988, Reprint of the 1962 original. MR **92d:**46066

[41] B. Hollenbeck and I. Verbitsky, *Best constants for the Riesz projection*, J. Funct. Anal. **175** (2000), no. 2, 370–392. MR **2001i:**42010

[42] S.V. Hruščev, *The problem of simultaneous approximation and of removal of the singularities of Cauchy type integrals*, Trudy Mat. Inst. Steklov. **130** (1978), 124–195, 223, Spectral theory of functions and operators. MR **80j:**30055

[43] S.V. Hruščev and S.A. Vinogradov, *Free interpolation in the space of uniformly convergent Taylor series*, Complex analysis and spectral theory (Leningrad, 1979/1980), Springer, Berlin, 1981, pp. 171–213. MR **83b:**30032

[44] ______, *Inner functions and multipliers of Cauchy type integrals*, Ark. Mat. **19** (1981), no. 1, 23–42. MR **83c:**30027

[45] R. Hunt, B. Muckenhoupt, and R. Wheeden, *Weighted norm inequalities for the conjugate function and Hilbert transform*, Trans. Amer. Math. Soc. **176** (1973), 227–251. MR 47 #701

[46] J.-P. Kahane, *Best approximation in $L^1(T)$*, Bull. Amer. Math. Soc. **80** (1974), 788–804. MR 52 #3845

[47] S. Kakutani, *Concrete representation of abstract (L)-spaces and the mean ergodic theorem*, Ann. of Math. (2) **42** (1941), 523–537. MR 2,318d

[48] Y. Katznelson, *An introduction to harmonic analysis*, corrected ed., Dover Publications Inc., New York, 1976. MR 54 #10976

[49] S.V. Kisljakov, *The Dunford-Pettis, Pełczyński and Grothendieck conditions*, Dokl. Akad. Nauk SSSR **225** (1975), no. 6, 1252–1255. MR 53 #1241

[50] A.N. Kolmogorov, *Sur les fonctions harmoniques conjuquées et les séries de Fourier*, Fund. Math. **7** (1925), 24–29.

[51] P. Koosis, *Introduction to H_p spaces*, second ed., Cambridge University Press, Cambridge, 1998. MR **2000b:**30052

[52] B.I. Korenblum, *Closed ideals of the ring A^n*, Funkcional. Anal. i Priložen. **6** (1972), no. 3, 38–52. MR 48 #2776

[53] P.I. Kuznetsov and E.D. Solomentsev, *Ivan Ivanovich Privalov (on the ninetieth anniversary of his birth)*, Uspekhi Mat. Nauk **37** (1982), no. 4(226), 193–211 (1 plate). MR **84b**:01049

[54] A.I. Markushevich, *Theory of functions of a complex variable. Vol. I, II, III*, English ed., Chelsea Publishing Co., New York, 1977. MR 56 #3258

[55] L.A. Markushevich and G.Ts. Tumarkin, *On a class of functions that can be represented in a domain by an integral of Cauchy-Stieltjes type*, Uspekhi Mat. Nauk **52** (1997), no. 3(315), 169–170. MR **98j**:30045

[56] V.G. Maz′ya and T.O. Shaposhnikova, *Theory of multipliers in spaces of differentiable functions*, Pitman (Advanced Publishing Program), Boston, MA, 1985. MR **87j**:46074

[57] M. Mooney, *A theorem on bounded analytic functions*, Pacific J. Math. **43** (1972), 457–463. MR 47 #2374

[58] N.I. Muskhelishvili, *Singular integral equations*, Dover Publications Inc., New York, 1992, Boundary problems of function theory and their application to mathematical physics, Translated from the second (1946) Russian edition and with a preface by J. R. M. Radok, Corrected reprint of the 1953 English translation. MR **94a**:45001

[59] D.J. Newman, *The nonexistence of projections from L^1 to H^1*, Proc. Amer. Math. Soc. **12** (1961), 98–99. MR 22 #11276

[60] N.K. Nikol′skiĭ, *Treatise on the shift operator*, Springer-Verlag, Berlin, 1986. MR **87i**:47042

[61] H. Pajot, *Analytic capacity, rectifiability, Menger curvature and the Cauchy integral*, Lecture Notes in Mathematics, vol. 1799, Springer-Verlag, Berlin, 2002. MR 1 952 175

[62] A. Pełczyński, *Banach spaces of analytic functions and absolutely summing operators*, American Mathematical Society, Providence, R.I., 1977, Expository lectures from the CBMS Regional Conference held at Kent State University, Kent, Ohio, July 11–16, 1976, Conference Board of the Mathematical Sciences Regional Conference Series in Mathematics, No. 30. MR 58 #23526

[63] S.K. Pichorides, *On the best values of the constants in the theorems of M. Riesz, Zygmund and Kolmogorov*, Studia Math. **44** (1972), 165–179. (errata insert), Collection of articles honoring the completion by Antoni Zygmund of 50 years of scientific activity, II. MR 47 #702

[64] J. Plemelj, *Ein Ergänzungssatz zur Cauchyschen Integraldarstellung analytischer Functionen, Randwerte betreffend*, Monatshefte für Math. u. Phys. **XIX** (1908), 205–210.

[65] A. Poltoratski, *Boundary behavior of pseudocontinuable functions*, Algebra i Analiz **5** (1993), no. 2, 189–210. MR **94k**:30090

[66] ———, *On the distributions of boundary values of Cauchy integrals*, Proc. Amer. Math. Soc. **124** (1996), no. 8, 2455–2463. MR **96j**:30057

[67] ———, *Kreĭn's spectral shift and perturbations of spectra of rank one*, Algebra i Analiz **10** (1998), no. 5, 143–183. MR **2000d**:47028

[68] ———, *Maximal properties of the normalized Cauchy transform*, J. Amer. Math. Soc. **16** (2003), no. 1, 1–17 (electronic). MR **2003j**:30056

[69] I.I. Privalov, *Sur les fonctions conjuguées*, Bull. Soc. Math. France **44** (1916), 100–103.

[70] ———, *Intégral de Cauchy*, Ph.D. thesis, Saratov, 1919.

[71] ———, *Randeigenschaften analytischer Funktionen*, VEB Deutscher Verlag der Wissenschaften, Berlin, 1956. MR 18,727f

[72] F. Riesz and M. Riesz, *Über die Randwerte einer analytischen Function*, IV Skand. Math. Kongr., Mittag-Leffler, Uppsala (1920), 27–44.

[73] M. Riesz, *Sur les fonctions conjugées*, Math. Z. **27** (1927), 218–244.

[74] W.W. Rogosinski and H.S. Shapiro, *On certain extremum problems for analytic functions*, Acta Math. **90** (1953), 287–318. MR 15,516a

[75] W.T. Ross and H.S. Shapiro, *Generalized analytic continuation*, University Lecture Series, vol. 25, American Mathematical Society, Providence, RI, 2002. MR **2003h:**30003

[76] W. Rudin, *The closed ideals in an algebra of analytic functions*, Canad. J. Math. **9** (1957), 426–434. MR 19,641c

[77] ———, *Real and complex analysis*, third ed., McGraw-Hill Book Co., New York, 1987. MR **88k:**00002

[78] ———, *Functional analysis*, second ed., McGraw-Hill Inc., New York, 1991. MR **92k:**46001

[79] A. Rybkin, *On an analogue of Cauchy's formula for $H^p, 1/2 \leq p < 1$, and the Cauchy type integral of a singular measure*, Complex Variables Theory Appl. **43** (2000), no. 2, 139–149. MR **2001i:**30042

[80] D. Sarason, *Functions of vanishing mean oscillation*, Trans. Amer. Math. Soc. **207** (1975), 391–405. MR 51 #13690

[81] ———, *Composition operators as integral operators*, Analysis and partial differential equations, Dekker, New York, 1990, pp. 545–565. MR **92a:**47040

[82] ———, *Sub-Hardy Hilbert spaces in the unit disk*, University of Arkansas Lecture Notes in the Mathematical Sciences, 10, John Wiley & Sons Inc., New York, 1994, A Wiley-Interscience Publication. MR **96k:**46039

[83] J. Shapiro and C. Sundberg, *Compact composition operators on L^1*, Proc. Amer. Math. Soc. **108** (1990), no. 2, 443–449. MR **90d:**47035

[84] N.A. Shirokov, *Analytic functions smooth up to the boundary*, Springer-Verlag, Berlin, 1988. MR **90h:**30087

[85] V.I. Smirnov, *Sur les valeurs limites des fonctions, régulières à l'intérieur d'un cercle*, Journal de la Société Phys.-Math. de Léningrade **2** (1929), 22–37.

[86] F. Smithies, *Cauchy and the creation of complex function theory*, Cambridge University Press, Cambridge, 1997. MR **99b:**01013

[87] Yu.B. Sokhotskii, *On definite integrals and functions using series expansions*, Ph.D. thesis, St. Petersburg, 1873.

[88] S. Spanne, *Sur l'interpolation entre les espaces $\mathcal{L}_k{}^{p\Phi}$*, Ann. Scuola Norm. Sup. Pisa (3) **20** (1966), 625–648. MR 35 #728

[89] D. Stegenga, *Multipliers of the Dirichlet space*, Illinois J. Math. **24** (1980), no. 1, 113–139. MR **81a:**30027

[90] E.M. Stein, *Singular integrals, harmonic functions, and differentiability properties of functions of several variables*, Singular integrals (Proc. Sympos. Pure Math., Chicago, Ill., 1966), Amer. Math. Soc., Providence, R.I., 1967, pp. 316–335. MR 58 #2467

[91] ———, *Singular integrals and differentiability properties of functions*, Princeton Mathematical Series, No. 30, Princeton University Press, Princeton, N.J., 1970. MR 44 #7280

[92] E.M. Stein and G. Weiss, *An extension of a theorem of Marcinkiewicz and some of its applications*, J. Math. Mech. **8** (1959), 263–284. MR 21 #5888

[93] O.D. Tsereteli, *Remarks on the theorems of Kolmogorov and of F. and M. Riesz*, Proceedings of the Symposium on Continuum Mechanics and Related Problems of Analysis (Tbilisi, 1971), Vol. 1 (Russian), Izdat. "Mecniereba", Tbilisi, 1973, pp. 241–254. MR 52 #6306

[94] ———, *Conjugate functions*, Mat. Zametki **22** (1977), no. 5, 771–783. MR 58 #12166

[95] G.C. Tumarkin, *On integrals of Cauchy-Stieltjes type*, Uspehi Mat. Nauk (N.S.) **11** (1956), no. 4(70), 163–166. MR 18,725f

[96] P.L. Ul′yanov, *On the A-Cauchy integral. I*, Uspehi Mat. Nauk (N.S.) **11** (1956), no. 5(71), 223–229. MR 18,726a

[97] S.A. Vinogradov, *Properties of multipliers of integrals of Cauchy-Stieltjes type, and some problems of factorization of analytic functions*, Mathematical programming and related questions (Proc. Seventh Winter School, Drogobych, 1974), Theory of functions and functional analysis (Russian), Central Èkonom.-Mat. Inst. Akad. Nauk SSSR, Moscow, 1976, pp. 5–39. MR 58 #28518

[98] S.A. Vinogradov, M.G. Goluzina, and V.P. Havin, *Multipliers and divisors of Cauchy-Stieltjes type integrals*, Zap. Naučn. Sem. Leningrad. Otdel. Mat. Inst. Steklov. (LOMI) **19** (1970), 55–78. MR 45 #562

[99] P. Wojtaszczyk, *Banach spaces for analysts*, Cambridge University Press, Cambridge, 1991. MR **93d:**46001

[100] A. Zygmund, *Sur les fonctions conjugées*, Fund. Math. **13** (1929), 284–303.

[101] ———, *Trigonometric series. 2nd ed. Vols. I, II*, Cambridge University Press, New York, 1959. MR 21 #6498

Joseph A. Cima
Department of Mathematics
University of North Carolina
Chapel Hill, North Carolina 27599, USA
e-mail: `cima@math.unc.edu`

Alec Matheson
Department of Mathematics
Lamar University
Beaumont, Texas 77710, USA
e-mail: `matheson@math.lamar.edu`

William T. Ross
Department of Mathematics and Computer Science
University of Richmond
Richmond, Virginia 23173, USA
e-mail: `wross@richmond.edu`

Operator Theory:
Advances and Applications, Vol. 156, 113–129

Quadrature Domains and Fluid Dynamics

Darren Crowdy

Abstract. Few physical scientists interested in the mathematical description of fluid flows will know what a quadrature domain is; just as few mathematicians interested in quadrature domain theory would profess to know much about fluid dynamics. And yet, recent research has shown that a surprisingly large number of the by-now classic exact solutions of two-dimensional fluid dynamics can be understood within the context of quadrature domain theory. This article surveys a number of different physical applications of quadrature domain theory arising in the general field of fluid dynamics.

Mathematics Subject Classification (2000). Primary 99Z99; Secondary 00A00.
Keywords. quadrature domains, complex analysis, fluid dynamics.

1. Introduction

The simplest example of a quadrature domain is a circular disc. Let $z = x + iy$ and suppose the disc is centred at the origin $z = 0$ with radius r. The well-known "mean value theorem" says that, if $h(z)$ is any function analytic in the disc D, then

$$\int\int_D h(z)dxdy = \pi r^2 h(0). \tag{1}$$

(1) is a simple example of a *quadrature identity*. The idea of quadrature domain theory is to consider more complicated domains satisfying more complicated quadrature identities. Shapiro [1] gives an illuminating introduction to quadrature domain theory. See also Sakai [2].

Perhaps the first connection between quadrature domain theory and applications was made by Richardson [3] who was interested in understanding the motion of the free boundaries of blobs of fluid trapped between two plates in a Hele-Shaw cell. When the flow is driven by a distribution of sources and/or sinks and surface tension effects on the free boundaries are ignored, this free boundary problem admits wide classes of "exact solution", i.e., initial fluid domains can be found whose evolution under the dynamics of the physical problem can be computed by tracking a finite set of time-evolving parameters. This constitutes a remarkable simplification of the problem.

It is important for our general message to point out that the exact solutions found by Richardson had, in fact, been found many years before by Polubarinova-Kochina [4] and Kufarev [5] who were interested in the motion of the interface between oil and water in porous media (mathematically, this problem is identical to the Hele-Shaw problem). But even if the solutions were already known, Richardson's 1972 paper introduced a crucial new theoretical ingredient: an understanding of the problem within the framework of quadrature domain theory.

Varchenko and Etingof [6] provide a comprehensive review of the impact of this new perspective in the context of flows in porous media and Hele-Shaw flows. The purpose of this article is different. The goal is to describe a broad spectrum of distinct physical problems, all emanating from the field of fluid dynamics, which can usefully be interpreted within the context of quadrature domain theory. The history of the Hele-Shaw problem illustrates the power of rephrasing a well-known problem in a new mathematical language. Here we describe the relevance of quadrature domain theory to a variety of different physical applications and describe its associated impact.

2. Quadrature domains

First, some background on quadrature domains. Consider a planar domain D. Let $h(z)$ be any function that is analytic in D and integrable over it. Suppose that

$$\int\int_D h(z)dxdy = \sum_{k=1}^{N} \sum_{j=0}^{n_k-1} c_{jk} h^{(j)}(z_k) \tag{2}$$

where $\{z_k \in \mathbb{C}\}$ is a set of points strictly inside D, $\{c_{jk} \in \mathbb{C}\}$ and $h^{(j)}(z)$ denotes the j-th derivative of h. Here, N and $\{n_k \geq 1\}$ are integers. Then D is very special and is known as a *quadrature domain* because of the remarkable fact, embodied in (2), that the two-dimensional integral on the left hand side of (2) in fact requires only the sum of a finite number of terms given on the right hand side for its evaluation. The quadrature identity (2) generalizes (1).

An alternative way to understand quadrature domains [6] is to consider their *Cauchy transforms* $C(z)$ defined as

$$C(z) = \frac{1}{\pi} \int\int_D \frac{dx'dy'}{z' - z}, \quad z \notin D. \tag{3}$$

This function is well-defined if D is bounded and is analytic for $z \notin D$. By Green's theorem, we also have

$$C(z) = \frac{1}{2i\pi} \oint_{\partial D} \frac{d\bar{z}'}{z' - z}, \quad z \notin D \tag{4}$$

and this form can be used to define Cauchy transforms for unbounded domains. Choosing $h(z) = z^n$ for $n \geq 0$ in the left hand side of (2) defines the geometrical *moments* M_n of a domain D, i.e.,

$$M_n = \frac{1}{\pi} \int\int_D z^n dxdy. \tag{5}$$

$C(z)$ is a generating function for these moments because its Laurent expansion coefficients valid as $|z| \to \infty$ are the moments (5). If the moments of a domain encode information concerning its shape, then so does $C(z)$. In physical problems, it is usually the evolution of the Cauchy transform which can be established most directly from the problem statement. The physical problem then reduces to that of reconstructing the domain from knowledge of its Cauchy transform. Mathematically, this is identical to the inverse problem of two-dimensional potential theory and such a viewpoint offers a helpful perspective. There are deep theoretical connections [1] between quadrature domain theory, potential theory and the concept of "balayage", and the theory of the Schwarz function [7].

Quadrature domains come in a variety of flavours. Basically, they are domains where the continuation of $C(z)$ into the domain has a special set of singularities. The most common quadrature domains satisfy quadrature identities of the type (2) and have Cauchy transforms that have a finite set of poles so that $C(z)$ is a rational function. Varchenko and Etingof [6] call these *algebraic domains*. They also introduced an *abelian domain* to be one where $C'(z)$, rather than $C(z)$, is rational. An ellipse is a quadrature domain, but it is neither algebraic nor abelian. Instead, it has a Cauchy transform with two square-root branch points at the foci. For example, an ellipse D with major and minor axes a and b respectively and with foci at ± 1 satisfies the quadrature identity [1]

$$\int\int_D h(z)dxdy = 2ab \int_{-1}^{1} h(x)(1 - x^2)^{1/2} dx. \tag{6}$$

A different class of domains, called *quadrature domains for arclength* [9], satisfy identities of the form

$$\oint_{\partial D} h(z)|dz| = \sum_{k=1}^{N} \sum_{j=0}^{n_k - 1} c_{jk} h^{(j)}(z_k). \tag{7}$$

Remarkably, all these classes of domain have been found to arise in applications, each having very different physics.

3. Constructing quadrature domains

Suppose $C(z)$ is known, then it remains to reconstruct the associated quadrature domain. There are various ways to do this. In a number of applications, Crowdy has made pragmatic use of the fact that the boundaries of quadrature domains are algebraic curves [10] [11]. Conformal maps from a pre-image ζ-plane (say) to the domain can also be used [12] [14]. Here we briefly describe the approach to reconstructing quadrature domains presented by Crowdy and Marshall [12] (who also survey the various other known methods of construction). It is known [6] [8] that the conformal mappings to a bounded $(g+1)$-connected quadrature domain is a meromorphic function on a Riemann surface of genus g. One model of such functions uses a mapping from a pre-image region in a ζ-plane consisting of the interior unit ζ-disc with g smaller interior discs excised. This is the Schottky model. The associated mappings must be invariant with respect to a group of Mobius transformations associated with this pre-image region. This group (Θ, say) is a classical *Schottky group*. Mumford, Series and Wright [15] give a very accessible and modern discussion of Schottky groups and their applications.

Given a Schottky group Θ the *Schottky-Klein prime function* is defined as [16]

$$\omega(\zeta,\gamma) = (\zeta-\gamma)\prod_{\Theta'}\{\zeta,\gamma/\gamma_i,\zeta_i\} \tag{8}$$

where Θ' denotes all transformations in Θ excluding the identity and all inverses while ζ_i and γ_i denote images of ζ and γ respectively under the i-th map ($\theta_i(\zeta)$, say) in this set. $\{\zeta,\gamma/\gamma_i,\zeta_i\}$ denotes a cross-ratio. Then one representation for a function invariant with respect to transformations in Θ is a ratio of products of Schottky-Klein prime functions, i.e.,

$$z(\zeta) = R\,\frac{\prod_{j=1}^{N}\omega(\zeta,\beta_j)}{\prod_{j=1}^{N}\omega(\zeta,\alpha_j)} \tag{9}$$

where the N poles $\{\alpha_j|j=1,\ldots,N\}$ and N zeros $\{\beta_j|j=1,\ldots,N\}$ satisfy the g conditions

$$\prod_{j=1}^{N}\prod_{\theta_i\in\Theta_k}\frac{(\beta_j-\theta_i(B_k))}{(\beta_j-\theta_i(A_k))}\Big/\frac{(\alpha_j-\theta_i(B_k))}{(\alpha_j-\theta_i(A_k))} = 1,\ k=1,\ldots,g. \tag{10}$$

A_k and B_k are the two fixed points of the k-th Mobius map generating the group and Θ_k is another subset of Θ (see [12] for a precise definition). Figure 1 illustrates various multiply-connected square packings of near-circular discs all constructed using mappings of the general form (9).

When the Schottky group is trivial, the associated prime function is $\omega(\zeta,\gamma) = (\zeta-\gamma)$ and the functions (9) are just the rational functions. When Θ is generated by the single Mobius map $\theta_1(\zeta)=\rho^2\zeta$ then $\omega(\zeta,\gamma)\propto P(\zeta/\gamma,\rho)$ where

$$P(\zeta,\rho) \equiv (1-\zeta)\prod_{k=1}^{\infty}(1-\rho^{2k}\zeta)(1-\rho^{2k}/\zeta) \tag{11}$$

which is closely related to the classical Jacobi theta functions. With the choice (11), (9) then yields the class of *loxodromic functions* [13]. Such functions have been used to construct explicit solutions to the rotating Hele-Shaw problem [18], the viscous sintering problem [19] [20], the problem of finding vortical equilibria of the Euler equation [21] and the problem of free surface Euler flows with surface tension [24].

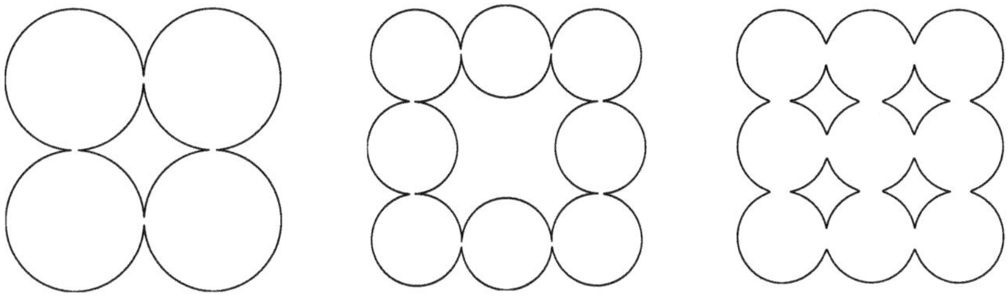

FIGURE 1. Three distinct square packings of near-circular particles constructed using conformal mappings based on the Schottky-Klein prime function (9). All are quadrature domains. The first two are doubly-connected, the third is quintuply-connected.

4. Applications in fluid dynamics

This section describes a series of physical problems where quadrature domains arise.

4.1. Hele-Shaw flows and flows in porous media

This class of problems is where use of quadrature domains and Cauchy transforms is best known and so we review it only briefly. Consider the two-dimensional flow-field $\mathbf{u}$ of a blob of fluid $D(t)$ of viscosity μ sandwiched between two plates of glass separated by b. Under appropriate assumptions, the flow is modelled by $\mathbf{u} = \nabla\phi$ where

$$\nabla^2\phi = 0, \text{ in } D(t). \tag{12}$$

If there is no surface tension then $\phi =$ constant on each free surface, while it must also be true that

$$V_n = \nabla\phi.\mathbf{n} \tag{13}$$

where V_n denotes the normal velocity of the free surface and $\mathbf{n}$ is the normal to the boundary. Richardson [3] studied the case of flows driven by sources of strength $Q_j(t)$ at positions $z = z_j$.

The above is a canonical example of a "laplacian growth problem". An extensive array of analytic results are known (see a website by Howison [25] for a comprehensive list of related references). These equations describe a number of different physical situations including, for example, flow in porous media, electrodeposition and the slow solidification in a supercooled liquid. Entov, Etingof and Kleinbock [26] discuss a number of variants of the Hele-Shaw problem for which there exist exact solutions. These include flow in a rotating Hele-Shaw cell where the flow is driven by centrifugal effects, Hele-Shaw flows with gravity and "squeeze flow" in a Hele-Shaw cell where the plates making up the cell are moved together (or apart). Extensions to flows in non-planar cells, multiply-connected fluid regions and to three dimensions have all been made (see [25] for references).

As an example of how useful it can be to understand such problems in terms of quadrature domains, we examine the case of Hele-Shaw flows in rotating cells. Recent experiments have investigated the various interfacial instabilities that can occur in an initial concentric annulus of fluid placed in a rotating cell [17]. The time-evolving concentric annulus is a trivial exact solution to this problem, but it fails to exhibit any of the nonlinear phenomena observed in the experiments. Knowledgeable of the fact (see [26]) that simply-connected quadrature domains are preserved in a rotating cell, Crowdy [18] generalized this result to doubly-connected domains relevant to the experiments involving an annulus. Having derived a general class of solutions, to mimic the experiments of [17] it was necessary to construct an initial quadrature domain that is close to a concentric annulus. Gustafsson [8] showed that multiply-connected quadrature domains are dense (in an appropriate sense) in the general class of multiply-connected domains so the existence of such a quadrature domain was guaranteed. Indeed, in [18], it is shown that conformal maps from the annulus $\rho < |\zeta| < 1$ of the form

$$z(\zeta) = \zeta \frac{P_N(\zeta \rho^{2/N} a^{-1}, \rho)}{P_N(\zeta a^{-1}, \rho)} \tag{14}$$

where $1 < a < \rho^{-1}$ (chosen so that the map in univalent) give images that are quadrature domains getting arbitrarily close to the annulus $\rho < |z| < 1$ as $N \to \infty$. Here, $P_N(\zeta, \rho)$ is equivalent to a product of the functions $P(\zeta, \rho)$ and is defined by

$$P_N(\zeta, \rho) = (1 - \zeta^N) \prod_{k=1}^{\infty} (1 - \rho^{2kN} \zeta^N)(1 - \rho^{2kN} \zeta^{-N}). \tag{15}$$

Under evolution, the map takes the form

$$z(\zeta, t) = R(t) \zeta \frac{P_N(\zeta \rho(t)^{2/N} a(t)^{-1}, \rho(t))}{P_N(\zeta a(t)^{-1}, \rho(t))}, \tag{16}$$

the parameters R, a and ρ evolving in time according to a coupled system of ordinary differential equations. In [18], this class of exact solutions is studied and compared to the qualitative results of the experiments in [17].

4.2. Rotating vortex arrays

A famous exact solution in vortex dynamics is the celebrated Kirchhoff elliptical vortex patch [27] which rotates, under the dynamics of the Euler equation, at constant angular velocity without changing its shape. The interior of an ellipse is a generalized quadrature domain satisfying the identity (1), while the exterior can be viewed as an unbounded quadrature domain with a Cauchy transform $C(z)$ which is a linear polynomial. By considering generalizations of these facts and using ideas involving the Schwarz function, broad new classes of exact solution have been found [28] for rotating vortex arrays with finite area cores having distributed vorticity. These solutions generalize the classic 19th century investigations of polygonal vortex arrays by Thomson [27]. Here, the flow $\mathbf{u}$ is given by

$\mathbf{u} = (\psi_y, -\psi_x)$ where ψ is a *streamfunction* governed by the steady nonlinear Euler equation for a two-dimensional incompressible fluid of constant density which takes the form

$$\frac{\partial(\psi, \nabla^2\psi)}{\partial(x,y)} = 0. \tag{17}$$

The key observation is that, in a frame of reference co-rotating with the configuration, carefully-chosen streamfunctions ψ having the form of *modified Schwarz potentials* [1], i.e.,

$$\psi(x,y) = -\frac{\omega}{4}\left(z\bar{z} - \int^{z} S(z')dz' - \int^{\bar{z}} \overline{S}(z')dz'\right) \tag{18}$$

in the fluid region D, where $S(z)$ is the Schwarz function of the boundary ∂D (and $\bar{S}(z)$ its conjugate function), can represent dynamically-consistent equilibrium solutions of the Euler equation (17). (Such potentials also play an important role in the general theory of quadrature domains [1]). The constant ω is the magnitude of the uniform vorticity in the fluid. This new perspective has led to the discovery of a wide range of new exact solutions of the steady Euler equations, including those for rotating vortex configurations involving multiple interacting vortex patches. Some typical configurations are shown in Figure 2. The fluid regions exterior to the five co-rotating vortex patches in Figure 2 are unbounded, quintuply-connected quadrature domains constructed using conformal mappings based on the Schottky-Klein prime function. The pre-image region consists of the unit ζ-circle with four smaller discs of radius q excised. Figure 2 shows six different vortical configurations for various values of q.

4.3. Multipolar vortices in the plane and on the sphere

Motivated by the observation, experiments and numerical investigations of a class of coherent structures known collectively as *multipolar vortices* (e.g., [30]), quadrature domain theory has been used [29] to construct a class of exact stationary equilibrium solutions of the Euler equations displaying all the qualitative properties of the multipolar vortices observed in practice. The essence of the approach in [29] is to reappraise the classical circular vortex patch known as the Rankine vortex [27]. If one thinks of it instead as the simplest form of quadrature domain (i.e., a circular disc) then the multipolar vortex solutions correspond to generalized quadrature domains (subject, of course, to the physical constraints of the Helmholtz laws of vortex motion [27]). Again, streamfunctions of the form (18) turn out to be significant. These ideas have proven to be generalizable in a number of directions including finding multipolar vortices in annular arrays [21], vortices with more complicated topology [10] as well as finding equilbrium regions of distributed vorticity on surfaces with non-zero curvature [31]. Examples of a triangular (or quadrupolar) vortices in equilibrium on the plane and on a sphere are shown in Figure 3. The right-hand diagram in Figure 3 has an interpretation as a quadrature domain on the surface of a sphere.

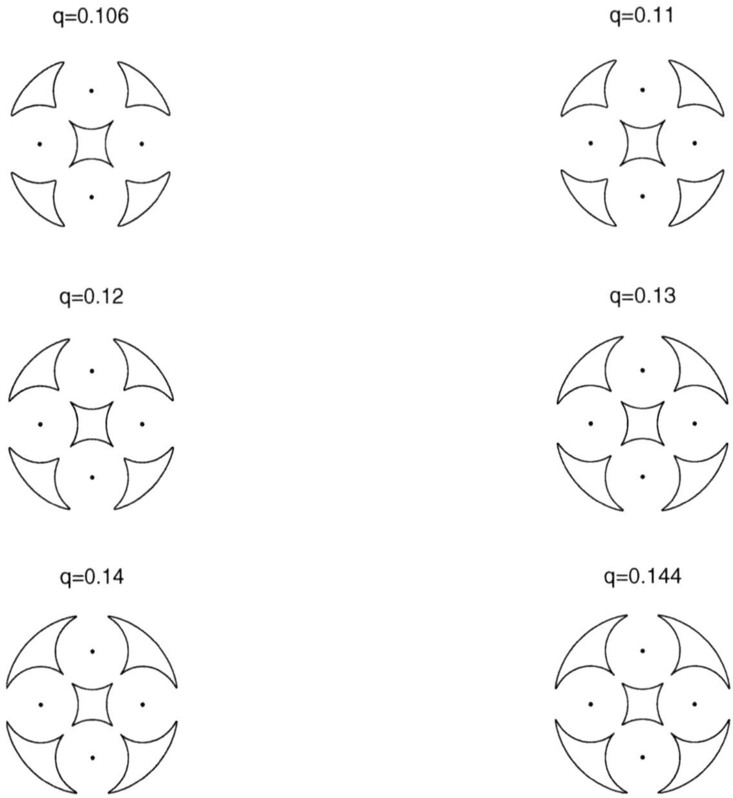

FIGURE 2. Steadily rotating vortex configuration consisting of 5 vortex patches (one central and four satellite patches) and 4 point vortices. The point vortices correspond to the singularities of the global Schwarz function.

Shapiro [32] introduced the notion of a *special point* of a quadrature domain. The boundary of a bounded quadrature domain is known to be given by all the continuous, non-isolated solutions of

$$P(z, \bar{z}) = 0 \tag{19}$$

where $P(z, w) = 0$ is an algebraic curve whose order equals that of the quadrature identity. There are often a number of isolated solutions of (19) occurring inside the domain and these have been dubbed *special points* of the quadrature domain. It is interesting to remark that, in the context of steady vortical flows of the Euler equation, these special points have a physical interpretation; they are precisely the stagnation points of the flow [10].

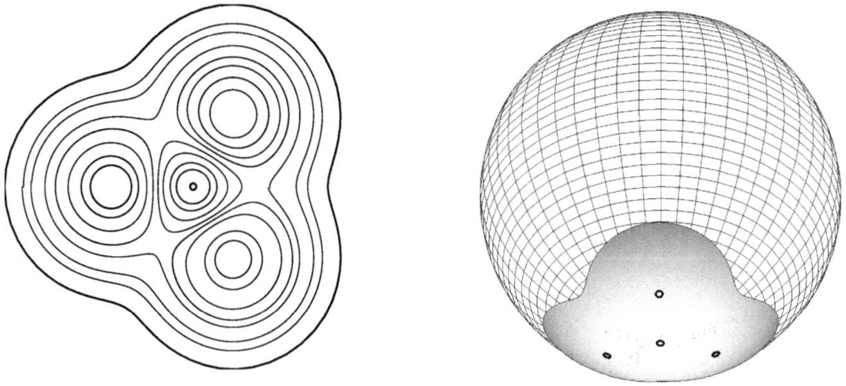

FIGURE 3. Streamlines and shape of a triangular vortex in equilibrium on the plane [29] (left). A triangular vortex in equilibrium on the surface of a sphere [31] (right). These are quadrature domains on the plane and on the sphere.

4.4. Free surface Euler flows with capillarity

Another famous exact solution known to fluid dynamicists is that for steady deep-water capillary waves found by Crapper in 1957 [22]. By reappraising this solution [33], broad new classes of exact solution for equilibrium configurations of free surface irrotational Euler flows with interfacial tension on the free boundaries have been identified. In the fluid region, the incompressible velocity field is given by $\mathbf{u} = \nabla\phi$ where

$$\nabla^2\phi = 0. \tag{20}$$

In equilibrium, any free boundary must be a streamline. If there is uniform surface tension T on the free boundary, the fluid pressure must balance the capillary forces. Using a well-known theorem due to Bernoulli, we can write

$$T\kappa + \Gamma = |\nabla\phi|^2 \tag{21}$$

where κ is the surface curvature and Γ is the Bernoulli constant.

Crowdy [34] has found new exact solutions for the shape deformations of both a bubble placed in an ambient circulatory flow of circulation γ and of a blob of fluid with internal circulation modelled by a contained line vortex singularity of strength γ. Non-trivial equilibrium shapes of a bubble in a circulatory flow are shown in Figure 4. In the case of both a bubble and a blob, the conformal mappings $z(\zeta)$ from a unit ζ-circle to the equilibrium shapes are such that $z(\zeta)$ and $\sqrt{z_\zeta(\zeta)}$ are rational functions. This means that the equilibrium shapes are simultaneously quadrature domains in the sense of satisfying identities of the form

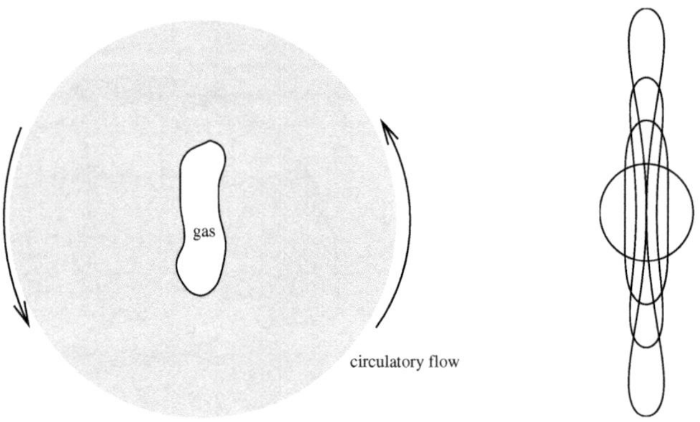

FIGURE 4. Schematic illustrating the problem of a bubble with capillarity in an ambient circulatory flow (left). Equilibrium bubble shapes, computed using exact solutions, for different values of the circulation [34] (right). At a critical circulation, the bubble is found to pinch.

(2) *and* quadrature domains for arclength in the sense of satisfying identities of the form (7). Indeed, using the formulae in [34], the fluid domains D exterior to the equilibrium bubble configurations shown in Figure 4 satisfy both

$$\oint_{\partial D} h(z)|dz| = -\frac{2\pi}{\Gamma}h(z_1) - \frac{2\pi}{\Gamma}h(-z_1) + \frac{\pi}{\Gamma}(1+\sqrt{1+2\Gamma\gamma^2})h(\infty). \tag{22}$$

where $h(z)$ is some function analytic in the fluid domain D, and

$$\int\int_D h(z)dxdy = \Gamma^2 h(z_1) + \Gamma^2 h(-z_1) \tag{23}$$

where $h(z)$ is some function analytic in the fluid domain D and decaying sufficiently fast at infinity. In (22) and (23), $z_1 = z_1(\Gamma,\gamma)$ is some algebraic function of the physical parameters Γ and γ.

This new understanding has led to a range of new mathematical results for this class of flows (e.g., [24]) including a simplified representation of the classic exact solutions of Kinnersley [23] for waves on fluid sheets. Moreover, all these new results are automatically applicable to a quite separate physical problem in electrophysics involving the shaping of conducting metal jets using electric fields [35] which has identical governing equations.

4.5. Steady Hele-Shaw flows with surface tension

Exact solutions for the equilibrium shapes of simply-connected blobs and bubbles in a Hele-Shaw flow where there is non-zero interfacial tension and the flow is driven by quadrupoles (or higher order poles) have been found by Entov *et al.* [26]. This problem also admits equilibrium shapes which are images of the unit ζ-circle under conformal mappings where $\sqrt{z_\zeta(\zeta)}$ is a rational function so that the equilibria can be interpreted as quadrature domains for arclength (but are *not*, in general, also quadrature domains in the usual sense). The solutions of Entov *et al* have been generalized in various directions [36] (for example, to doubly-connected fluid configurations). The mathematical similarities and differences between the two physically-distinct problems of §4.4 and 4.5 have also been discussed [36].

4.6. Viscous sintering

Hopper [37] found a remarkable exact solution for the surface tension-driven coalescence of two near-circular viscous fluid blobs. This is the planar analogue of the two-sphere coalescence "unit problem" which is an important microscale model of an industrially-important manufacturing process known as *viscous sintering.* Howison [25] has also compiled a comprehensive list of related references.

Hopper's mathematical model is as follows. In a time-evolving region $D(t)$, of incompressible fluid of viscosity μ, a streamfunction ψ satisfies

$$\nabla^4\psi = 0 \text{ in } D(t), \tag{24}$$

so that $\psi = \text{Im}[\bar{z}f(z,t) + g(z,t)]$ for some $f(z,t)$ and $g(z,t)$ analytic in $D(t)$. On the boundary of the fluid,

$$-pn_i + 2\mu e_{ij}n_j = \kappa n_i \text{ and } V_n = \mathbf{n}.\nabla^{\perp}\psi \text{ on } \partial D(t) \tag{25}$$

where κ is the boundary curvature, p is the fluid pressure and e_{ij} is the fluid rate-of-strain tensor. V_n denotes the normal velocity of the boundary. Crowdy [38] [44] has shown explicitly that the dynamics of these equations can, in certain circumstances, preserve quadrature domains. Reappraising Hopper's work within this framework has led to generalizations of his solutions.

As an initial sinter compact of touching particles is heated, the particles coalesce and the compact densifies as the interparticulate pores close up under the effects of surface tension. In certain circumstances, the above mathematical problem admits exact solutions in the form of time-evolving quadrature domains. Figures 5 and 6 show time sequences of the sintering of two doubly-connected packings of near-circular viscous blobs computed using the methodology presented in [20]. The sequences are shown up to the time at which the central pore has closed up. Figures 5 and 6 are calculated using conformal mappings, dependent on just a finite set of time-evolving parameters, based on the Schottky-Klein prime function representation (9) [20]. Indeed, Figure 5 was computed using conformal mappings of precisely the form (16) with $N = 4$, thus by virtue of the association with quadrature domains, the full dynamics of the problem is reduced to the solution of just three ordinary differential equations for $a(t), R(t)$ and $\rho(t)$. The evolution in Figure 5 has also been computed using elliptic function theory by Richardson [40] and using purely numerical methods by Van de Vorst [39].

FIGURE 5. Viscous sintering of four near-cylindrical viscous blobs in an initially doubly-connected configuration up to the time of pore closure. Times shown are $t = 0,\ 0.2,\ 0.4,\ 0.6,\ 0.8,\ 1.14$.

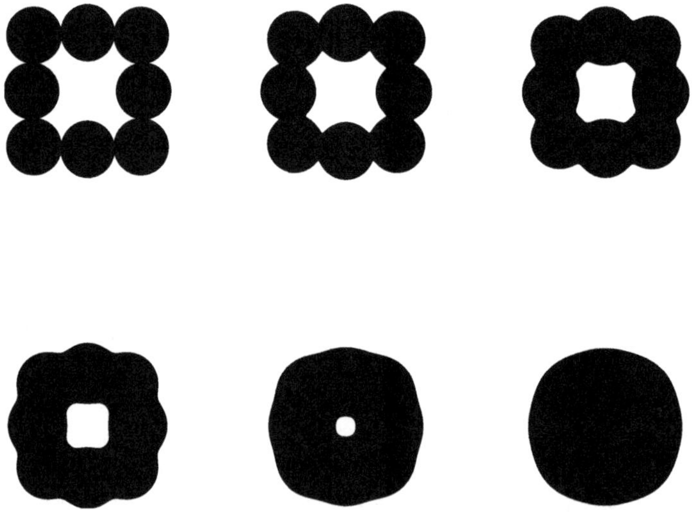

FIGURE 6. Viscous sintering of a looser square packing of eight near-circular viscous blobs up to the time of pore closure. Times shown are $t = 0, 0.2, 0.4, 0.6, 0.8$ and 1.36.

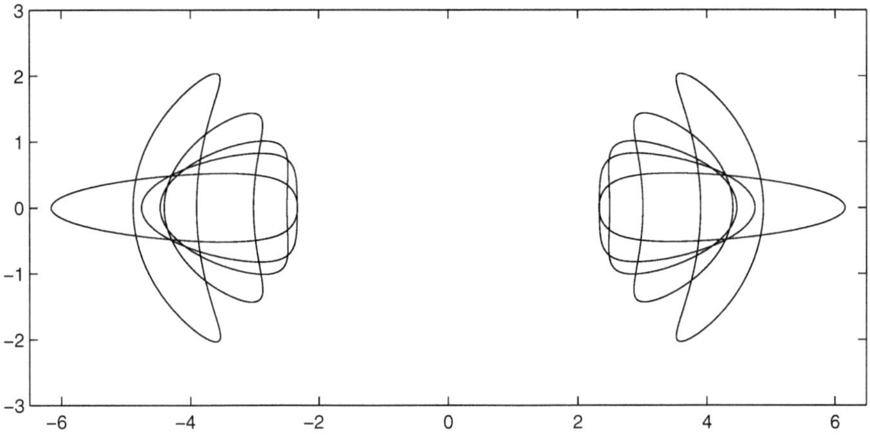

FIGURE 7. Configurations of two steady bubbles in an ambient Stokes flow of the form (26). Different shapes correspond to different choices of the far-field parameters. The domains exterior to the two bubbles are unbounded doubly-connected quadrature domains with Cauchy transforms given by (27).

4.7. Bubbles in Stokes flows

Quadrature domain theory can also be used to understand a range of exact solutions for bubbles in ambient Stokes flows. An example is the work of Tanveer and Vasconcelos [41] who consider time-evolving bubbles in ambient straining and shear flows (among others). The latter exact solutions can be generalized to the case of compressible bubbles [42] and to steady two-bubble configurations [43]. Figure 7 shows various steady two-bubble configurations placed in an ambient flow where the Goursat functions have the far-field form

$$\begin{aligned} f(z) &\sim f_3 z^3 + f_1 z + O(z^{-1}), \\ g'(z) &\sim g_4 z^4 + g_2 z^2 + O(z^{-1}) \end{aligned} \tag{26}$$

where f_3, f_1, g_4 and g_2 are some parameters dictated by the imposed far-field conditions. The fluid domains exterior to the various two bubble configurations in Figure 7 are doubly-connected, unbounded quadrature domains corresponding to Cauchy transforms of the form

$$C(z) = A_\infty z + \frac{A_0}{z} \tag{27}$$

where the parameters A_∞ and A_0 depend on the far-field parameters in (26). The domains in Figure 7 were constructed using conformal maps based on the Schottky-Klein prime function (11) for the genus 1 case.

5. Discussion and future directions

It has been seen that quadrature domain theory arises in a surprisingly broad range of distinct physical contexts, even just within the field of fluid dynamics. It is to be expected that the same will be true of other disciplines (e.g., plane elasticity, electrostatics). As a result of this interpretation in terms of quadrature domain theory, new mathematical results have been found. The abstraction of quadrature domain theory can immediately give invaluable insight into the scope of what is possible mathematically.

There is no doubt that there are many other areas where quadrature domain theory will be found, in future, to have relevance. One unifying observation is that *all* of the time-evolving free boundary problems admitting exact solutions just described, whatever the governing physics, have Cauchy transforms obeying a partial differential equation of the form

$$\frac{\partial C(z,t)}{\partial t} + \frac{\partial I(z,t)}{\partial z} + \sigma_2(z,t)C(z,t) = R(z,t), \; z \notin D(t) \tag{28}$$

where

$$I(z,t) = \int\int_{D(t)} \frac{\sigma_1(z',t)dx'dy'}{z'-z}, \tag{29}$$

for various choices of $\sigma_1(z,t)$ and $\sigma_2(z,t)$ which are analytic in $D(t)$ and $R(z,t)$ which is meromorphic in $D(t)$. Crowdy [44] gives details as to why such an equation can be expected to preserve the rational character of $C(z,t)$. The choice $\sigma_1(z,t) = \sigma_2(z,t) = 0$ and $R(z,t) = \sum_{j=1}^{M} \frac{Q_j(t)}{z-z_j}$ gives the case of Hele-Shaw flow driven by sources/sinks; $\sigma_1(z,t) = \frac{b^2\rho g}{12\mu}, \sigma_2(z,t) = 0 = R(z,t)$ gives the case with gravity; $\sigma_1(z,t) = \frac{b^2\rho\omega^2}{12\mu}z$, $\sigma_2(z,t) = R(z,t) = 0$ gives flows in a Hele-Shaw cell rotating with angular velocity ω; the choice $\sigma_1(z,t) = R(z,t) = 0$ and $\sigma_2(z,t) = \frac{\dot{b}}{b}$ gives the case of "squeeze flow" in Hele-Shaw cell where $b(t)$ is the separation of the plates; $\sigma_1(z,t) = -2f(z,t)$, $\sigma_2(z,t) = R(z,t) = 0$ gives the case of viscous sintering; $\sigma_1(z,t) = -2f(z,t)$, $\sigma_2(z,t) = 0$ and

$$R(z,t) = \frac{1}{2\pi i}\oint_{\partial D(t)} \frac{2g'(z',t)}{z'-z}dz' \tag{30}$$

corresponds to the case of bubbles placed in singular Stokes flows.

Intriguingly, (28) has recently been found to give rise to a tantalizing theoretical link between fixed and free boundary problems. Fokas [45] has introduced a flexible new transform method that is applicable not only to mixed linear boundary value problems in fixed domains but also to integrable nonlinear problems. The method involves the introduction of a differential 1-form which is closed if and only if the governing field equation holds in the domain. The closure of this form implies what Fokas calls a "global relation". If Fokas's algorithmic method is applied to the free boundary problems described herein, the corresponding global relation turns out to be precisely (28). Thus, considering the evolution of $C(z,t)$ in the solution of these problems is no longer arbitrary but becomes a natural *consequence* of a general algorithmic method that applies not only to free boundary problems but to a broad range of applied mathematical problems. This perspective might very well prove to be valuable in future.

Another connection to integrable systems theory has recently arisen in the work of Wiegmann and Zabrodin [46]. They find a theoretical connection between the problem of reconstructing a domain from its harmonic moments and dispersionless integrable hierarchies. The connection applies to very general domains but it is conceivable that quadrature domains have some interpretation as special solutions or reductions of these hierarchies. A tempting connection of quadrature domains with "finite-gap solutions" of nonlinear integrable systems is irresistible but still largely intuitive at present.

Concerning three-dimensional results, there is broad scope for future results there too. These will be far more important for realistic application. The analysts are already paving the way for possible applications by determining what is mathematically possible in higher dimensions. See Shapiro [1] for a discussion of this. As an example, Dritschel and co-workers [47] have recently found a fascinating practical application to the problem of modelling three-dimensional multi-vortex interactions in geostrophic flows of the fact that the exterior potential generated by a uniform ellipsoid is equivalent to that induced by a non-uniform two-dimensional "focal ellipse" [1] [48]. That is, an ellipsoid satisfies a higher-dimensional analogue of (6). Such ideas lie at the heart of quadrature domain theory.

Acknowledgment

The author wishes to sincerely thank Professor Harold Shapiro for his continued interest and support for the author's work over the last few years.

References

[1] H.S. Shapiro, The Schwarz functions and its generalization to higher dimension, Wiley, New York, (1992).

[2] M. Sakai, Quadrature domains, Lecture notes in mathematics, **934**, Springer-Verlag, (1982).

[3] S. Richardson, Hele-Shaw flows with a free boundary produced by the injection of fluid into a narrow channel, *J. Fluid Mech.*, **56**, 609–618, (1972).

[4] P.Ya. Polubarinova-Kochina, On the motion of the oil contour, *Dokl. Akad. Nauk. SSSR*, **47**, 254–257, (1945).

[5] P.P Kufarev, The oil contour problem for the circle with any number of wells, *Dokl. Akad. Nauk. SSSR*, **75**, 507–510, (1950).

[6] A.N. Varchenko and P.I. Etingof, Why the boundary of a round drop becomes a curve of order four, American Mathematical Society *University Lecture Series*, **3**, (1994).

[7] P. Davis, The Schwarz function and its applications, Carus Mathematical Monographs 17, Math. Assoc. of America, (1974).

[8] B. Gustafsson, Quadrature identities and the Schottky double, *Acta. Appl. Math.*, **1**, 209–240, (1983).

[9] B. Gustafsson, Applications of half-order differentials on Riemann surfaces to quadrature domains for arc-length, *J. d'Analyse Math.*, **49**, 54–89, (1987).

[10] D.G. Crowdy, Multipolar vortices and algebraic curves, *Proc. Roy. Soc. A*, **457**, 2337–2359, (2001).

[11] D.G. Crowdy and H. Kang, Squeeze flow of multiply-connected fluid domains in a Hele-Shaw cell, *J. Nonlin. Sci.*, **11**, 279–304, (2001).

[12] D.G. Crowdy and J.S. Marshall, Constructing multiply-connected quadrature domains, *SIAM J. Appl. Math.*, **64**, 1334–1359, (2004).

[13] G. Valiron, Cours d'Analyse Mathematique, Theorie des fonctions, 2nd Edition, Masson et Cie, Paris (1947).

[14] S. Richardson, Hele-Shaw flows with time-dependent free boundaries involving a multiply-connected fluid region, *Eur. J. Appl. Math.*, **12**, 571–599, (2002).

[15] D. Mumford, C. Series and D. Wright, Indra's Pearls: the vision of Felix Klein, Cambridge University Press, (2002).

[16] H. Baker, Abelian functions, Cambridge University Press, Cambridge, (1995).

[17] L. Carrillo, J. Soriano and J. Ortin, Radial displacement of a fluid annulus in a rotating Hele-Shaw cell, *Phys. Fluids*, **11**, 778, (1999).

[18] D.G. Crowdy, Theory of exact solutions for the evolution of a fluid annulus in a rotating Hele-Shaw cell, *Q. Appl. Math*, **LX**(1), 11–36, (2002).

[19] D.G. Crowdy and S. Tanveer, A theory of exact solutions for annular viscous blobs, *J. Nonlinear Sci.*, **8**, 375–400, (1998). Erratum, **11**, 237, (2001).

[20] D.G. Crowdy, Viscous sintering of unimodal and bimodal cylindrical packings with shrinking pores, *Eur. J. Appl. Math.*, **14**, 421–445, (2003).

[21] D.G. Crowdy, On the construction of exact multipolar equilibria of the 2D Euler equations, *Phys. Fluids*, **14**(1), (2002), 257–267.

[22] G.D. Crapper, An exact solution for progressive capillary waves of arbitrary amplitude, *J. Fluid Mech.*, **2**, 532, (1957).

[23] W. Kinnersley, Exact large amplitude capillary waves on sheets of fluid, *J. Fluid Mech.*, **76**, 229–241, (1977).

[24] D.G. Crowdy, Steady nonlinear capillary waves on curved sheets, *Eur. J. Appl. Math.*, **12**, (2001), 689–708.

[25] S. Howison, `www.maths.ox.ac.uk/howison/Hele-Shaw`

[26] V.M. Entov, P.I. Etingof and D. Ya Kleinbock, On nonlinear interface dynamics in Hele-Shaw flows, *Eur. J. Appl. Math.*, **6**, 399–420, (1995).

[27] P.G. Saffman, Vortex dynamics, Cambridge University Press, Cambridge, (1992).

[28] D.G. Crowdy, Exact solutions for rotating vortex arrays with finite-area cores, *J. Fluid Mech.*, **469**, 209–235, (2002).

[29] D.G. Crowdy, A class of exact multipolar vortices, *Phys. Fluids*, **11**(9), 2556–2564, (1999).

[30] C.F. Carnevale and R.C. Kloosterziel, Emergence and evolution of triangular vortices, *J. Fluid Mech.*, **259**, 305–331, (1994).

[31] D.G. Crowdy and M. Cloke, Analytical solutions for distributed multipolar vortex equilibria on a sphere, *Phys. Fluids*, **15**, 22–34, (2002).

[32] H.S. Shapiro, Unbounded quadrature domains, in Complex Analysis I, Proceedings, University of Maryland 1985–1986, C.A. Berenstein (ed.), Lecture Notes in Mathematics, **1275**, Springer-Verlag, Berlin, pp. 287–331, (1987).

[33] D.G. Crowdy, A new approach to free surface Euler flows with surface tension, *Stud. Appl. Math.*, **105**, 35–58, (2000).

[34] D.G. Crowdy, Circulation-induced shape deformations of drops and bubbles: exact two-dimensional models, *Phys. Fluids*, **11**(10), 2836–2845, (1999).

[35] N.M Zubarev, Exact solution of the problem of the equilibrium configuration of the charged surface of a liquid metal, *J.E.T.P.*, **89**(6), 1078–1085, (1999).

[36] D.G. Crowdy, Hele-Shaw flows and water waves, *J. Fluid Mech.*, **409**, 223–242, (2000).

[37] R.W. Hopper, Plane Stokes flow driven by capillarity on a free surface, *J. Fluid Mech.*, **213**, 349–375, (1990).

[38] D.G. Crowdy, A note on viscous sintering and quadrature identities, *Eur. J. Appl. Math.*, **10**, 623–634, (1999).

[39] G.A.L. Van de Vorst, Integral method for a two-dimensional Stokes flow with shrinking holes applied to viscous sintering, *J. Fluid Mech.*, **257**, 667–689, (1993).

[40] S. Richardson, Plane Stokes flow with time-dependent free boundaries in which the fluid occupies a doubly-connected region, *Eur. J. Appl. Math.*, **11** 249–269, (2000).

[41] S. Tanveer and G.L. Vasconcelos, Time-evolving bubbles in two-dimensional Stokes flow, *J. Fluid Mech.*, **301**, 325–344, (1995).

[42] D.G. Crowdy, Compressible bubbles in Stokes flow, *J. Fluid Mech.*, **476**, 345–356, (2003).

[43] D.G. Crowdy, Exact solutions for two bubbles in the flow-field of a four-roller mill, *J. Eng. Math.*, **44**, 311–330, (2002).

[44] D.G. Crowdy, On a class of geometry-driven free boundary problems, *SIAM J. Appl. Math.*, **62**(2), 945–954, (2002).

[45] A.S. Fokas, On the integrability of linear and nonlinear PDE's, *J. Math. Phys.*, **41**, 4188, (2000).

[46] P.B. Wiegmann and A. Zabrodin, Conformal maps and dispersionless integrable hierarchies, *Comm. Math. Phys.*, **213**, 523–538, (2000).

[47] D.G. Dritschel, J.N. Reinaud and W.J. McKiver, The quasi-geostrophic ellipsoidal vortex model, *J. Fluid Mech.*, **505**, 201–223, (2004).

[48] D. Khavinson and H.S. Shapiro, The Schwarz potential in $\mathbb{R}^n$ and Cauchy's problem for the Laplace equation, TRITA-MAT-1989-36, Royal Institute of Technology research report, (1989).

Darren Crowdy
Department of Mathematics
Imperial College of Science, Technology and Medicine
180 Queen's Gate
London, SW7 2AZ
United Kingdom
e-mail: d.crowdy@imperial.ac.uk

Operator Theory:
Advances and Applications, Vol. 156, 131–150

On Uniformly Discrete Sequences in the Disk

Peter Duren, Alexander Schuster and Dragan Vukotić

Dedicated to Harold Shapiro on the Occasion of his 75th Birthday

Abstract. We present an overview of uniformly discrete sequences and their relationship to other important types of sets, such as uniformly separated sequences, sets of interpolation and sampling, and zero-sets for Bergman spaces.

Mathematics Subject Classification (2000). Primary 32A36; Secondary 30H05.

Keywords. Bergman spaces, pseudohyperbolic metric, uniformly discrete sequences, densities, sets of sampling and interpolation, zero-sets.

Introduction

A sequence $\Gamma = \{z_k\}$ in the unit disk $\mathbb{D}$ is said to be *uniformly discrete* if

$$\rho(z_j, z_k) = \left| \frac{z_j - z_k}{1 - \overline{z_j} z_k} \right| \geq \delta$$

for some fixed $\delta > 0$ and all $j \neq k$. The supremum of all such δ is called the *separation constant* of Γ and is denoted by $\delta(\Gamma)$. For $0 < p < \infty$ the *Bergman space* A^p consists of those functions f analytic in $\mathbb{D}$ for which the area integral

$$\|f\|_p^p = \int_{\mathbb{D}} |f(z)|^p \, d\sigma(z)$$

is finite. Here $d\sigma$ denotes normalized area measure, with $\sigma(\mathbb{D}) = 1$. An analytic function f belongs to the *Hardy space* H^p if

$$\|f\|_{H^p}^p = \sup_{r<1} \frac{1}{2\pi} \int_0^{2\pi} |f(re^{i\theta})|^p \, d\theta < \infty .$$

It is clear that $H^p \subset A^p$.

Uniformly discrete sequences play a fundamental role in the theory of Bergman spaces. For instance, they occur in the description of sampling and interpola-

The first and the third author are both supported by MCyT grant BFM2003-07294-C02-01, Spain.

tion sets as given by Seip [24, 25]. It is well known that all Hardy spaces H^p have the same zero-sets, characterized by the Blaschke condition $\sum_{k=1}^{\infty}(1 - |z_k|) < \infty$. However, the problem of describing the zero-sets of A^p functions still remains open. A sequence $\Gamma = \{z_k\}$ is said to be an A^p *zero-set* if some function in A^p vanishes *precisely* on Γ with the prescribed multiplicity. Horowitz [10] showed that the A^p zero-sets depend on p and that every subset of an A^p zero-set is an A^p zero-set. We will see in this paper that uniformly discrete sequences satisfy exactly the same Blaschke-type condition found by Horowitz [11] for zero-sets of Bergman spaces. In fact, every uniformly discrete sequence is an A^p zero-set for all p less than an explicitly given constant depending only on the separation constant $\delta(\Gamma)$. Further information about Bergman spaces can be found in the recent books [9] and [5].

One purpose of this paper is to collect a variety of results about uniformly discrete sequences, especially as they relate to Bergman spaces. We begin with basic properties of the pseudohyperbolic metric. Then comes a discussion of some geometric properties of uniformly discrete sequences and their relation to uniformly separated and exponential sequences. After a brief account of Carleson measures for Bergman spaces, we present a theorem that characterizes finite unions of uniformly discrete sequences in several equivalent ways. This is an analogue of a result of Horowitz [11] and others on finite unions of uniformly separated sequences. Finally, we give relatively simple sufficient conditions, expressed in terms of the pseudohyperbolic metric, for a sequence to be interpolating or sampling for an A^p space. These criteria are derived from Seip's theorems but have the advantage of avoiding the difficult calculations of densities required for direct application of his theorems.

1. The pseudohyperbolic metric

Uniformly discrete sequences are defined in terms of the pseudohyperbolic metric, a concept more natural than the Euclidean metric for problems in the unit disk. The *pseudohyperbolic distance* between two points z and α in $\mathbb{D}$ is defined by

$$\rho(z,\alpha) = |\varphi_\alpha(z)|, \qquad \text{where } \varphi_\alpha(z) = \frac{\alpha - z}{1 - \overline{\alpha} z}.$$

The function φ_α is a Möbius transformation, a conformal automorphism of the disk with $\varphi_\alpha(\alpha) = 0$. Note that $\rho(z,0) = |z|$.

It is not hard to verify that ρ is a true metric on $\mathbb{D}$. It is clear that $0 \leq \rho(z,\alpha) < 1$, with $\rho(z,\alpha) = 0$ if and only if $z = \alpha$. The symmetry property $\rho(\alpha,z) = \rho(z,\alpha)$ is immediate. The triangle inequality

$$\rho(z,\zeta) \leq \rho(z,\alpha) + \rho(\alpha,\zeta)$$

is less obvious, but it follows from the stronger inequality

$$\frac{|\rho(z,\alpha) - \rho(\alpha,\zeta)|}{1 - \rho(z,\alpha)\rho(\alpha,\zeta)} \leq \rho(z,\zeta) \leq \frac{\rho(z,\alpha) + \rho(\alpha,\zeta)}{1 + \rho(z,\alpha)\rho(\alpha,\zeta)}.$$

Direct calculations show that the function φ_α has the following additional properties.

(i) φ_α is an involution: $\varphi_\alpha(\varphi_\alpha(z)) = z$ for all $z \in \mathbb{D}$, or $\varphi_\alpha^{-1} = \varphi_\alpha$.

(ii) φ_α is an isometry with respect to the pseudohyperbolic metric:

$$\rho(\varphi_\alpha(z), \varphi_\alpha(\zeta)) = \rho(z, \zeta) \qquad \text{for all } z, \zeta \in \mathbb{D}.$$

(iii) The quantities $1 - |z|^2$ and $1 - \rho(z, \alpha)^2$ are related by the expression

$$1 - \rho(z, \alpha)^2 = \frac{(1 - |\alpha|^2)(1 - |z|^2)}{|1 - \overline{\alpha} z|^2} = (1 - |z|^2)|\varphi_\alpha'(z)|. \tag{1}$$

Property (ii) says equivalently that ρ is Möbius-invariant.

The *pseudohyperbolic disk* with *center* α and *radius* r $(0 < r < 1)$ is

$$\Delta(\alpha, r) = \{z : \rho(z, \alpha) < r\}.$$

It is clear that $\Delta(\alpha, r) \subset \mathbb{D}$. Moreover, $\Delta(\alpha, r) = \varphi_\alpha(\Delta(0, r))$ by the Möbius invariance of ρ. Hence, $\Delta(\alpha, r)$ is a true Euclidean disk, since linear fractional transformations preserve circles. However, its Euclidean center and radius are different from α and r unless $\alpha = 0$. A simple calculation shows that the (normalized) area is

$$\sigma(\Delta(\alpha, r)) = \left(\frac{r(1 - |\alpha|^2)}{1 - r^2|\alpha|^2} \right)^2.$$

Consequently, the Euclidean radius of the pseudohyperbolic disk $\Delta(\alpha, r)$ is

$$R = \frac{r(1 - |\alpha|^2)}{1 - r^2|\alpha|^2},$$

because its area is equal to R^2. The Euclidean center of $\Delta(\alpha, r)$ is

$$\gamma = \frac{(1 - r^2)\alpha}{1 - r^2|\alpha|^2}.$$

The *hyperbolic area* of a measurable subset $\Omega \subset \mathbb{D}$ is defined by

$$a(\Omega) = \int_\Omega \frac{d\sigma(z)}{(1 - |z|^2)^2}.$$

It is easily seen to be Möbius invariant. Indeed, if $w = \varphi_\alpha(z)$, then

$$\int_{\varphi_\alpha(\Omega)} \frac{d\sigma(w)}{(1 - |w|^2)^2} = \int_\Omega \frac{|\varphi_\alpha'(z)|^2 \, d\sigma(z)}{(1 - |\varphi_\alpha(z)|^2)^2} = \int_\Omega \frac{d\sigma(z)}{(1 - |z|^2)^2},$$

by (1). As a consequence, a pseudohyperbolic disk $\Delta(\alpha, r)$ has hyperbolic area

$$a(\Delta(\alpha, r)) = a(\Delta(0, r)) = \frac{1}{\pi} \int_0^r \int_0^{2\pi} \frac{s}{(1 - s^2)^2} \, d\theta \, ds = \frac{r^2}{1 - r^2}.$$

Thus the hyperbolic area of $\Delta(\alpha, r)$ depends only on the radius r, not on the center α.

2. Geometric Properties of Uniformly Discrete Sequences

Uniformly discrete sequences often play the same role for Bergman spaces that uniformly separated sequences play for Hardy spaces. A sequence $\{z_k\}$ in $\mathbb{D}$ is called a *Blaschke sequence* if $\sum(1-|z_k|)<\infty$. A Blaschke sequence is said to be *uniformly separated* if

$$\prod_{j\neq k}\left|\frac{z_j-z_k}{1-\overline{z_j}z_k}\right|\geq\delta\,,\qquad k=1,2,\dots\,,$$

for some constant $\delta>0$ independent of k. It is easy to see that every uniformly separated sequence is uniformly discrete, but the converse is not true. A sequence $\{z_k\}$ in the disk is said to be an *exponential sequence* if

$$1-|z_{k+1}|\leq r(1-|z_k|)$$

for some $0<r<1$. By a geometric series argument, every exponential sequence is a Blaschke sequence. There is a simple description of uniformly separated sequences that lie on a single ray.

Proposition 1. *Let $\{z_k\}$ be an infinite sequence such that $0\leq z_1<z_2<\ldots<1$. Then the following statements are equivalent:*

(a) *$\{z_k\}$ is exponential.* (b) *$\{z_k\}$ is uniformly separated.*
(c) *$\{z_k\}$ is uniformly discrete.*

Proof. It is well known that (a) implies (b) in general, and that (b) implies (a) for radial sequences. (See for instance [3], Theorem 9.2.) Since (b) trivially implies (c), it suffices to show that (c) implies (a) for radial sequences. To this end, suppose that $0\leq z_1<z_2<\ldots<1$ and

$$\frac{z_{k+1}-z_k}{1-z_kz_{k+1}}\geq\delta>0\,.$$

By simple algebra, we have

$$z_{k+1}\geq\frac{\delta+z_k}{1+\delta z_k}\,,$$

and therefore

$$1-z_{k+1}\leq 1-\frac{\delta+z_k}{1+\delta z_k}=\frac{1-\delta}{1+\delta z_k}\cdot(1-z_k)<(1-\delta)(1-z_k)\,,$$

which proves the statement. □

The proof suggests a specific example. For $0<r<1$, let the sequence $\{r_k\}$ be defined inductively by

$$r_1=r\,,\quad r_{k+1}=\frac{r+r_k}{1+r\,r_k}\,.$$

Then it is easy to check that $0<r_1<r_2<\ldots<1$ and $r_k\to 1$ as $k\to\infty$. Also,

$$\rho(r_k,r_{k+1})=\frac{r_{k+1}-r_k}{1-r_k\,r_{k+1}}=r\,,\qquad k=1,2,\dots\,.$$

Thus $\{r_k\}$ is uniformly discrete, hence exponential.

Proposition 1 has a partial generalization to Stolz angles, i.e., regions of the form

$$\{z \in \mathbb{D} : |1-z| \le a(1-|z|), |1-z| < b\}, \quad a, b > 0.$$

A result of Tse [28] shows that a sequence in a Stolz angle is uniformly discrete if and only if it is uniformly separated. However, the equivalence of the exponential condition fails for Stolz angles. Here is a simple example.

Proposition 2. *There exists a uniformly discrete sequence in a Stolz angle that is not exponential.*

Proof. Let L denote the interval $(1, i)$ in the plane and $\overline{L} = \{\overline{z} : z \in L\}$ the interval obtained by reflecting L across the real axis. Then for any two points $z \in L$, $w \in \overline{L}$ we have $\rho(z,w) \ge 1/\sqrt{2}$. Indeed, if $z = 1-r+ri$ and $w = 1-s-si$, where $0 < r, s < 1$, then a brief computation shows that

$$\begin{aligned}\rho(z,w)^2 &= \frac{(r-s)^2+(r+s)^2}{(r+s)^2+[r(1-s)+s(1-r)]^2} \\ &\ge \frac{(r-s)^2+(r+s)^2}{(r+s)^2+(r+s)^2} = \frac{r^2+s^2}{(r+s)^2} \ge \tfrac{1}{2}.\end{aligned}$$

It now suffices to place a uniformly discrete sequence on L. Take for instance $z_k \in L$ with $1-|z_k| = 1/2^k$. Then the sequence $\{z_k\} \cup \{\overline{z}_k\}$ is uniformly separated but far from exponential because $1-|z_k| = 1-|\overline{z}_k|$. □

It is shown in a forthcoming paper [7] that a uniformly separated sequence $\{z_k\}$ in a Stolz angle (and hence a finite union of such sequences) is a finite union of exponential sequences. Further equivalent characterizations are also given there.

For later applications we now collect some basic properties of uniformly discrete sequences. One aim is to characterize finite unions of uniformly discrete sequences in several equivalent ways. We begin with an estimate of the *counting function* $n(\Gamma, \alpha, r)$, defined as the number of points in a sequence $\Gamma = \{z_k\}$ that lie in the pseudohyperbolic disk $\Delta(\alpha, r)$.

Lemma 1. *Let $\Gamma = \{z_k\}$ be a uniformly discrete sequence with separation constant $\delta = \delta(\Gamma)$. Then its counting function satisfies*

$$n(\Gamma, \alpha, r) \le \left(\tfrac{2}{\delta}+1\right)^2 \frac{1}{1-r^2}$$

for every point $\alpha \in \mathbb{D}$ and $0 < r < 1$. In particular, $n(\Gamma, \alpha, r) = O\left(\frac{1}{1-r}\right)$ as $r \to 1$.

Proof. By the Möbius invariance of uniformly discrete sequences, there is no loss of generality in taking $\alpha = 0$. Uniform discreteness and the triangle inequality imply that the disks $\Delta(z_k, \frac{\delta}{2})$ are pairwise disjoint. By the strong form of the triangle inequality, each point $\zeta \in \Delta(0, r)$ satisfies

$$\Delta(\zeta, \tfrac{\delta}{2}) \subset \Delta(0, R), \qquad \text{where } R = \frac{r+\frac{\delta}{2}}{1+\frac{\delta r}{2}}.$$

Indeed, if $\rho(0,\zeta) < r$ and $\rho(z,\zeta) < \frac{\delta}{2}$, then

$$\rho(0,z) \le \frac{\rho(0,\zeta)+\rho(\zeta,z)}{1+\rho(0,\zeta)\rho(\zeta,z)} < \frac{r+\frac{\delta}{2}}{1+\frac{\delta r}{2}} = R\,.$$

Recall now that a pseudohyperbolic disk $\Delta(\zeta, s)$ has hyperbolic area

$$a(\Delta(\zeta,s)) = \frac{s^2}{1-s^2}\,, \qquad \zeta \in \mathbb{D}\,,\ 0 < s < 1\,.$$

Because the disks $\Delta(z_k, \frac{\delta}{2})$ are disjoint, hyperbolic areas can be compared to give

$$\sum_{z_k \in \Delta(0,r)} a(\Delta(z_k, \tfrac{\delta}{2})) \le a(\Delta(0,R))\,,$$

or

$$n(\Gamma,0,r)\,\frac{(\frac{\delta}{2})^2}{1-(\frac{\delta}{2})^2} \le \frac{R^2}{1-R^2}\,,$$

which reduces to the inequality

$$n(\Gamma,0,r) \le \left(\frac{2r+\delta}{\delta}\right)^2 \frac{1}{1-r^2} \le \left(\tfrac{2}{\delta}+1\right)^2 \frac{1}{1-r^2}\,. \qquad \square$$

The next result is a partial converse.

Lemma 2. *Let $\Gamma = \{z_k\}$ be a sequence of distinct points in the unit disk such that for some fixed radius $r > 0$, each pseudohyperbolic disk $\Delta(\alpha, r)$ contains at most N points z_k. Then Γ is the disjoint union of at most N uniformly discrete sequences.*

Proof. Consider first the disk $\Delta(z_1, r)$. By hypothesis, it contains at most N points in the sequence Γ, including z_1. Let those points be assigned to n different subsets, $\Gamma_1, \Gamma_2, \dots, \Gamma_n$, where $n \le N$. Let z_{k_1} be the first point of Γ not already assigned. Then $\rho(z_{k_1}, z_1) \ge r$, so z_{k_1} is placed into the set Γ_j containing z_1. Now proceed inductively. Suppose a finite number of points have been assigned to subsets $\Gamma_1, \Gamma_2, \dots, \Gamma_m$ with $m \le N$, and that $\rho(z,\zeta) \ge r$ for all points $z, \zeta \in \Gamma_j$, where $j = 1, \dots, m$. Let z^* be the first point of Γ not already assigned to a subset Γ_j. By hypothesis, the disk $\Delta(z^*, r)$ contains at most $N-1$ points of Γ that have already been assigned, and they represent at most $N-1$ different subsets Γ_j, so the point z^* can be assigned to some subset Γ_k not represented in this list. It is clear by construction that $\rho(z^*, \zeta) \ge r$ for all points $\zeta \in \Gamma$ already assigned to Γ_k. This inductive process therefore divides the given set Γ into disjoint subsets $\Gamma_1, \dots, \Gamma_m$ with $m \le N$ and $\rho(z,\zeta) \ge r$ for all $z, \zeta \in \Gamma_j$, $j = 1, \dots, m$. $\square$

3. The Bergman space and uniformly discrete sequences

Uniformly discrete sequences yield important examples of Carleson measures for the Bergman space. These are finite positive Borel measures μ on $\mathbb{D}$ with the property that $\int_{\mathbb{D}} |f(z)|^p \, d\mu(z) < \infty$ for every function f in the Bergman space

A^p, where $0 < p < \infty$. If μ is any such measure, it follows from the closed graph theorem that

$$\int_{\mathbb{D}} |f(z)|^p \, d\mu(z) \le K \, \|f\|_p^p \,, \qquad f \in A^p \,,$$

for some constant K depending only on p. A measure μ with this property is called a *Carleson measure for the Bergman space* A^p.

Hastings [8] showed that the Carleson measures for A^p are characterized by the property $\mu(S) \le C\sigma(S)$ for all Carleson squares

$$S = \{re^{i\theta} : 1 - h \le r < 1, \ |\theta - \theta_0| \le h\}, \qquad h > 0 \,.$$

Oleĭnik [19] found the same description in a broader context. The next theorem includes a version of this result, due to Luecking [12, 13], that brings into play the pseudohyperbolic metric. The *Berezin transform* $B\mu$ of a measure μ is defined by

$$B\mu(\zeta) = \int_{\mathbb{D}} |\varphi_\zeta'(z)|^2 d\mu(z) = \int_{\mathbb{D}} \frac{(1 - |\zeta|^2)^2}{|1 - \overline{\zeta} z|^4} d\mu(z) \,, \qquad \zeta \in \mathbb{D} \,.$$

Theorem A. *Let μ be a positive Borel measure on $\mathbb{D}$. Then for each fixed p with $0 < p < \infty$, the following four statements are equivalent.*

(i) *The inequality $\int_{\mathbb{D}} |f(z)|^p \, d\mu \le K \, \|f\|_p^p$ holds for some constant K and all $f \in A^p$.*

(ii) *An inequality $\mu(\Delta(\alpha, r)) \le C \, \sigma(\Delta(\alpha, r))$ holds for each r $(0 < r < 1)$, for some constant C depending only on r, and for all pseudohyperbolic disks $\Delta(\alpha, r)$, $\alpha \in \mathbb{D}$.*

(iii) *An inequality $\mu(\Delta(\alpha, r)) \le C \, \sigma(\Delta(\alpha, r))$ holds for some r $(0 < r < 1)$, for some constant C, and for all pseudohyperbolic disks $\Delta(\alpha, r)$.*

(iv) *The Berezin transform $B\mu$ is a bounded function on $\mathbb{D}$.*

The theorem implies in particular that Carleson measures are independent of p. For a proof of Theorem A, see [5], Section 2.10.

The next lemma shows how uniformly discrete sequences generate Carleson measures. The result appears in a paper by Zhu [30].

Lemma 3. *If $\{z_k\}$ is uniformly discrete, then*

$$\sum_{k=1}^{\infty} (1 - |z_k|^2)^2 |f(z_k)|^p \le C \, \|f\|_p^p \qquad \text{for every } f \in A^p \,,$$

where $0 < p < \infty$. Equivalently, the discrete measure μ that places mass $(1 - |z_k|^2)^2$ at z_k is a Carleson measure for the Bergman space A^p.

Proof. If $\rho(z_j, z_k) \ge \delta$, then the pseudohyperbolic disks $\Delta(z_j, \frac{\delta}{2})$ and $\Delta(z_k, \frac{\delta}{2})$ are disjoint for $j \ne k$, by the triangle inequality. By the sub-mean value property of $|g|^p$ for any analytic function g, we have

$$|g(0)|^p \le \frac{4}{\delta^2} \int_{\Delta(0, \frac{\delta}{2})} |g(\zeta)|^p \, d\sigma(\zeta) \,.$$

Now let $g(\zeta) = f(\varphi_{z_k}(\zeta))\varphi'_{z_k}(\zeta)^{2/p}$ to deduce that

$$\begin{aligned}(1-|z_k|^2)^2|f(z_k)|^p &\le \frac{4}{\delta^2}\int_{\Delta(0,\frac{\delta}{2})} |f(\varphi_{z_k}(\zeta))|^p|\varphi'_{z_k}(\zeta)|^2\,d\sigma(\zeta)\\ &= \frac{4}{\delta^2}\int_{\Delta(z_k,\frac{\delta}{2})} |f(z)|^p\,d\sigma(z)\,,\end{aligned}$$

by a change of variables. Since the disks $\Delta(z_k, \frac{\delta}{2})$ are disjoint, we conclude that

$$\sum_{k=1}^{\infty}(1-|z_k|^2)^2|f(z_k)|^p \le \frac{4}{\delta^2}\int_{\mathbb{D}} |f(z)|^p\,d\sigma(z)\,. \qquad \square$$

Corollary. *If $\{z_k\}$ is uniformly discrete and $\rho(z_j, z_k) \ge \delta > 0$ for $j \ne k$, then*

$$\sum_{k=1}^{\infty}(1-|z_k|^2)^2 \le \frac{4}{\delta^2}\,.$$

Lemma 3 may be compared with the well known fact, first found by Shapiro and Shields [26], that uniformly separated sequences generate Carleson measures for Hardy spaces in the sense that

$$\sum_{k=1}^{\infty}(1-|z_k|^2)|f(z_k)|^p \le C\,\|f\|^p_{H^p} \qquad \text{for every} \quad f \in H^p\,.$$

The following theorem combines results of Horowitz [11], McDonald and Sundberg [17], and others to characterize finite unions of uniformly separated sequences in a number of equivalent ways.

Theorem B. *For a Blaschke sequence Γ of points z_k in $\mathbb{D}$, the following six statements are equivalent.*

(i) *Γ is a finite union of uniformly separated sequences.*

(ii) *For some $p \in (0,\infty)$, there exists a constant C such that*

$$\sum_{k=1}^{\infty}(1-|z_k|^2)|f(z_k)|^p \le C\,\|f\|^p_{H^p} \qquad \textit{for every} \quad f \in H^p\,.$$

(iii) *For each $p \in (0,\infty)$, there exists a constant C such that*

$$\sum_{k=1}^{\infty}(1-|z_k|^2)|f(z_k)|^p \le C\,\|f\|^p_{H^p} \qquad \textit{for every} \quad f \in H^p\,.$$

(iv) *For some $p \in (0,\infty)$, the Blaschke product B with zeros z_k is a universal divisor of A^p in the sense that $f/B \in A^p$ for every $f \in A^p$ that vanishes on Γ.*

(v) *For each $p \in (0,\infty)$, the Blaschke product B is a universal divisor of A^p.*

(vi)
$$\sup_{\alpha\in\mathbb{D}}\sum_{k=1}^{\infty}(1-|\varphi_\alpha(z_k)|) < \infty\,.$$

For a streamlined version of the proof, see [4] or [5]. Also see [16]. We can now give a similar description of finite unions of uniformly discrete sequences. The next theorem combines the previous lemmas with one further condition. Again let $n(\Gamma, \alpha, r)$ denote the number of points of a given sequence $\Gamma = \{z_k\}$ that lie in the disk $\Delta(\alpha, r)$.

Theorem 1. *For a sequence Γ of distinct points $z_k \in \mathbb{D}$, the following six statements are equivalent.*

(i) *Γ is a finite union of uniformly discrete sequences.*
(ii) $\sup_{\alpha\in\mathbb{D}} n(\Gamma, \alpha, r) < \infty$ *for some* $r \in (0, 1)$.
(iii) $\sup_{\alpha\in\mathbb{D}} n(\Gamma, \alpha, r) < \infty$ *for each* $r \in (0, 1)$.
(iv) *For some $p \in (0, \infty)$, there exists a constant C such that*

$$\sum_{k=1}^{\infty}(1 - |z_k|^2)^2 |f(z_k)|^p \le C\,\|f\|_p^p \qquad \textit{for every } f \in A^p .$$

(v) *For each $p \in (0, \infty)$, there exists a constant C such that*

$$\sum_{k=1}^{\infty}(1 - |z_k|^2)^2 |f(z_k)|^p \le C\,\|f\|_p^p \qquad \textit{for every } f \in A^p .$$

(vi)
$$\sup_{\alpha\in\mathbb{D}} \sum_{k=1}^{\infty}(1 - |\varphi_\alpha(z_k)|^2)^2 < \infty .$$

Proof. Combining Lemmas 1 and 2, we see that (i) $\iff$ (ii) $\iff$ (iii). Theorem A shows that (iv) $\iff$ (v), since the Carleson measures for A^p are independent of p. Lemma 3 asserts that (i) $\implies$ (v). Thus it will be enough to show that (v) $\implies$ (vi) $\implies$ (iii).

(v) $\implies$ (vi). For an arbitrary point $\alpha \in \mathbb{D}$, let

$$f_\alpha(z) = \left\{ \frac{1 - |\alpha|^2}{(1 - \overline{\alpha} z)^2} \right\}^{2/p} .$$

Then $f_\alpha \in A^p$ and $\|f_\alpha\|_p = 1$, so we can use (1) to conclude from (v) that

$$\begin{aligned}\sum_{k=1}^{\infty}(1 - |\varphi_\alpha(z_k)|^2)^2 &= \sum_{k=1}^{\infty} \frac{(1 - |z_k|^2)^2 (1 - |\alpha|^2)^2}{|1 - \overline{\alpha} z_k|^4} \\ &= \sum_{k=1}^{\infty}(1 - |z_k|^2)^2 |f_\alpha(z_k)|^p \le C\,\|f_\alpha\|_p^p = C .\end{aligned}$$

(vi) $\implies$ (iii). Since $|\varphi_\alpha(z_k)| < r$ for $n(\Gamma, \alpha, r)$ points z_k, it follows from (vi) that

$$n(\Gamma, \alpha, r)(1 - r^2)^2 \le \sum_{k=1}^{\infty}(1 - |\varphi_\alpha(z_k)|^2)^2 \le C , \qquad \alpha \in \mathbb{D} ,$$

which shows that $n(\Gamma, \alpha, r)$ is bounded for each fixed r. □

The equivalence of (vi) with (ii) or (iii) is actually contained in Theorem A, since the sum in (vi) is the Berezin transform of the discrete measure μ that places mass $(1-|z_k|^2)^2$ at the point z_k.

We have seen that the sum $\sum(1-|z_k|^2)^2$ converges for every uniformly discrete sequence. In fact, a much stronger statement can be made. The following theorem says that uniformly discrete sequences are "almost" Blaschke sequences.

Theorem 2. *A uniformly discrete sequence* $\{z_k\}$ *satisfies*

$$\sum_{k=1}^{\infty}(1-|z_k|)\left(\log\frac{1}{1-|z_k|}\right)^{-(1+\varepsilon)}<\infty \tag{2}$$

for each $\varepsilon>0$. *The statement is false when* $\varepsilon=0$.

Horowitz [10] found that the zero-sets of Bergman spaces have exactly the same property (2) for $\varepsilon>0$, and that the series may diverge when $\varepsilon=0$. This result is also implicit in earlier work of Shapiro and Shields [27]; *cf.* [5], Section 4.1. As we shall see, the coincidence is partially explained by the fact that every uniformly discrete sequence is an A^p zero-set for some p. On the other hand, an A^p zero-set (even a Blaschke sequence) need not be uniformly discrete.

Proof of Theorem 2. Let δ be the separation constant. Draw a disk of pseudohyperbolic radius $\delta/2$ around each z_k. Then $\Delta(z_j,\delta/2)\cap\Delta(z_k,\delta/2)\neq\emptyset$ for $j\neq k$, by the uniform discreteness assumption. Let $r_j=1-2^{-j}$, and let N_j denote the number of points of our sequence in the annulus $R_j=\{z\in\mathbb{D}:r_j\le|z|<r_{j+1}\}$. We first obtain an estimate on N_j. This is done by taking slightly larger annuli $R_j^*\supset R_j$ defined by $R_j^*=\{z:s_j<|z|<S_j\}$, where

$$s_j=\frac{r_j-\delta/2}{1-r_j\delta/2},\qquad S_j=\frac{r_{j+1}+\delta/2}{1+r_{j+1}\delta/2}.$$

Then for each point $\zeta\in R_j$, the strong form of the triangle inequality for the pseudohyperbolic metric ensures that $\Delta(\zeta,\delta/2)\subset R_j^*$. The hyperbolic area of R_j^* is

$$\begin{aligned}a(R_j^*)&=\frac{S_j^2}{1-S_j^2}-\frac{s_j^2}{1-s_j^2}<\frac{S_j-s_j}{(1-s_j)(1-S_j)}\\&\le C\,\frac{S_j-s_j}{(1-r_j)(1-r_{j+1})}\le\frac{C}{1-r_{j+1}}=C\,2^j,\end{aligned}$$

where C is a constant, not necessarily the same at each occurrence, depending only on δ. Since

$$a(\Delta(z_k,\delta/2))=\frac{\delta^2}{4-\delta^2},\qquad k=1,2,\dots,$$

and the disks $\Delta(z_k,\delta/2)$ are nonoverlapping, we conclude that

$$\frac{\delta^2}{4-\delta^2}N_j=\sum_{z_k\in R_j}a(\Delta(z_k,\delta/2))\le a(R_j^*)\le C\,2^j,$$

whence $N_j \le C\, 2^j$. Now we have

$$\begin{aligned}\sum_{k=1}^{\infty}(1-|z_k|)\left(\log\frac{1}{1-|z_k|}\right)^{-(1+\varepsilon)} &= \sum_{j=0}^{\infty}\sum_{z_k\in R_j}(1-|z_k|)\left(\log\frac{1}{1-|z_k|}\right)^{-(1+\varepsilon)}\\ &\le \sum_{j=0}^{\infty} N_j(1-r_j)\left(\log\frac{1}{1-r_j}\right)^{-(1+\varepsilon)}\\ &\le C\sum_{j=0}^{\infty} 2^j 2^{-j} j^{-(1+\varepsilon)}\\ &= C\sum_{j=0}^{\infty} j^{-(1+\varepsilon)} < \infty,\end{aligned}$$

which is what we needed to prove.

An alternate proof can be based on Lemma 1. The sum is expressed as a Stieltjes integral with respect to the counting function $n(\Gamma, 0, r)$, whereupon an integration by parts allows the estimate $n(\Gamma, 0, r) = O(\frac{1}{1-r})$ to be introduced.

Finally, an example shows that the conclusion fails when $\varepsilon = 0$. Let the sequence Γ consist of 2^n points equidistributed on the circle of radius $1 - 2^{-n}$ centered at the origin. Such a sequence is uniformly discrete (as we will show presently) and we have

$$\sum_{k=1}^{\infty}(1-|z_k|)\left(\log\frac{1}{1-|z_k|}\right)^{-1} = \sum_{n=1}^{\infty} 2^{-n}\, 2^n\, (n\log 2)^{-1} = \infty. \qquad \square$$

The last example requires further discussion to justify our claim that the given sequence is uniformly discrete. Without much additional effort, we can carry out a more general construction. This will provide a family of uniformly discrete sequences that are regularly distributed in the unit disk.

Theorem 3. *Given an integer $k \ge 2$, let Γ_k be the set consisting of k^n points equidistributed on the circle $|z| = 1 - k^{-n}$, chosen to include the points $1 - k^{-n}$, where $n = 1, 2, \ldots$. Then Γ_k is uniformly discrete with separation constant*

$$\delta(\Gamma_k) \ge \min\left\{\frac{k-1}{k+1}, \frac{2}{\sqrt{5}}\frac{k-1}{\sqrt{k^2+1}}\right\}.$$

Proof. We need to consider two types of situations. A pair of points may lie on different circles, or they may lie on the same circle $|z| = r_n = 1 - k^{-n}$. Suppose first that the points $a, b \in \Gamma_k$ lie on two different circles: $|a| = r_n$ and $|b| = r_m$, where $m > n$. Then a simple inequality for Möbius mappings (*cf.* [3], page 154) gives

$$\rho(a,b) \ge \frac{r_m - r_n}{1 - r_m r_n} = \frac{k^{-n} - k^{-m}}{k^{-n} + k^{-m} - k^{-n-m}} \ge \frac{k^{m-n}-1}{k^{m-n}+1} \ge \frac{k-1}{k+1}.$$

Next suppose that $|a| = |b| = r_n$. The least pseudohyperbolic distance is attained for adjacent points, so we need only consider $a = r_n$ and $b = r_n e^{2\pi i/k^n}$. Then

$$\begin{aligned} \rho(a,b) &= r_n \left|1 - e^{2\pi i/k^n}\right| \left|1 - r_n^2\, e^{2\pi i/k^n}\right|^{-1} \\ &= 2r_n \sin(\pi/k^n) \left\{(1-r_n^2)^2 + 4r_n^2 \sin^2(\pi/k^n)\right\}^{-1/2} \\ &= \left\{1 + \left[(1-r_n^2)/(2r_n \sin(\pi/k^n))\right]^2\right\}^{-1/2}. \end{aligned}$$

But $\sin(\pi/k^n) \geq (2/\pi)(\pi/k^n) = 2/k^n$, so we deduce that

$$\begin{aligned} \rho(a,b) &\geq \left\{1 + \left[(1-r_n^2)k^n/(4r_n)\right]^2\right\}^{-1/2} = \left\{1 + \left[(1+r_n)/(4r_n)\right]^2\right\}^{-1/2} \\ &\geq \left\{1 + \left[(1+r_1)/(4r_1)\right]^2\right\}^{-1/2} = 4(k-1)\left\{20k^2 - 36k + 17\right\}^{-1/2} \\ &\geq \frac{2}{\sqrt{5}} \frac{k-1}{\sqrt{k^2+1}}. \end{aligned}$$

Combining the two lower bounds, we obtain the stated result. It may be remarked that the minimum is equal to $(k-1)/(k+1)$ for $k \leq 7$ but is equal to the second term for $k \geq 8$. In particular, $\delta(\Gamma_2) \geq \frac{1}{3}$. □

We have already remarked that a Blaschke sequence need not be uniformly discrete. In fact, it need not be a finite union of uniformly discrete sequences.

Proposition 3. *There exists a Blaschke sequence that is not a finite union of uniformly discrete sequences.*

Proof. A sequence $\Gamma = \{z_k\}$ will be constructed to place n distinct points z_k on the circle $|z| = r_n = 1 - 2^{-n}$, for $n = 1, 2, \ldots$. Then

$$\sum_{k=1}^{\infty}(1 - |z_k|) = \sum_{n=1}^{\infty} n(1 - r_n) = \sum_{n=1}^{\infty} n\, 2^{-n} < \infty,$$

which shows that Γ will be a Blaschke sequence. Now observe that $\{r_n\}$ is an exponential sequence, and is therefore uniformly discrete, so that $\rho(r_n, r_m) \geq \delta > 0$ for $n \neq m$. Thus by the triangle inequality, the pseudohyperbolic disks $\Delta(r_n, \delta/2)$ are pairwise disjoint. Now construct the sequence Γ by placing n distinct points z_k on the circle $|z| = r_n$ and inside the disk $\Delta(r_n, \delta/(2n))$, for $n = 1, 2, \ldots$. Any pair of points z_j and z_k in the nth disk must satisfy $\rho(z_j, z_k) < \delta/n$, which tends to zero as $n \to \infty$. Since Γ contains an unbounded number of points with arbitrarily small separation, it cannot be a finite union of uniformly discrete sequences. □

4. Interpolation and sampling

A sequence $\{z_k\}$ in $\mathbb{D}$ is said to be a *set of interpolation for* A^p if for each sequence of complex numbers w_k such that

$$\sum_{k=1}^{\infty} (1-|z_k|^2)^2\,|w_k|^p < \infty\,,$$

there exists a function $f \in A^p$ with the property $f(z_k) = w_k$ for all $k \in \mathbb{N}$. This definition is analogous to the definition given by Shapiro and Shields [26] for the Hardy spaces H^p (*cf.* [3], Chapter 9 for further details).

A sequence $\{z_k\}$ in $\mathbb{D}$ is said to be a *set of sampling for* A^p if

$$m\|f\|_p^p \leq \sum_{k=1}^{\infty}(1-|z_k|^2)^2|f(z_k)|^p \leq M\|f\|_p^p\,, \qquad f \in A^p\,,$$

for some positive constants m and M depending only on p. This concept has no immediate analogue in Hardy spaces, but was instead motivated by the theory of frames (in the case $p = 2$), closely related to the Fourier and wavelet analysis. (See [5], Section 6.1 for further discussion.)

Interpolation and sampling sets are Möbius invariant. For instance, if $\{z_k\}$ is a set of interpolation for A^p, then so is the sequence $\{\varphi_\alpha(z_k)\}$ for each $\alpha \in \mathbb{D}$. It is easy to see that no set can be both sampling and interpolating for the same space A^p. Indeed, an interpolation set is always an A^p zero-set, whereas a sampling set can never be an A^p zero-set.

Sampling and interpolation sequences for the Bergman space were characterized by Seip [25]. To discuss his results, we first need to define certain notions of density. Let Γ be a sequence of points in $\mathbb{D}$, and let $n(\Gamma,\zeta,s)$ be its pseudohyperbolic counting function. The quantity

$$E(\Gamma,\zeta,r) = \frac{\int_0^r n(\Gamma,\zeta,s)\,ds}{2\int_0^r a(\Delta(\zeta,s))\,ds}$$

is, roughly speaking, a measure of the average number of points of Γ contained in the disks $\Delta(\zeta,s)$, per unit of hyperbolic area, for $0 < s < r$.

The *lower uniform density* of Γ is defined to be

$$D^-(\Gamma) = \liminf_{r\to 1}\ \inf_{\zeta\in\mathbb{D}} E(\Gamma,\zeta,r)\,,$$

and the *upper uniform density* is

$$D^+(\Gamma) = \limsup_{r\to 1}\ \sup_{\zeta\in\mathbb{D}} E(\Gamma,\zeta,r)\,.$$

A simple calculation shows that

$$2\int_0^r a(\Delta(\zeta,s))\,ds = 2\int_0^r \frac{s^2}{1-s^2}\,ds = \log\frac{1+r}{1-r} - 2r\,,$$

so the densities have the more explicit expressions

$$D^-(\Gamma) = \liminf_{r\to 1} \left(\log \frac{1}{1-r}\right)^{-1} \inf_{\zeta\in\mathbb{D}} \int_0^r n(\Gamma,\zeta,s)\,ds$$

$$D^+(\Gamma) = \limsup_{r\to 1} \left(\log \frac{1}{1-r}\right)^{-1} \sup_{\zeta\in\mathbb{D}} \int_0^r n(\Gamma,\zeta,s)\,ds\,.$$

It is clear from the definitions that $0 \le D^-(\Gamma) \le D^+(\Gamma) \le \infty$. By Lemma 1, the counting function of a uniformly discrete sequence Γ satisfies

$$n(\Gamma,\zeta,s) = O\left(\frac{1}{1-s}\right),$$

which implies that $D^+(\Gamma) < \infty$ if Γ is uniformly discrete. In particular, every uniformly discrete sequence is an interpolation set, hence a zero-set, for some space A^p. It can be shown that $D^+(\Gamma) = 0$ if Γ is uniformly separated (*cf.* [5], page 175). This has the nontrivial implication that a uniformly separated sequence is interpolating for every Bergman space A^p.

Having defined the densities, we can now state the main theorems that characterize sampling and interpolation sets.

Theorem C. *A sequence Γ of distinct points in the disk is an interpolating sequence for A^p if and only if it is uniformly discrete and $D^+(\Gamma) < 1/p$.*

Theorem D. *A sequence Γ of distinct points in the disk is a sampling sequence for A^p if and only if it is a finite union of uniformly discrete sequences and it contains a uniformly discrete subsequence Γ' with the property $D^-(\Gamma') > 1/p$.*

Theorems C and D say that interpolation sequences are sparse everywhere in the unit disk, whereas sampling sequences must be sufficiently dense. The precise conditions are expressed by upper and lower uniform densities, but these densities can be quite difficult to compute. Our goal is now to develop sufficient conditions for sampling and interpolation, directly based on the pseudohyperbolic metric, that are relatively easy to verify.

For $0 < \varepsilon < 1$, a sequence $\Gamma = \{z_k\}$ of points in the unit disk $\mathbb{D}$ is said to be an *ε-net* if each point $z \in \mathbb{D}$ has the property $\rho(z, z_k) < \varepsilon$ for some z_k in Γ. An equivalent statement is that

$$\mathbb{D} = \bigcup_{k=1}^{\infty} \Delta(z_k, \varepsilon)\,,$$

where $\Delta(z_k, \varepsilon)$ again denotes a pseudohyperbolic disk. Luecking [14] showed that every uniformly discrete ε-net with sufficiently small ε is a sampling set for A^p. With the aid of Theorem D, this principle can be expressed in quantitative form.

Theorem 4. *Let $0 < p < \infty$. If Γ is a uniformly discrete ε-net with*

$$\varepsilon < \left(1 + \sqrt{\tfrac{2}{p}}\right)^{-1},$$

then Γ is a sampling set for A^p.

In particular, all uniformly discrete ε-nets with $\varepsilon < \frac{1}{2}$ are sampling sets for A^2. Another consequence of the theorem is that every ε-net is a sampling set for some Bergman space A^p.

Theorem 4 is a direct consequence of Theorem D and the following lemma.

Lemma 4. *Let* $0 < \varepsilon < 1$. *If* Γ *is an* ε*-net, then*

$$D^-(\Gamma) \geq \frac{(1-\varepsilon)^2}{2\varepsilon^2}.$$

Deduction of Theorem 4. Since the function

$$f(x) = \frac{(1-x)^2}{2x^2}$$

is decreasing for $0 < x < 1$, Lemma 4 implies that

$$D^-(\Gamma) \geq f(\varepsilon) > f\left(\frac{1}{1+\sqrt{\frac{2}{p}}}\right) = \frac{1}{p},$$

and the result follows from Theorem D. □

Before passing to the proof of Lemma 4, we note the following corollary.

Corollary. *A sequence* Γ *has lower density* $D^-(\Gamma) > 0$ *if and only if* Γ *is an* ε*-net for some* $\varepsilon < 1$.

In light of Theorem D, the corollary shows that a uniformly discrete set Γ is a sampling set for some A^p space if and only if Γ is an ε-net for some ε. This is analogous to the fact that Γ is an interpolation set for some A^p space if and only if Γ is uniformly discrete.

Proof of the Corollary. It needs only be observed that Γ is an ε-net if $D^-(\Gamma) > 0$. Suppose, on the contrary, that Γ is *not* an ε-net for any $\varepsilon < 1$. Then for each radius $r < 1$, the pseudohyperbolic counting function $n(\Gamma, \alpha, r) = 0$ for some $\alpha \in \mathbb{D}$ depending on r. Consequently, $n(\Gamma, \alpha, s) = 0$ for all $s < r$, so that $\int_0^r n(\Gamma, \alpha, s)\, ds = 0$. In particular,

$$\inf_{\zeta \in \mathbb{D}} \int_0^r n(\Gamma, \zeta, s)\, ds = 0$$

for each $r < 1$, which implies that $D^-(\Gamma) = 0$ by the definition of lower density. □

Proof of Lemma 4. Let $\zeta \in \mathbb{D}$ and define

$$h(s) = \frac{s-\varepsilon}{1-s\varepsilon}, \qquad \varepsilon < s < 1.$$

It follows from the strong form of the triangle inequality for the pseudohyperbolic metric that

$$\Delta(\zeta, h(s)) \subset \bigcup_{z_k \in \Delta(\zeta, s)} \Delta(z_k, \varepsilon). \tag{3}$$

Indeed, since Γ is an ε-net, each point $w \in \mathbb{D}$ belongs to some disk $\Delta(z_k, \varepsilon)$. If $w \in \Delta(\zeta, h(s))$, then

$$\rho(z_k, \zeta) \le \frac{\rho(z_k, w) + \rho(w, \zeta)}{1 + \rho(z_k, w)\rho(w, \zeta)}.$$

But $\rho(z_k, w) < \varepsilon$ and $\rho(w, \zeta) < h(s)$, so

$$\rho(z_k, \zeta) < \frac{\varepsilon + h(s)}{1 + \varepsilon h(s)} = s.$$

This shows that $z_k \in \Delta(\zeta, s)$, which proves (3). Recall now that a pseudohyperbolic disk $\Delta(\alpha, r)$ has hyperbolic area

$$a(\Delta(\alpha, r)) = \frac{r^2}{1 - r^2}.$$

Thus it follows from (3) that

$$n(\Gamma, \zeta, s) \ge \frac{a(\Delta(\zeta, h(s)))}{a(\Delta(z_k, \varepsilon))} = \frac{h(s)^2}{1 - h(s)^2} \frac{1 - \varepsilon^2}{\varepsilon^2} = \frac{(s - \varepsilon)^2}{\varepsilon^2(1 - s^2)}.$$

Consequently, for $r > \varepsilon$,

$$\int_0^r n(\Gamma, \zeta, s)\, ds \ge \int_\varepsilon^r n(\Gamma, \zeta, s)\, ds \ge \frac{1}{\varepsilon^2} \int_\varepsilon^r \frac{(s - \varepsilon)^2}{1 - s^2}\, ds.$$

But a straightforward calculation gives

$$\int_\varepsilon^r \frac{(s - \varepsilon)^2}{1 - s^2}\, ds = -r + \tfrac{1}{2}(1 + \varepsilon)^2 \log(1 + r) - \tfrac{1}{2}(1 - \varepsilon)^2 \log(1 - r) + C(\varepsilon),$$

where $C(\varepsilon)$ denotes a constant depending only on ε. Therefore,

$$D^-(\Gamma) \ge \lim_{r \to 1} \left(\log \frac{1}{1 - r}\right)^{-1} \frac{1}{\varepsilon^2} \int_\varepsilon^r \frac{(s - \varepsilon)^2}{1 - s^2}\, ds = \frac{(1 - \varepsilon)^2}{2\varepsilon^2},$$

which is the desired inequality. □

There is a similar criterion for interpolation. If a sequence $\Gamma = \{z_k\}$ is uniformly discrete with separation constant

$$\delta = \delta(\Gamma) = \inf_{j \ne k} \rho(z_j, z_k)$$

sufficiently close to 1, then Γ is an interpolation set for A^p. This principle was discovered by Amar [1] and Rochberg [20]. The following lemma, when combined with Theorem C, gives a quantitative version for $0 < p < \infty$.

Lemma 5. *If Γ is uniformly discrete with separation constant δ, where $0 < \delta < 1$, then*

$$D^+(\Gamma) \le (2\pi + 1) \frac{\sqrt{1 - \delta}}{(1 - \sqrt{1 - \delta})^2}.$$

Proof. It is convenient to define

$$\varepsilon = 1 - \sqrt{1-\delta}\,, \qquad \text{so that} \quad \delta = \frac{2\varepsilon}{1+\varepsilon^2} \quad \text{and} \quad 0 < \varepsilon < \delta\,.$$

Begin by writing

$$\int_0^r n(\Gamma,\zeta,s)\,ds = \int_0^\varepsilon n(\Gamma,\zeta,s)\,ds + \int_\varepsilon^r n(\Gamma,\zeta,s)\,ds$$

for $\varepsilon < r < 1$. Note that

$$\int_0^\varepsilon n(\Gamma,\zeta,s)\,ds \le n(\Gamma,\zeta,\varepsilon) \le \left(\frac{2}{\delta}+1\right)^2 \frac{1}{1-\varepsilon^2}\,,$$

by Lemma 1. Next write

$$\int_\varepsilon^r n(\Gamma,\zeta,s)\,ds = \int_\varepsilon^r n(\Gamma,\zeta,h(s))\,ds + \int_\varepsilon^r n(\Gamma,\zeta,h(s),s)\,ds\,,$$

where

$$h(s) = \frac{s-\varepsilon}{1-s\varepsilon} < s$$

and $n(\Gamma,\zeta,r_1,r_2)$ is the number of points of Γ lying in the pseudohyperbolic annulus $\Omega(\zeta,r_1,r_2)$.

To deal with the first integral on the right-hand side above, apply the strong form of the triangle inequality to see that $\Delta(z_j,\varepsilon)\cap\Delta(z_k,\varepsilon)=\emptyset$ for $j\neq k$, since $\rho(z_j,z_k)\ge\delta$. Moreover, the triangle inequality shows that if $z_k\in\Delta(\zeta,h(s))$, then $\Delta(z_k,\varepsilon)\subset\Delta(\zeta,s)$. Therefore,

$$n(\Gamma,\zeta,h(s)) \le \frac{a(\Delta(\zeta,s))}{a(\Delta(z_k,\varepsilon))} = \frac{(1-\varepsilon^2)s^2}{\varepsilon^2(1-s^2)}\,,$$

and so

$$\begin{aligned}\int_\varepsilon^r n(\Gamma,\zeta,h(s))\,ds &\le \frac{1-\varepsilon^2}{\varepsilon^2}\int_0^r \frac{s^2}{1-s^2}\,ds \\ &= \frac{1-\varepsilon^2}{2\varepsilon^2}\left(\log\frac{1+r}{1-r} - 2r\right).\end{aligned}$$

Our next task is to estimate $\int_\varepsilon^r n(\Gamma,\zeta,h(s),s)\,ds$. Since the pseudohyperbolic metric is invariant under Möbius self-mappings of the disk, there is no loss of generality in taking $\zeta=0$. Then $\Omega(0,h(s),s)$ is a Euclidean annulus bounded by the circles $|z|=s$ and $|z|=h(s)$. Use radial segments of angular separation θ to divide this annulus into a minimum number of congruent cells of pseudohyperbolic diameter smaller than δ. Then there can be at most one point of Γ in each cell. To determine this minimum number of cells, suppose first that θ is chosen to make cells of precise diameter δ, so that $\rho(h(s),se^{i\theta})=\delta$. Using a standard identity for the pseudohyperbolic metric, one computes

$$\begin{aligned} 1-\rho(h(s),se^{i\theta})^2 &= \frac{(1-s^2)(1-h(s)^2)}{|1-h(s)se^{i\theta}|^2} \\ &= \frac{(1-s^2)^2(1-\varepsilon^2)}{(1-s^2)^2+2s(s-\varepsilon)(1-s\varepsilon)(1-\cos\theta)}. \end{aligned}$$

Setting this expression equal to $1-\delta^2$, we find after some manipulation

$$\theta^2 \geq 2(1-\cos\theta) = \frac{\delta^2-\varepsilon^2}{1-\delta^2}\frac{(1-s^2)^2}{s(s-\varepsilon)(1-s\varepsilon)} \geq \frac{\delta^2-\varepsilon^2}{1-\delta^2}\frac{(1-s^2)^2}{s^2(1-\varepsilon^2)}.$$

Further calculation gives

$$\frac{\delta^2-\varepsilon^2}{(1-\delta^2)(1-\varepsilon^2)} = \frac{\varepsilon^2(\varepsilon^2+3)}{(1-\varepsilon^2)^2} \geq \frac{\varepsilon^4}{(1-\varepsilon^2)^2},$$

so we have

$$\theta \geq \frac{\varepsilon^2}{1-\varepsilon^2}\frac{1-s^2}{s}.$$

Consequently,

$$n(\Gamma,\zeta,h(s),s) \leq \frac{2\pi}{\theta} \leq \frac{2\pi(1-\varepsilon^2)}{\varepsilon^2}\frac{s}{1-s^2},$$

and

$$\int_\varepsilon^r n(\Gamma,\zeta,h(s),s)\,ds \leq \frac{2\pi(1-\varepsilon^2)}{\varepsilon^2}\int_0^r \frac{s}{1-s^2}\,ds = \frac{\pi(1-\varepsilon^2)}{\varepsilon^2}\log\frac{1}{1-r^2}.$$

Putting everything together, we conclude that

$$\begin{aligned} D^+(\Gamma) &\leq \lim_{r\to 1}\left(\log\frac{1}{1-r}\right)^{-1}\left\{\frac{1-\varepsilon^2}{2\varepsilon^2}\log\frac{1+r}{1-r} + \frac{\pi(1-\varepsilon^2)}{\varepsilon^2}\log\frac{1}{1-r^2}\right\} \\ &= (\pi+\tfrac{1}{2})\frac{1-\varepsilon^2}{\varepsilon^2} < (2\pi+1)\frac{\sqrt{1-\delta}}{(1-\sqrt{1-\delta})^2}. \end{aligned}$$

□

With an appeal to Theorem C, Lemma 5 yields the following theorem.

Theorem 5. *Let $0<p<\infty$. If Γ is a uniformly discrete sequence with separation constant δ satisfying*

$$(2\pi+1)\frac{\sqrt{1-\delta}}{(1-\sqrt{1-\delta})^2} < \frac{1}{p},$$

then Γ is an interpolation set for A^p.

Schuster and Varolin [23] have recently improved the condition of Theorem 5 to $\delta > \sqrt{p(p+2)}/(p+1)$. They have also improved Theorem 4 by showing that every uniformly discrete ε-net with $\varepsilon < \sqrt{p/(p+2)}$ is a sampling set for A^p.

References

[1] E. Amar, *Suites d'interpolation pour les classes de Bergman de la boule et du polydisque de* $\mathbb{C}^n$. Canad. J. Math **30** (1978), 711–737.

[2] R.R. Coifman and R. Rochberg, *Representation theorems for holomorphic and harmonic functions in* L^p. *Representation theorems for Hardy spaces.* Astérisque **77**, 11–66, Soc. Math. France, 1980.

[3] P.L. Duren, *Theory of* H^p *Spaces.* Academic Press, 1970. Reprinted with supplement by Dover Publications, 2000.

[4] P. Duren and A. Schuster, *Finite unions of interpolation sequences.* Proc. Amer. Math. Soc. **130** (2002), 2609–2615.

[5] P. Duren and A. Schuster, *Bergman Spaces.* American Mathematical Society, 2004.

[6] P. Duren, A.P. Schuster, and K. Seip, *Uniform densities of regular sequences in the unit disk.* Trans. Amer. Math. Soc. **352** (2000), 3971–3980.

[7] D. Girela, J.A. Peláez, and D. Vukotić, *Interpolating Blaschke products: Stolz and tangential approach regions.* Preprint.

[8] W.W. Hastings, *A Carleson measure theorem for Bergman spaces*, Proc. Amer. Math. Soc. **52** (1975), 237–241.

[9] H. Hedenmalm, B. Korenblum, and K. Zhu, *Theory of Bergman Spaces.* Springer, 2000.

[10] C. Horowitz, *Zeros of functions in the Bergman space.* Duke Math. J. **41** (1974), 693–710.

[11] C. Horowitz, *Factorization theorems for functions in the Bergman space.* Duke Math. J. **44** (1977), 201–213.

[12] D.H. Luecking, *Inequalities on Bergman spaces.* Illinois J. Math. **25** (1981), 1–11.

[13] D.H. Luecking, *A technique for characterizing Carleson measures on Bergman spaces.* Proc. Amer. Math. Soc. **87** (1983), 656–660.

[14] D.H. Luecking, *Forward and reverse Carleson inequalities for functions in Bergman spaces and their derivatives.* Amer. J. Math. **107** (1985), 85–111.

[15] D.H. Luecking, *Zero sequences for Bergman spaces.* Complex Variables Theory Appl. **30** (1996), 345–362.

[16] D.H. Luecking, *Finite unions of interpolating sequences.* To appear.

[17] G. McDonald and C. Sundberg, *Toeplitz operators on the disc.* Indiana Univ. Math. J. **28** (1979), 595–611.

[18] P.J. McKenna, *Discrete Carleson measures and some interpolation problems.* Michigan Math. J. **24** (1977), 311–319.

[19] V. Oleĭnik, *Imbedding theorems for weighted classes of harmonic and analytic functions.* (Russian) Zap. Naučn. Sem. Leningrad. Otdel. Mat. Inst. Steklov. (LOMI) **47** (1974), 120–137, 187, 192–193.

[20] R. Rochberg, *Interpolation by functions in Bergman spaces.* Michigan Math. J. **29** (1982), 229–236.

[21] A.P. Schuster, *Sets of sampling and interpolation in Bergman spaces.* Proc. Amer. Math. Soc. **125** (1997), 1717–1725.

[22] A.P. Schuster, *On Seip's description of sampling sequences for Bergman spaces.* Complex Variables Theory Appl. **42** (2000), 347–367.

[23] A. Schuster and D. Varolin, *Sampling and interpolation on Riemann surfaces.* Preprint.

[24] K. Seip, *Regular sets of sampling and interpolation for weighted Bergman spaces.* Proc. Amer. Math. Soc. **117** (1993), 213–220.

[25] K. Seip, *Beurling type density theorems in the unit disk.* Invent. Math. **113** (1993), 21–39.

[26] H.S. Shapiro and A.L. Shields, *On some interpolation problems for analytic functions.* Amer. J. Math. **83** (1961), 513–532.

[27] H.S. Shapiro and A.L. Shields, *On the zeros of functions with finite Dirichlet integral and some related function spaces.* Math. Z. **80** (1962), 217–229.

[28] K.-F. Tse, *Nontangential interpolating sequences and interpolation by normal functions.* Proc. Amer. Math. Soc. **29** (1971), 351–354.

[29] K. Zhu, *Operator Theory on Function Spaces.* Marcel Dekker 1990.

[30] K. Zhu, *Evaluation operators on Bergman spaces.* Math. Proc. Cambridge Philos. Soc. **117** (1995), 513–523.

Peter Duren
Department of Mathematics
University of Michigan
Ann Arbor, Michigan 48109–1109, USA
e-mail: `duren@umich.edu`

Alexander Schuster
Department of Mathematics
San Francisco State University
San Francisco, California 94132, USA
e-mail: `schuster@sfsu.edu`

Dragan Vukotić
Departamento de Matemáticas
Universidad Autónoma de Madrid
E-28049 Madrid, Spain
e-mail: `dragan.vukotic@uam.es`

Operator Theory:
Advances and Applications, Vol. 156, 151–172

Algebraic Aspects of the Dirichlet Problem

P. Ebenfelt, D. Khavinson and H.S. Shapiro

1. Introduction

Our aim in this paper is to make some remarks about certain algebraic aspects of the classical Dirichlet problem for the Laplace operator

$$\Delta := \sum_{k=1}^{n} \frac{\partial^2}{\partial x_k^2}$$

in the unit ball $\mathbb{B}^n$ of $\mathbb{R}^n$, $n \geq 2$. Thus, letting $\mathbb{S}^{n-1} := \partial\mathbb{B}^n$ denote the unit sphere, we pose the Dirichlet problem

$$\begin{cases} \Delta u = 0, & \text{in } \mathbb{B}^n \\ u = f, & \text{on } \mathbb{S}^{n-1}, \end{cases} \tag{1.1}$$

where the Dirichlet data f is a continuous function on the sphere $\mathbb{S}^{n-1}$. It will be convenient for us to assume, without loss of generality of course, that the data function f is defined in an open neighborhood of $\mathbb{S}^{n-1}$ in $\mathbb{R}^n$. The solution u is given by the Poisson integral

$$u(x) = \int_{\mathbb{S}^{n-1}} f(y) K_n(y,x) dS_y, \tag{1.2}$$

where dS_y is the surface area measure of the sphere and $K_n(x)$ is the Poisson kernel

$$K_n(x) = \omega_n \frac{1 - \|x\|^2}{\|y - x\|^n} \tag{1.3}$$

and ω_n is the reciprocal of the surface area of the sphere $\mathbb{S}^{n-1}$. Many classical analytic properties of the solution u (such as, e.g., regularity up to the boundary) can be gleaned from this formula. In this paper, we shall address some questions of algebraic nature which do not seem to follow as easily from (1.2).

The first and second authors were partially supported by the NSF grants DMS-0100110 and DMS-0139008 respectively.

Mathematics Subject Classification (2000). 31A25, 31B20.

We take as our starting point the following well known result: *If $f(x)$ is a polynomial of degree m, then the solution $u(x)$ to the Dirichlet problem* (1.1) *is a polynomial of degree at most m.* The precise origin of this result is difficult to trace (see, e.g., [F] for the result in $\mathbb{R}^3$), but a simple and elementary proof based on linear algebra can be found in, e.g., [K], [KS], and [Sh] (where the conclusion is in fact proved for the Dirichlet problem in an ellipsoid). In light of this basic result, it seems natural to ask if the solution operator preserves wider classes of algebraic functions:

Question A. Assume that the data $f(x)$ is a *rational* function without poles on $\mathbb{S}^{n-1}$. Is then the solution $u(x)$ of the Dirichlet problem (1.1) necessarily rational?

Professor W. Ross' student Mr. T. Fergusson (private communication) has answered Question A in the affirmative for $n = 2$. (This also follows from a more general result due to the first author [E]; a very short and simple proof is given in Section 3.) In the general form Question A was posed to us by Fergusson and Ross (written communication). More generally, one may ask the following.

Question B. Assume that the data $f(x)$ is an *algebraic* function which is analytic in a neighborhood of $\mathbb{S}^{n-1}$ in $\mathbb{R}^n$. Is then the solution $u(x)$ of the Dirichlet problem (1.1) necessarily algebraic?

Somewhat surprisingly perhaps, even though the answer to Question A is affirmative for $n = 2$, it is negative for all odd $n \geq 3$ and, at least, for all even n with $4 \leq n \leq 270$ (see Theorem 2.2 and the remark following it). The answer to Question B is negative in all dimensions. Our main results are examples of rational data functions f in $\mathbb{R}^n$, for all odd $n \geq 3$ (Theorem 2.1) and even $n \geq 4$ such that an additional condition is satisfied (Theorem 2.2), which yield non-rational – even non-algebraic – solutions to the Dirichlet problem (1.1), but also some positive results (Theorems 2.5 and 5.1) on Question A when f belongs to certain subclasses of rational functions in $\mathbb{R}^n$ with n *even*. We also give examples (Theorem 2.4) of algebraic data functions in $\mathbb{R}^2$ for which the solutions are non-algebraic.

The paper is organized as follows. In Section 2, we state our main results more precisely. In Section 3, we discuss the Dirichlet problem in two dimensions, its relation to the problem of Laurent decomposition, and give a proof of Theorem 2.4. The proofs of the remaining Theorems are then given in Sections 4 and 5. As concluding remarks, we discuss in Section 6 a connection with ultraspherical polynomials and the "Nehari transform".

2. Main results

Our first two results provide very simple rational functions $f(x)$ in $\mathbb{R}^n$, $n \geq 3$ (moreover if n is even the further arithmetical condition (2.2) is assumed), for which the solutions to (1.1) are not rational, or even algebraic, showing that the answers to both Questions A and B above are negative for these values of $n \geq 3$.

Theorem 2.1. *Let $|a| > 1$ and $n = 2k + 1 \geq 3$ be odd. Then the solution $u(x)$ to the Dirichlet problem (1.1) in $\mathbb{R}^n$ with data*

$$f(x) = \frac{1}{(x_1 - a)^{2k-1}} \tag{2.1}$$

is not algebraic.

When $n = 4$, the solution to (1.1) with data given by (2.1) turns out to be rational (cf. Theorem 2.5 below). However, we have the following for the even dimensional Dirichlet problem.

Theorem 2.2. *Let $n = 2k \geq 4$ be even and assume that*

$$\sum_{\substack{0 \leq j \leq k-1 \\ k+j \text{ odd}}} \sum_{l=j}^{k-1} \left(-\frac{1}{2}\right)^l \binom{k-1}{l} \binom{l+k-1}{l} \binom{l}{j} (-i)^j \neq 0, \tag{2.2}$$

Then the solution $u(x)$ to the Dirichlet problem (1.1) in $\mathbb{R}^n$ with data

$$f(x) = \frac{1}{(1 + x_1^2 + x_2^2)^{k-1}} \tag{2.3}$$

is not algebraic.

Remark 2.3. The authors suspect, but have been unable to prove, that (2.2) holds for all $k \geq 2$. At least, we have verified, using the software package MatLab, that (2.2) holds for all $2 \leq k \leq 135$. The modulus of the left hand side grows steadily, although not quite monotonically, as k grows in this range. For $k = 2$, the left hand side is i, and for $k = 135$ it is of the order 10^{80}. Thus, the solution in $\mathbb{R}^n$ with data (2.3) is not algebraic, at least for all even n with $4 \leq n \leq 270$.

As mentioned above, the answer to Question A when $n = 2$ is affirmative (see, e.g., Section 3 below for a simple proof). However, the answer to Question B is negative, as is shown by the following result. We shall identify $\mathbb{R}^2$ with the complex plane $\mathbb{C}$ in the usual way via $z = x_1 + ix_2$.

Theorem 2.4. *Let $w \in \mathbb{C} \setminus (\mathbb{S}^1 \cup \{0\})$. Then, the solution $u(x_1, x_2)$ to the Dirichlet problem (1.1) in $\mathbb{R}^2 \cong \mathbb{C}$ with data*

$$f(x_1, x_2) = |z - w|, \quad z = x_1 + ix_2, \tag{2.4}$$

is not algebraic.

Theorem 2.4 will follow from a more general result about Laurent decompositions in Section 3.

We shall conclude this section by giving a positive result on Question A for axially symmetric rational data functions in $\mathbb{R}^4$. Recall that a function $f(x)$ in $\mathbb{R}^n$ is called *axially symmetric* if there is a unit vector $v \in \mathbb{R}^n$ such that

$$f(Ax) = f(x), \quad \forall A \in O_v(\mathbb{R}^n), \tag{2.5}$$

for all $x \in \mathbb{R}^n$ where both sides are defined. Here, $O_v(\mathbb{R}^n)$ denotes the subgroup of the orthogonal linear group $O(\mathbb{R}^n)$ which leaves the vector v invariant; the line spanned by v is called the axis of symmetry.

Theorem 2.5. *Consider the Dirichlet problem* (1.1) *in* $\mathbb{R}^4$. *If the data* $f(x)$ *is an axially symmetric rational function, then the solution is a rational (axially symmetric) function.*

Theorem 2.5 will follow from a more general, but slightly more technical result (Theorem 5.1 in Section 5 below) for axially symmetric rational data functions in $\mathbb{R}^n$ for even n.

3. The two dimensional case

We identify $\mathbb{R}^2$ with $\mathbb{C}$ in the usual way via $z = x_1 + ix_2$, so that $\Delta = 4\,\partial^2/\partial z\partial\bar{z}$. Hence, any harmonic (not necessarily real-valued) function $u(x) = u(z,\bar{z})$ in a simply connected domain $\Omega \subset \mathbb{C}$ is of the form $u(z,\bar{z}) = h(z) + \overline{g(z)}$, where $h, g \in \mathcal{O}(\Omega)$ and $\mathcal{O}(\Omega)$ denotes the space of analytic functions in Ω. If $f = f(z,\bar{z})$ is a (possibly complex-valued) real-analytic function in a neighborhood of $\mathbb{S}^1$ and u is the solution to the Dirichlet problem (1.1), with $n = 2$, then it is well known that u extends real-analytically to an open neighborhood of the closed unit disk $\overline{\mathbb{D}}$; we prefer to use the notation $\mathbb{D}$ rather than $\mathbb{B}^2$ for the open unit disk in $\mathbb{C}$. Thus, we have $u(z,\bar{z}) = h(z) + \overline{g(z)}$ for $h, g \in \mathcal{O}(\overline{\mathbb{D}})$ (i.e., h, g are analytic in some neighborhood of the closed unit disk $\overline{\mathbb{D}}$) and we have

$$f(z,\bar{z}) = h(z) + \overline{g(z)}, \quad z \in \mathbb{S}^1. \tag{3.1}$$

The functions h and g will be uniquely determined if we require, e.g., $g(0) = 0$. Observe that on $\mathbb{S}^1$ we have $\bar{z} = 1/z$. Thus, if we write

$$F(z) := f(z, 1/z), \tag{3.2}$$

then $F(z)$ is analytic in some annular neighborhood of the unit circle $\mathbb{S}^1$ and, in view of (3.1), we have the identity

$$F(z) = h(z) + \bar{g}(1/z) \tag{3.3}$$

in some open neighborhood of $\mathbb{S}^1$, where $\bar{g}(w)$ is the analytic function

$$\bar{g}(w) := \overline{g(\bar{w})}.$$

This illustrates the connection between the Dirichlet problem in two dimensions and the problem of *Laurent splitting* of analytic functions. Recall that if $F(z)$ is an analytic function in a neighborhood of $\mathbb{S}^1$, then it has a unique series expansion

$$F(z) = \sum_{k=0}^{\infty} a_k z^k + \sum_{k=1}^{\infty} b_k \left(\frac{1}{z}\right)^k, \tag{3.4}$$

where the first series converges in a neighborhood of the closed unit disk $\overline{\mathbb{D}}$ and the second series converges in a neighborhood of the closed complement $\mathbb{C} \setminus \mathbb{D}$ (with the limit 0 at ∞). Hence, if we set

$$h(z) := \sum_{k=0}^{\infty} a_k z^k, \quad g(w) = \sum_{k=1}^{\infty} \bar{b}_k w^k, \tag{3.5}$$

then we obtain a unique splitting (3.3) of $F(z)$, for z in some neighborhood of $\mathbb{S}^1$, with $h, g \in \mathcal{O}(\overline{\mathbb{D}})$ and $g(0) = 0$; we shall refer to (h, g) as the *Laurent pair* of F. Thus, we have proved the following.

Proposition 3.1. *Let $F(z)$ be an analytic function in some open neighborhood of the unit circle $\mathbb{S}^1$ and $f(z, \bar{z})$ a (possibly complex-valued) real-analytic function such that (3.2) holds. If $h, g \in \mathcal{O}(\overline{\mathbb{D}})$ with $g(0) = 0$, then (h, g) is the Laurent pair of F (i.e., (3.3) holds) if and only if*

$$u(z, \bar{z}) := h(z) + \bar{g}(\bar{z})$$

is the solution of the Dirichlet problem (1.1) with $n = 2$ and data $f = f(z, \bar{z})$.

Thus, we can, and we shall, reformulate and prove Theorem 2.4 regarding the two-dimensional Dirichlet problem in terms of Laurent splittings. First, however, let us state the following proposition which is equivalent to the statement in the introduction that the solution to the Dirichlet problem with rational data is rational. We shall also give a very short and simple proof. The proof by T. Fergusson, also not difficult, was based on the Cauchy residues theorem [Fe, written communication]. See also [GS] for an algorithmic approach.

Proposition 3.2. *Let F be analytic in a neighborhood of $\mathbb{S}^1$ and let (h, g) be its Laurent pair (as defined above). If F is rational, then so are h and g.*

Proof. Since F is rational (or, equivalently, a meromorphic function on the Riemann sphere $\hat{\mathbb{C}}$) and g is analytic in a neighborhood of $\overline{\mathbb{D}}$, we can extend h, originally analytic in a neighborhood of $\overline{\mathbb{D}}$, as a meromorphic function on the Riemann sphere $\hat{\mathbb{C}}$ by defining it for $z \in \hat{\mathbb{C}} \setminus \overline{\mathbb{D}}$ by

$$h(z) = F(z) - \bar{g}(1/z),$$

where of course $z = \infty$ corresponds to $1/z = 0$. This implies that h is rational, and hence also g, being the difference between two rational functions, is rational. □

In order to prove Theorem 2.4, we observe that the data $f(z, \bar{z}) = |z - w| = \sqrt{(z - w)(\bar{z} - \bar{w})}$, with $w \in \mathbb{C} \setminus (\mathbb{S}^1 \cup \{0\})$, corresponds to

$$F(z) = f(z, 1/z) = \sqrt{\frac{(z - w)(1 - z\bar{w})}{z}}. \tag{3.6}$$

Thus, in view of Proposition 3.1, Theorem 2.4 follows from the following more general theorem about Laurent splittings.

Theorem 3.3. *Let F be analytic in a neighborhood of $\mathbb{S}^1$ and assume that F is of the form*

$$F(z) = \sqrt{R(z)}, \tag{3.7}$$

where R is a rational function with at least one simple zero in $\mathbb{D}$ and at least one simple zero in $\mathbb{C} \setminus \overline{\mathbb{D}}$. If (h, g) is the Laurent pair of F, then neither h nor g is algebraic.

Remark 3.4. Observe that the assumption that F is analytic in a neighborhood of $\mathbb{S}^1$ restricts the possible choices of rational functions R. For instance, R cannot have poles or simple zeros on $\mathbb{S}^1$. Moreover, R must have an even number of zeros and poles (counted with multiplicities) in $\mathbb{D}$.

Proof. Let $a \in \mathbb{D}$ and $b \in \mathbb{C} \setminus \overline{\mathbb{D}}$ be simple zeros of R, and let L be a smooth curve from a to b which does not intersect any other poles or zeros of R. We shall, as we may, take L to be the shortest arc of a circle with large radius through a and b. Let U be a small disk centered at the point of intersection between L and $\mathbb{S}^1$. We shall choose U so small that F, h, and g are all analytic in U. We shall also let $U_1 \subset \mathbb{D}$ and $U_2 \subset \mathbb{C} \setminus \overline{\mathbb{D}}$ be disks centered at a and b respectively, so small that R has no poles or zeros, except a and b, inside U_1 and U_2. Now define the closed curve γ to be the piece L' of L going from the point of intersection $L \cap \partial U_1$ to the point $L \cap \partial U_2$, around the circle ∂U_2 in the positive direction, back along $-L'$, and finally around the circle ∂U_1 in the positive direction. Observe that in a neighborhood of γ, by the construction, there is an analytic function $A(z)$ such that

$$F(z) = \sqrt{(z-a)(z-b)A(z)}. \tag{3.8}$$

Let us start in U and continue the function $h(z)$ analytically along the portion of γ contained in $\mathbb{C} \setminus \overline{\mathbb{D}}$ until we return to U by using the identity $h(z) = F(z) - \bar{g}(1/z)$. Note that the analytic continuation of $F(z)$ changes sign as we go around b whereas $1/z$ stays inside $\overline{\mathbb{D}}$ where $\bar{g}$ is analytic. Thus, if we denote by $\tilde{h}(z)$ the analytic function in U obtained by this analytic continuation, then we conclude that $\tilde{h}(z) = -F(z) + \bar{g}(1/z)$. Hence, we have

$$\tilde{h}(z) = h(z) - 2F(z)$$

in U. If we use this identity to continue the new branch $\tilde{h}(z)$ in U along the remaining portion of γ inside $\mathbb{D}$ (where $h(z)$ is analytic) until we come back to U – and have completed a full tour of γ – then $F(z)$ changes sign again (since we go around a). The conclusion is that if we denote by $h_1(z)$ the analytic function in U obtained by analytically continuing $h(z)$ along the closed curve γ, then

$$h_1(z) = h(z) + 2F(z). \tag{3.9}$$

Also, as we have already seen, analytic continuation of the function $F(z)$ in U around γ leads to the same function $F(z)$ (since we change sign twice). Thus, it follows from (3.9) that the analytic continuation $h_k(z)$ of $h(z)$ around $k\gamma$ (i.e., k times around γ) satisfies

$$h_k(z) = h(z) + 2kF(z).$$

In particular, repeated analytic continuation of $h(z)$ yields an *infinite number* of different branches over U. This means, of course, that $h(z)$ cannot be an algebraic function. Consequently, g can also not be algebraic. This completes the proof. □

4. Proof of Theorems 2.1 and 2.2

The strategy in proving Theorems 2.1 and 2.2 will be to first show that the solutions to the corresponding Dirichlet problems, restricted to the line $x = (s, 0, \dots, 0)$, can be expressed by integrals of a certain type. The proofs then reduce to showing that these integrals cannot be algebraic functions of s. Our starting point will be the Poisson integral (1.2) for the solution of (1.1) in $\mathbb{R}^n$. If we assume that the data function f is a function of x_1 alone, as in Theorem 2.1, then by using cylindrical coordinates $y = (t, r\xi)$, where $\xi \in \mathbb{S}^{n-2}$, $t \in [-1, 1]$ and $r = \sqrt{1-t^2}$, we obtain

$$u(x) = \omega_n(1 - \|x\|^2) \int_{-1}^{1} g(t) r^{n-3} \left(\int_{\mathbb{S}^{n-2}} \frac{dS_\xi}{((t-x_1)^2 + \sum_{j=2}^{n} (r\xi_j - x_j)^2)^{n/2}} \right) dt, \tag{4.1}$$

where $g = g(t)$ is the function of one variable such that $f(x) = g(x_1)$. As mentioned above, to prove Theorem 2.1 we shall show that the restriction of $u(x)$ to the axis of symmetry $x = (s, 0, \dots, 0)$ is not algebraic. The function inside the integral over $\mathbb{S}^{n-2}$ then reduces to

$$\begin{aligned} \frac{1}{((t-s)^2 + \sum_{j=2}^{n} (r\xi_j)^2)^{n/2}} &= \frac{1}{((t-s)^2 + r^2)^{n/2}} \\ &= \frac{1}{(2s)^{n/2}} \frac{1}{(\phi(s) - t)^{n/2}}, \end{aligned} \tag{4.2}$$

where in the first step we used that $\xi \in \mathbb{S}^{n-2}$, and in the second that $r^2 = 1 - t^2$; we also use the notation

$$\phi(s) = \frac{1}{2}\left(s + \frac{1}{s}\right). \tag{4.3}$$

Observe that the expression in (4.2), which appears as an integrand in (4.1), is in fact independent of ξ. Thus, we arrive at the following result.

Proposition 4.1. *Let $u(x)$ be the solution in $\mathbb{R}^n$ to the Dirichlet problem (1.1) in which the data function f is a function of x_1 alone, i.e., $f(x) = g(x_1)$. If we set $v(s) := u(s, 0, \dots, 0)$, with $s \in (0, 1)$, then we have, for some constant c_n depending only on n,*

$$v(s) = c_n \frac{1 - s^2}{s^{n/2}} P_n(g)(\phi(s)), \tag{4.4}$$

where ϕ is given by (4.3) and P_n is the integral operator

$$P_n(g)(z) := \int_{-1}^{1} g(t)(1 - t^2)^{(n-3)/2} \frac{dt}{(z-t)^{n/2}}. \tag{4.5}$$

To show that $u(x)$ is not algebraic, it suffices to show that $P_n(g)(z)$ is not an algebraic function of z. Before proceeding to do this, we shall derive a similar, but slightly more complicated, integral formula for the solution $u(x)$ when the data function, as in Theorem 2.2, is a function of $x_1^2 + x_2^2$. Thus, assume that $f(x) = g(x_1^2 + x_2^2)$, where $g = g(u)$ is a function of one variable; we shall also assume here that $n \geq 4$. This time we use, in the Poisson integral (1.2), coordinates $y = (t_1, t_2, \rho\xi)$, where $(t_1, t_2) \in \mathbb{B}^2$, $\xi \in \mathbb{S}^{n-3}$, and $\rho = \sqrt{1 - (t_1^2 + t_2^2)}$. We then obtain

$$u(x) = \omega_n(1 - \|x\|^2) \int_{\mathbb{B}^2} g(t_1^2 + t_2^2)\rho^{n-4} \times \left(\int_{\mathbb{S}^{n-3}} \frac{dS_\xi}{((t_1 - x_1)^2 + (t_2 - x_2)^2 + \sum_{j=3}^{n}(\rho\xi_j - x_j)^2)^{n/2}} \right) dt, \tag{4.6}$$

where dt denotes $dt_1 dt_2$. Again restricting to the line $x = (s, 0, \dots, 0)$ and essentially repeating the computation in (4.2), we obtain

$$u(s, 0 \dots, 0) = c_n \frac{1 - s^2}{s^{n/2}} \int_{\mathbb{B}^2} g(t_1^2 + t_2^2)\rho^{n-4} \frac{1}{(\phi(s) - t_1)^{n/2}} dt, \tag{4.7}$$

where c_n is some constant depending only on n; in what follows, we shall, as is customary, use c_n to denote such a constant and the reader should be warned that the precise value of c_n may be different from formula to formula. By using polar coordinates $t_1 = r\cos\theta$, $t_2 = r\sin\theta$, we arrive at

$$u(s, 0 \dots, 0) = c_n \frac{1 - s^2}{s^{n/2}} \int_0^1 g(r^2)(1 - r^2)^{(n-4)/2}\, r \left(\int_0^{2\pi} \frac{1}{(\phi(s) - r\cos\theta)^{n/2}} d\theta \right) dr. \tag{4.8}$$

We shall compute the inner integral over the circle for even $n = 2k$ with $k \geq 2$. We shall, initially, be interested in the solution $u(s, 0, \dots, 0)$ with $s \in (0, 1)$. Observe that this corresponds to $z := \phi(s)$ real-valued with $z > 1$. Thus, in the computation that follows, we shall let z be real with $z > 1$ and $0 \leq r \leq 1$. We introduce $\zeta = e^{i\theta}$ and obtain

$$\int_0^{2\pi} \frac{d\theta}{(z - r\cos\theta)^k} = -i\left(-\frac{2}{r}\right)^k \int_\gamma \frac{\zeta^{k-1} d\zeta}{(\zeta^2 - (2z/r)\zeta + 1)^k}, \tag{4.9}$$

where γ denotes the positively oriented unit circle. The latter integral can be computed by residues. If we denote

$$h(\zeta) := \frac{\zeta^{k-1}}{(\zeta^2 - (2z/r)\zeta + 1)^k} \tag{4.10}$$

and observe that the denominator has two distinct real zeros (of multiplicity k) at the points

$$\zeta_0 := \frac{1}{r}(z - \sqrt{z^2 - r^2}), \quad \zeta_1 := \frac{1}{r}(z + \sqrt{z^2 - r^2}), \tag{4.11}$$

then we conclude, since $0 < \zeta_0 < 1 < \zeta_1$, that

$$\begin{aligned}\int_0^{2\pi} \frac{d\theta}{(z - r\cos\theta)^k} &= 2\pi\left(-\frac{2}{r}\right)^k \operatorname{Res}_{\zeta_0} h \\ &= 2\pi\left(-\frac{2}{r}\right)^k \frac{1}{(k-1)!}\left[\left(\frac{d}{d\zeta}\right)^{k-1} \frac{\zeta^{k-1}}{(\zeta - \zeta_1)^k}\right]\Bigg|_{\zeta=\zeta_0}. \end{aligned} \tag{4.12}$$

If we write

$$\frac{\zeta^{k-1}}{(\zeta-\zeta_1)^k} = \frac{1}{\zeta - \zeta_1}\left(1 + \frac{\zeta_1}{\zeta-\zeta_1}\right)^{k-1} = \sum_{l=0}^{k-1}\binom{k-1}{l}\frac{\zeta_1^l}{(\zeta-\zeta_1)^{l+1}},$$

then, by carrying out the differentiation in (4.12) term by term,

$$\begin{aligned}\int_0^{2\pi} \frac{d\theta}{(z - r\cos\theta)^k} &= 2\pi\left(-\frac{2}{r}\right)^k \frac{1}{(k-1)!} \\ &\quad \times \sum_{l=0}^{k-1}\binom{k-1}{l}(-1)^{k-1}(k-1)!\binom{l+k-1}{l}\frac{\zeta_1^l}{(\zeta_0-\zeta_1)^{l+k}} \\ &= -2\pi\left(\frac{2}{r}\right)^k \sum_{l=0}^{k-1}\binom{k-1}{l}\binom{l+k-1}{l}\frac{\zeta_1^l}{(\zeta_0-\zeta_1)^{l+k}}. \end{aligned} \tag{4.13}$$

Now, $\zeta_0 - \zeta_1 = -2\sqrt{z^2-r^2}/r$ and hence we can rewrite (4.13) as follows

$$\begin{aligned}&\int_0^{2\pi} \frac{d\theta}{(z - r\cos\theta)^k} \\ &= -2\pi\left(\frac{2}{r}\right)^k \sum_{l=0}^{k-1}\binom{k-1}{l}\binom{l+k-1}{l}\left(-\frac{1}{2}\right)^{l+k}\frac{r^k(z+\sqrt{z^2-r^2})^l}{(z^2-r^2)^{(l+k)/2}}. \end{aligned} \tag{4.14}$$

By expanding $(z + \sqrt{z^2-r^2})^l$ and simplifying, we obtain

$$\begin{aligned}&\int_0^{2\pi} \frac{d\theta}{(z - r\cos\theta)^k} \\ &= (-1)^{k+1}2\pi\sum_{l=0}^{k-1}\sum_{j=0}^{l}\left(-\frac{1}{2}\right)^l\binom{k-1}{l}\binom{l+k-1}{l}\binom{l}{j}\frac{z^j}{(z^2-r^2)^{(k+j)/2}}. \end{aligned} \tag{4.15}$$

We can formulate this as follows.

Lemma 4.2. *For real z with $z > 1$, $0 \le r \le 1$, and $k \ge 2$, we have the following identity*

$$\int_0^{2\pi} \frac{d\theta}{(z - r\cos\theta)^k} = (-1)^{k+1}2\pi\sum_{j=0}^{k-1} B_{jk}\frac{z^j}{(z^2-r^2)^{(k+j)/2}}, \tag{4.16}$$

where

$$B_{jk} := \sum_{l=j}^{k-1} \left(-\frac{1}{2}\right)^l \binom{k-1}{l} \binom{l+k-1}{l} \binom{l}{j}. \tag{4.17}$$

Plugging this into (4.8), we obtain for the solution $u(x)$, with data $f(x) = g(x_1^2 + x_2^2)$ in $\mathbb{R}^n = \mathbb{R}^{2k}$,

$$u(s,0\ldots,0) = c_n \frac{1-s^2}{s^k} \sum_{j=0}^{k-1} B_{jk} z^j \int_0^1 g(r^2)(1-r^2)^{k-2} \frac{2rdr}{(z^2-r^2)^{(k+j)/2}}, \tag{4.18}$$

where B_{jk} is given by (4.17) above and $z = \phi(s)$ as given by (4.3). Thus, by making the substitution $u = r^2$, we obtain the following.

Proposition 4.3. *Let $u(x)$ be the solution in $\mathbb{R}^n$, with $n = 2k$ even and $k \geq 2$, to the Dirichlet problem (1.1) in which the data function f is a function of $x_1^2 + x_2^2$, i.e., $f(x) = g(x_1^2 + x_2^2)$. If we set $v(s) := u(s, 0, \ldots, 0)$, with $s \in (0, 1)$, then we have, for some constant c_n depending only on n,*

$$v(s) = c_n \frac{1-s^2}{s^k} \tilde{P}_k(g)(\phi(s)), \tag{4.19}$$

where ϕ is given by (4.3) and $\tilde{P}_k$ is the operator

$$\tilde{P}_k(g)(z) := \sum_{j=0}^{k-1} B_{jk} z^j \int_0^1 g(u)(1-u)^{k-2} \frac{du}{(z^2-u)^{(k+j)/2}} \tag{4.20}$$

and B_{jk} is given by (4.17).

Motivated by Propositions 4.1 and 4.3, we shall now proceed to study the integral, initially defined for real $w > 1$,

$$I(w) := \int_a^1 h(u) \frac{du}{(w-u)^{m/2}}, \tag{4.21}$$

in which m is an integer ≥ 2, a a real number ≤ 0, $h(u)$ is a rational function without poles on the real line segment $[a, \infty)$. We shall further assume that, as u goes to infinity in the complex plane, we have the following estimate

$$|h(u)| \leq C|u|^q, \tag{4.22}$$

for some constant $C > 0$ and some q with

$$q < \frac{m}{2} - 1. \tag{4.23}$$

Clearly, $I(w)$ extends as an analytic function in the region $\Omega := \mathbb{C} \setminus (-\infty, 1]$ by simply letting the square root in (4.24) be the principal branch. Moreover, $I(w)$ is analytically extendable across the segment $(-\infty, a)$, and the analytic continuation across this segment equals $(-1)^m I(w)$; in particular, if m is an even integer, we may continue $I(w)$ to an analytic function in the doubly connected region $\mathbb{C} \setminus [a, 1]$. Comparing the definition of $I(w)$ and the expressions for $v(s)$ in Propositions 4.1 and 4.3 in which w would correspond to $\phi(s)$ or $\phi(s)^2$, respectively, we realize that we should continue $I(w)$ analytically across the segment $[a,1]$ (which corresponds to continuing our solution $v(s)$ across the unit circle in the s-plane) before attempting to detect any non-algebraic singularities at infinity. Let us denote by $I_+(x)$ and $I_-(x)$ the boundary value of I at x, for $x \in (a,1)$, approaching from $\{\text{Im } w > 0\}$ and $\{\text{Im } w < 0\}$, respectively. Let Γ_+ and Γ_- denote smooth simple oriented curves from a to 1 such $\Gamma_\pm \setminus \{a,1\}$ is contained in $\{\pm\text{Im } w > 0\}$ and such that $h(w)$ has no singularities in the region Ω bounded by the closed simple oriented curve $\Gamma := \Gamma_- - \Gamma_+$. By deforming the contour of integration in the definition of $I(w)$ into $\Gamma_\pm$, we conclude that

$$I_+(x) = \int_{\Gamma_-} h(u)\frac{du}{(w-u)^{m/2}}, \quad I_-(x) = \int_{\Gamma_+} h(u)\frac{du}{(w-u)^{m/2}}, \tag{4.24}$$

where the branch of the square root in each case (when m is odd) is the principal branch. Let us now denote by $\tilde{I}(w)$ the analytic continuation of $I_+(x)$ into the half plane $\{\text{Im } w < 0\}$, or equivalently the analytic continuation of $I(w)$ across the segment $(a,1)$ from the upper half plane into the lower half plane. We then obtain, for $w = x - iy$ with $x \in (a,1)$ and $y > 0$, $(I_+ = I_- + I_+ - I_-)$

$$\tilde{I}(w) = I(w) + \int_{\Gamma_-} h(u)\frac{du}{(w-u)^{m/2}} - \int_{\Gamma_+} h(u)\frac{du}{(w-u)^{m/2}}, \tag{4.25}$$

where the square roots again are the principal branches. Note that when $m = 2p$ is even, we obtain

$$\tilde{I}(w) = I(w) + \int_{\Gamma} h(u)\frac{du}{(w-u)^p} = I(w) + H(w), \tag{4.26}$$

where again Γ is the closed simple contour $\Gamma_- - \Gamma_+$ and

$$H(w) = \frac{1}{(p-1)!}\left[\left(\frac{d}{du}\right)^{p-1} h(u)\right]\Bigg|_{u=w}. \tag{4.27}$$

In particular, if m is even, then $\tilde{I}(w)$ is equal to $I(w)$ modulo a rational function. When m is odd, we can compute the integrals over $\Gamma_\pm$ in (4.25), for $w \in (a,1)$, as follows. For $R > 0$ large, let γ_R be the closed oriented curve consisting of Γ_+, followed by Γ_-, the line segment $I_R := [1,R]$, the circle $C_R := \{|w| = R\}$ in the negative direction, and finally the line segment $-I_R$. For fixed $w \in (a,1)$, we continue the square root $\sqrt{w-u}$ to a continuous function of u along γ_R, starting from the principal branch along $-\Gamma_+ + \Gamma_-$. This defines an analytic branch of this square root, as a function of u, in the open domain V bounded by the closed curve γ_R. The orientation of γ_R, however, is opposite the positively oriented boundary ∂V. Thus, by the Residue Theorem, we have, for R large enough,

$$\int_{\gamma_R} h(u)\frac{du}{(w-u)^{m/2}} = -2\pi i \sum_{l=1}^{q} \text{Res}_{a_l}\left[\frac{h(u)}{(w-u)^{m/2}}\right]. \tag{4.28}$$

where $a_1, \ldots, a_q$ denote the poles of h and the residue is taken with respect to the variable u. (Recall that h was assumed not to have any poles on $[a,\infty)$ or inside the curve Γ.) The integral over the circle C_R in (4.28) tends to 0 as $R \to \infty$ in view of (4.22) and (4.23). Observe also that the integrals over the two oppositely oriented line segments I_R and $-I_R$ do not cancel due to the fact that the branches of the square root $\sqrt{w-u}$ differ by a sign in each integral. Thus, by letting $R \to \infty$, we conclude that

$$\begin{aligned}\int_{\Gamma_-} h(u)\frac{du}{(w-u)^{m/2}} - \int_{\Gamma_+} h(u)\frac{du}{(w-u)^{m/2}} + 2\int_1^\infty h(u)\frac{du}{(w-u)^{m/2}} \\ = -2\pi i \sum_{l=1}^{q} \mathrm{Res}_{a_l}\left[\frac{h(u)}{(w-u)^{m/2}}\right],\end{aligned} \tag{4.29}$$

where the square root of the negative quantity $w-u$ for $u \in [1,\infty)$ is $i\sqrt{u-w}$. By using (4.25), we conclude that for the analytic continuation $\tilde{I}(w)$ of $I(w)$ across $(a,1)$ from the upper half plane into the lower half plane, we have

$$\tilde{I}(w) = I(w) - 2i\int_1^\infty h(u)\frac{du}{(u-w)^{m/2}} - 2\pi i \sum_{l=1}^{q} \mathrm{Res}_{a_l}\left[\frac{h(u)}{(w-u)^{m/2}}\right]. \tag{4.30}$$

Proof of Theorem 2.1. Recall that $n = 2k+1$ with $k \geq 1$. As mentioned above, it suffices to show that the function $J(w) := P_n(g)(w)$, given by (4.5), is not algebraic with $g(t) = 1/(t+2)^{2k-1}$. Thus, to reach a contradiction, we shall assume that $J(w)$ is algebraic. Observe that $J(w)$ is equal to the integral $I(w)$ above with $m = 2k+1$, $a = -1$, and

$$h(u) = \frac{(1-u^2)^{k-1}}{(u+2)^{2k-1}}. \tag{4.31}$$

Hence by (4.30), if we denote by $\tilde{J}(w)$ the analytic continuation of $J(w)$ across $(-1,1)$ from the upper half plane to the lower half plane, then we have

$$\tilde{J}(w) = J(w) - 2i\int_1^\infty h(u)\frac{du}{(u-w)^{k+1/2}} - 2\pi i\,\mathrm{Res}_{-2}\left[\frac{h(u)}{(w-u)^{k+1/2}}\right], \tag{4.32}$$

where h is as in (4.31). Observe that the function

$$H(w) := -2\pi i\,\mathrm{Res}_{-2}\left[\frac{h(u)}{(w-u)^{k+1/2}}\right]$$

is algebraic, since it is a finite linear combination of derivatives, with respect to u, of $(w-u)^{-k-1/2}$ evaluated at $u=-2$. Now, if $J(w)$ were algebraic, then of course $\tilde{J}(w)$ would be algebraic. Thus, we conclude that if $J(w)$ were algebraic, then the integral

$$K(w) := \int_1^\infty \frac{(1-u^2)^{k-1}}{(u+2)^{2k-1}}\frac{du}{(u-w)^{k+1/2}} \tag{4.33}$$

would be an algebraic function of w. Observe that we may rewrite this integral as follows

$$\begin{aligned} K(w) &= \int_1^\infty \frac{(1-u^2)^{k-1}}{(u+2)^{2k-1}} \frac{du}{(u-w)^{k+1/2}} \\ &= \sum_{l=0}^{k-1} (-1)^l \binom{k-1}{l} \int_1^\infty \frac{u^{2l}}{(u+2)^{2k-1}} \frac{du}{(u-w)^{k+1/2}} \end{aligned} \tag{4.34}$$

To obtain a contradiction, we shall need the following lemma.

Lemma 4.4. *Let $b > -1$ and $p + \alpha > 1$. Then, as $x \to \infty$, we have*

$$\int_1^\infty \frac{du}{(u+b)^p (u+x)^\alpha} = \begin{cases} O(x^{-\alpha}), & \text{if } p > 1 \\ x^{-\alpha} \ln x + O(x^{-\alpha}), & \text{if } p = 1. \end{cases} \tag{4.35}$$

Proof. The statement for $p > 1$ is an immediate consequence of Lebesgue's dominated convergence theorem. (Simply multiply the integral by x^α and take the limit as $x \to \infty$.) Let us therefore assume that $p = 1$. Making the change of variables $u + b = t(x - b)$, we obtain, with $y = x - b$,

$$\begin{aligned} \int_1^\infty \frac{du}{(u+b)(u+x)^\alpha} &= y^{-\alpha} \int_{(1+b)/y}^\infty \frac{dt}{t(t+1)^\alpha} \\ &= y^{-\alpha} \left(\int_{(1+b)/y}^1 \frac{dt}{t(t+1)^\alpha} + \int_1^\infty \frac{dt}{t(t+1)^\alpha} \right). \end{aligned} \tag{4.36}$$

Observe that the function $1/(t(t+1)^\alpha)$ is integrable on $[1, \infty]$ by the assumption that $1 + \alpha > 1$. Thus, to prove the lemma it suffices to show that

$$\int_{(1+b)/y}^\infty \frac{dt}{t(t+1)^\alpha} = \ln y + O(1). \tag{4.37}$$

An integration by parts yields

$$\int_{(1+b)/y}^1 \frac{dt}{t(t+1)^\alpha} = \frac{1}{(1+(1+b)/y)^\alpha} \ln(y/(1+b)) + \alpha \int_{(1+b)/y}^1 \frac{\ln t \, dt}{(t+1)^{\alpha+1}}. \tag{4.38}$$

Since the function $f(t) = \ln t/(t+1)^{\alpha+1}$ is integrable on $[0, 1]$, the conclusion of Lemma 4.4 now follows. □

Now, observe that, for $u \in [1, \infty)$, we have $u < u + 2 < 4u$. Thus, if we set $w = -x$ in (4.34) with $x > 0$, then we conclude from (4.34) and Lemma 4.4 (with $b = 0$) that, as $x \to \infty$,

$$Ax^{-k-1/2} \ln x \le |K(-x)| \le Bx^{-k-1/2} \ln x, \tag{4.39}$$

for some constants $0 < A < B$. If $K(w)$ were algebraic, then $K(w)$ would have a Puiseux expansion at infinity, i.e., it would be given for large w by a convergent series

$$K(w) = \sum_{l=-r}^\infty b_l \left(\frac{1}{w}\right)^{l/p},$$

for some integers p and r. This clearly contradicts the estimate (4.39) and, hence, completes the proof of Theorem 2.1. $\square$

Proof of Theorem 2.2. The proof of Theorem 2.2 proceeds along the same lines as that of Theorem 2.1. This time, we have $n = 2k$ with $k \geq 2$, and we shall show, as is sufficient to prove Theorem 2.2, that $J(z) := \tilde{P}_k(g)(z)$, given by (4.20) with $g(u) = 1/(1+u)^{k-1}$, is not algebraic. Observe that we can write

$$J(z) = \sum_{j=0}^{k-1} B_{jk} z^j I_j(w),$$

where B_{jk} is given by (4.17), $w = z^2$, and $I_j(w)$ is as in (4.21) with $a = 0$, $m = k + j$, and

$$h(u) = \frac{(1-u)^{k-2}}{(1+u)^{k-1}}. \tag{4.40}$$

Thus, if we denote by $\tilde{J}(z)$ the analytic continuation of $J(z)$ across the segment $(0,1)$ from the upper half plane into the lower half plane, we obtain, by the arguments preceeding the proof of Theorem 2.1 above (since $w = z^2$ then also crosses $(0,1)$ from the upper half plane into the lower half plane),

$$\begin{aligned}\tilde{J}(z) = J(z) &+ \sum_{\substack{0\leq j\leq k-1\\ k+j\,\mathrm{even}}} z^j H_j(w)\\ &- \sum_{\substack{0\leq j\leq k-1\\ k+j\,\mathrm{odd}}} B_{jk} z^j \left(2i\int_1^\infty h(u)\frac{du}{(u-w)^{(k+j)/2}} + 2\pi i \,\mathrm{Res}_{-1}\left[\frac{h(u)}{(w-u)^{(k+j)/2}}\right]\right),\end{aligned} \tag{4.41}$$

where $H_j(w)$ is the rational function given by (4.27) with h is as in (4.40) and $p = (k+j)/2$. Hence, if we assume, in order to reach a contradiction, that $J(z)$ is algebraic, then we conclude (cf. the proof of Theorem 2.1 above) that the function

$$K(z) = \sum_{\substack{0\leq j\leq k-1\\ k+j\,\mathrm{odd}}} B_{jk} z^j \int_1^\infty \frac{(1-u)^{k-2}}{(1+u)^{k-1}} \frac{du}{(u-w)^{(k+j)/2}} \tag{4.42}$$

must be algebraic. We write

$$\frac{(1-u)^{k-2}}{(1+u)^{k-1}} = \frac{1}{1+u}\left(-1 + \frac{2}{1+u}\right)^{k-2}$$

and expand the right hand side to obtain

$$\begin{aligned}K(z) &= \sum_{\substack{0\leq j\leq k-1\\ k+j\,\mathrm{odd}}} B_{jk} z^j \int_1^\infty \frac{(1-u)^{k-2}}{(1+u)^{k-1}} \frac{du}{(u-w)^{(k+j)/2}} \\ &= \sum_{\substack{0\leq j\leq k-1\\ k+j\,\mathrm{odd}}} \sum_{l=0}^{k-2} (-1)^{k-2-l}\binom{k-2}{l} 2^l B_{jk} z^j \int_1^\infty \frac{1}{(1+u)^{l+1}} \frac{du}{(u-w)^{(k+j)/2}}.\end{aligned} \tag{4.43}$$

We set $z = -iy$, with $y > 0$, in $K(z)$ and let $y \to \infty$. By Lemma 4.4 (with $b = 1$), we then obtain

$$K(-iy) = (-1)^{k-2} \sum_{\substack{0 \le j \le k-1 \\ k+j \text{odd}}} B_{jk}(-iy)^j \left(y^{-(k+j)} \ln y + O(y^{-(k+j)})\right)$$
$$= My^{-k} \ln y + O(y^{-k}), \tag{4.44}$$

where

$$M = (-1)^{k-2} \sum_{\substack{0 \le j \le k-1 \\ k+j \text{odd}}} B_{jk}(-i)^j. \tag{4.45}$$

By using the definition (4.17) of B_{jk}, we obtain

$$M = (-1)^{k-2} \sum_{\substack{0 \le j \le k-1 \\ k+j \text{odd}}} \sum_{l=j}^{k-1} \left(-\frac{1}{2}\right)^l \binom{k-1}{l} \binom{l+k-1}{l} \binom{l}{j} (-i)^j, \tag{4.46}$$

Since $M \neq 0$ by the assumption (2.2), the estimate (4.44) contradicts the hypothesis that $K(w)$ is algebraic in precisely the same way as in the proof of Theorem 2.1 above (by considering the Puiseux expansion of K at infinity). □

5. Axially symmetric data in even dimensions

Let $f(x)$ be an axially symmetric rational function without singularities on the unit sphere $\mathbb{S}^{n-1}$ in $\mathbb{R}^n$. We shall assume that $n = 2k$ is even and $k \ge 2$. Since the Laplace operator commutes with rotations, we may assume that the axis of symmetry is the x_1-axis. If we write $r = \sqrt{x_2^2 + \ldots + x_n^2}$, then f is a function of x_1 and r. Since f is rational, f must in fact be a function of x_1 and r^2. Now, on $\mathbb{S}^{n-1}$ we have $x_1^2 + \ldots + x_n^2 = 1$, or equivalently $r^2 = 1 - x_1^2$. It follows that there is a rational function of one variable $g(t)$ such that $f(x) = g(x_1)$ on the sphere $\mathbb{S}^{n-1}$. Since f has no poles on $\mathbb{S}^{n-1}$, we deduce that g cannot have any poles on the segment $[-1, 1]$. Thus, we shall consider the Dirichlet problem (1.1) with data $f(x) = g(x_1)$, where $g(t)$ is a rational function without poles in $[-1, 1]$. We have the following result which contains, as a special case, Theorem 2.5.

Theorem 5.1. *Let $n = 2k \ge 4$ and $f(x) = g(x_1)$, where g is a rational function of one variable. Assume in addition, if $k \ge 3$, that g has $k-2$ rational integrals, i.e., there is a rational functions $G(t)$ such that $d^{k-2}G/dt^{k-2} = g$. Then, the solution $u(x)$ to (1.1) is an axially symmetric rational function.*

Proof. First, we observe that, for even $n = 2k \ge 4$ and $y \in \mathbb{R}^n \setminus \overline{\mathbb{B}^n}$, the harmonic potential

$$U_y(x) = \frac{1}{\|x-y\|^{n-2}} = \left(\sum_{j=1}^n (x_j - y_j)^2\right)^{-(k-1)} \tag{5.1}$$

is rational without poles in a neighborhood of $\overline{\mathbb{B}^n}$. Let us take $y = y(s) := (s, 0, \dots, 0)$ with $|s| > 1$. Then, for x on the sphere $\mathbb{S}^{n-1}$, we have (cf. (4.2))

$$U_{y(s)}(x) = \left((x_1 - s)^2 + \sum_{j=2}^{n} x_j^2\right)^{-(k-1)} = \frac{1}{(-2s)^{k-1}} \frac{1}{(x_1 - \phi(s))^{k-1}}, \tag{5.2}$$

where $\phi(s)$ is given in (4.3). Let us for simplicity denote $U_{y(s)}$ by U_s. Observe that ϕ is an analytic 2-to-1 mapping of $\mathbb{C} \setminus \{0\}$ onto $\mathbb{C}$ which sends both the punctured unit disk $\mathbb{D} \setminus \{0\}$ and $\mathbb{C} \setminus \overline{\mathbb{D}}$ conformally onto $\mathbb{C} \setminus [-1, 1]$. Hence, for $x \in S^{n-1}$, we may continue $U_s(x)$ as an analytic function of s to $\mathbb{C} \setminus \overline{\mathbb{D}}$ and, for any $w \in \mathbb{C} \setminus [-1, 1]$, there is a unique $s = s(w)$ with $s \in \mathbb{C} \setminus \overline{\mathbb{D}}$ such that $w = \phi(s)$ and hence

$$u(x) = (-2s)^{k-1} U_s(x)$$

is the solution to the Dirichlet problem (1.1) with

$$f(x) = \frac{1}{(x_1 - w)^{k-1}}.$$

Also, note that for any integer $m \geq 1$, the function

$$U_s^m(x) := \frac{\partial^m}{\partial s^m} U_s(x) \tag{5.3}$$

is harmonic away from the point $y = (s, 0, \dots, 0)$ and satisfies, for x on the sphere $\mathbb{S}^{n-1}$,

$$U_s^m(x) = \sum_{j=0}^{m} \psi_j(s) \frac{1}{(x_1 - \phi(s))^{k-1+j}}, \tag{5.4}$$

where $\psi_1(s), \dots, \psi_m(s)$ are rational functions of s and

$$\psi_m(s) = \frac{(k - 2 + m)!}{(k - 2)!} \frac{(-\phi'(s))^m}{(-2s)^{k-1}}.$$

In particular, for any $w \in \mathbb{C} \setminus [-1, 1]$, we have $\psi_m(s) \neq 0$ for $s = s(w)$ with $s \in \mathbb{C} \setminus \overline{\mathbb{D}}$ and $\phi(s) = w$. We conclude that there are coefficients $c_j \in \mathbb{C}$, $j = 0, \dots, m$, such that

$$u(x) := \sum_{j=0}^{m} c_j U_s^j(x), \tag{5.5}$$

where we use the notation $U_s^0 := U_s$, solves the Dirichlet problem (1.1) with data $f(x) = 1/(x_1 - w)^{k-1+m}$. In particular, the solution u to (1.1) with data $f(x) = 1/(x_1 - w)^l$, where $l \geq k - 1$, is rational. Recall that any rational function of one variable can be decomposed as a finite sum of singular parts

$$g(t) = p(t) + \sum_{j=1}^{p} \sum_{l=1}^{q_j} a_{jl} \frac{1}{(t - w_j)^l}, \tag{5.6}$$

where $p(t)$ is a polynomial, $w_1, \dots, w_p$ denote the poles of g in $\mathbb{C}$, and $q_1, \dots, q_p$ their orders. If we also recall that the solution of the Dirichlet problem (1.1) with polynomial data is polynomial, then it follows from the above discussion that the solution $u(x)$ to (1.1) with axially symmetric data $f(x) = g(x_1)$, where g is given by (5.6), is rational provided that the coefficients a_{jl} satisfy

$$a_{jl} = 0, \quad \forall l < k-1 \text{ and } j = 1, \ldots, p. \tag{5.7}$$

Clearly, this condition is vacuous when $k = 2$, and it is not difficult to see that for $k \geq 3$ the condition (5.7) is equivalent to $g(t)$ having $k-2$ rational integrals. Since, as is well known, the solution to the Dirichlet problem with axially symmetric data is axially symmetric, the proof of Theorem 5.1 is complete. □

We should point out that we do not know if the condition that $g(t)$ has $k-2$ rational integrals is necessary for the solution $u(x)$ to be rational in Theorem 5.1. In particular, we do not know if, e.g., the solution in, say, $\mathbb{R}^6$ with data $f(x) = 1/(x_1 - 2)$ is rational. However, we can say that the solution will be rational on the axis of symmetry. Indeed, by Proposition 4.1, the solution $u(x)$ to (1.1) with $f(x) = g(x_1)$, restricted to the axis of symmetry $x = (s, 0, \ldots, 0)$, is given by (4.4) where $P_n(g)(z)$ is given by (4.5). By an integration by parts, we have, using the notation $n = 2k$,

$$\begin{aligned}
P_n(g)(z) &= \int_{-1}^{1} g(t)(1-t^2)^{k-3/2} \frac{dt}{(z-t)^k} \\
&= \left[\frac{1}{k-1}\frac{g(t)(1-t^2)^{k-3/2}}{(z-t)^{k-1}}\right]_{-1}^{1} - \frac{1}{k-1}\int_{-1}^{1} \frac{d}{dt}\left(g(t)(1-t^2)^{k-3/2}\right)\frac{dt}{(z-t)^{k-1}} \\
&= -\frac{1}{k-1}\int_{-1}^{1} g'(t)(1-t^2)^{k-3/2}\frac{dt}{(z-t)^{k-1}} \\
&\qquad + \frac{2k-3}{k-1}\int_{-1}^{1} g(t)t(1-t^2)^{k-5/2}\frac{dt}{(z-t)^{k-1}} \\
&= -\frac{1}{k-1}P_{n-2}(T_1 g)(z) + \frac{2k-3}{k-1}P_{n-2}(T_2 g)(z) \\
&= P_{n-2}(V_k g)(z),
\end{aligned} \tag{5.8}$$

where

$$V_k g(t) = -\frac{1}{k-1}T_1 g(t) + \frac{2k-3}{k-1}T_2 g(t) \tag{5.9}$$

and

$$T_1 g(t) := g'(t)(1-t^2), \quad T_2 g(t) := t g(t). \tag{5.10}$$

Observe that if g is rational, then $T_1 g$ and $T_2 g$, and hence also $V_k g$, are rational. We conclude, by induction, that for any rational function $g(t)$ and $n = 2k \geq 6$, we have

$$P_n(g) = \sum_{l=1}^{2^{k-2}} a_l P_4(g_l), \tag{5.11}$$

where $a_l \in \mathbb{R}$ and g_l, for $l = 1, \ldots 2^{k-2}$, are rational functions. By Proposition 4.1, the solution to (1.1) restricted to $x = (s, \dots, 0)$ is given by

$$v(s) = \sum_{l=1}^{2^{k-2}} a'_l s^{k-2} v_l(s), \tag{5.12}$$

where $a'_l \in \mathbb{R}$ and $v_l(s)$, for $l = 1, \dots, 2^{k-2}$, is the restriction to the axis of symmetry of the solution to the Dirichlet problem (1.1) in $\mathbb{R}^4$ with data $f(x) = g_l(x_1)$. Hence, by Theorem 2.5, $v(s)$ is rational. We may formulate this as follows.

Theorem 5.2. *Let $n = 2k \geq 4$, $f(x) = g(x_1)$, where $g(t)$ is a rational function of one variable. Let $u(x)$ be the solution to the Dirichlet problem (1.1), and let $v(s) = u(s, 0 \dots, 0)$ be its restriction to the x_1-axis. Then, $v(s)$ is rational.*

6. Ultraspherical polynomials and the Nehari transform

We shall conclude this paper by exploring a relationship between the Dirichlet problem and series of ultraspherical polynomials. We shall follow closely the notations in [SW] and [S]. Thus, $P_m^{(\lambda)}(t)$ will denote the ultraspherical (or Gegenbauer) polynomial of degree m with parameter λ. Recall that the $P_m^{(\lambda)}(t)$ are given by the generating function

$$\frac{1}{(1 - 2tw + w^2)^\lambda} = \sum_{m=0}^{\infty} P_m^{(\lambda)}(t) w^m, \tag{6.1}$$

from which one easily deduces the recursion formula

$$\frac{dP_m^{(\lambda)}}{dt}(t) = 2\lambda P_{m-1}^{(\lambda+1)}(t), \quad m \geq 1. \tag{6.2}$$

We mention here that the ultraspherical polynomials are special cases of the Jacobi polynomials $P_m^{(\alpha,\beta)}(t)$ in the sense that $P_m^{(\lambda)}(t) = c(m, \lambda) P_m^{(\alpha,\beta)}(t)$ where $\alpha = \beta = \lambda - 1/2$ and $c(m, \lambda)$ is a normalization factor depending on m and λ. Furthermore, as special cases of the ultraspherical polynomials we have the Legendre polynomials $P_m(t)$ (normalized by $P_m(1) = 1$) via $P_m^{(1/2)}((t) = P_m(t)$, and the Chebyshev polynomials of the second kind $U_m(t)$ via $P_m^{(1)}(t) = U_m(t)$ (see, e.g., [S]). For our purposes, however, a more relevant connection is to the spherical harmonics in $\mathbb{R}^n$. For any $y \in S^{n-1} \subset \mathbb{R}^n$, the restriction of the function

$$f(x) := P_m^{(n-2)/2}(\langle x, y \rangle), \qquad \text{where} \qquad \langle x, y \rangle := \sum_{j=1}^{n} x_j y_j,$$

to the sphere S^{n-1} is a spherical harmonic of order m, i.e., the restriction to S^{n-1} of a homogeneous harmonic polynomial of degree m. More precisely, the function $H_m(x)$ defined in $\mathbb{R}^n$ by

$$H_m(x) := \|x\|^m P_m^{(n-2)/2} \left(\frac{\langle x, y \rangle}{\|x\|} \right) \tag{6.3}$$

is a homogeneous harmonic polynomial of degree m. As a consequence of Theorem 5.2, we obtain the following result.

Theorem 6.1. *Let $n \geq 4$ be even and $\{c_m\}_{m=0}^{\infty}$ a sequence of complex numbers such that*

$$\limsup_{m\to\infty} |c_m|^{1/m} < 1. \tag{6.4}$$

Then, the series

$$g(z) := \sum_{m=0}^{\infty} c_m P_m^{(n-2)/2}(z) \tag{6.5}$$

converges to a holomorphic function in an open neighborhood of the segment $[-1,1]$ in the complex plane. If g is rational, then the function $\tilde{g}(w)$ defined by the Taylor series

$$\tilde{g}(w) := \sum_{m=0}^{\infty} c_m P_m^{(n-2)/2}(1) w^n \tag{6.6}$$

is also rational.

Remark 6.2. The numbers $P_m^{\lambda}(1)$, which appear in the Taylor series (6.6) with $\lambda = (n-2)/2$, can be easily computed from the generating formula (6.1). One obtains

$$P_m^{\lambda}(1) = \frac{2\lambda(2\lambda+1)\dots(2\lambda+m-1)}{m!}. \tag{6.7}$$

For instance, if $n = 6$, so that $\lambda = 2$, then

$$P_m^{(2)}(1) = \binom{m-3}{m} = \frac{(m+1)(m+2)(m+3)}{3!}.$$

Thus, the conclusion of Theorem 6.1 with $n = 6$ implies that the Taylor series

$$\sum_{m=0}^{\infty} c_m w^{m+3}$$

has a rational third derivative provided that the corresponding ultraspherical series (6.5) is rational. Similarly, the conclusion of the theorem in general can be formulated as saying that the derivative of order $n-3$ of the Taylor series

$$\sum_{m=0}^{\infty} c_m w^{m+n-3}$$

is rational when the series (6.5) is.

Proof. The convergence of the series (6.5) defining g, subject to the condition (6.4), follows from well known estimates for Jacobi polynomials. If we now consider the Dirichlet problem (1.1) in $\mathbb{R}^n$ with data $f(x) = g(x_1)$, where g is assumed to be rational, then the solution $u(x)$ restricted to the x_1-axis is rational by Theorem 5.2. On the other hand, the solution $u(x)$ can be given explicitly by

$$u(x) = \sum_{m=0}^{\infty} c_m \|x\|^m P_m^{(n-2)/2}\left(\frac{x_1}{\|x\|}\right), \tag{6.8}$$

and hence the restriction of u to the x_1-axis is precisely $\tilde{g}(x_1)$. $\square$

The transformation of a sum of Jacobi polynomials to a Taylor series with the same coefficients was apparently first studied by Nehari [N] who demonstrated, under the assumption (6.4), a correspondence between the singular points of the series of Jacobi polynomials on its ellipse of convergence, and those of the associated Taylor series on its circle of convergence. However, results of the precision of Theorem 6.1 do not follow from Nehari's theorem. For further developments of Nehari's results, see also [EKS].

The conclusion of Theorem 6.1 is in general false for odd n. Let us demonstrate this for $n = 3$. The corresponding value of $\lambda = (n-2)/2$ is then $1/2$, and hence the relevant ultraspherical polynomials $P_m^{(1/2)}(t)$ are just the classical Legendre polynomials, which we denote, as is customary, by $P_m(t)$. (The integral operator P_n defined in (4.5) will not be used in this section, so there should be no risk of confusing the Legendre polynomials with these integral operators.) Since $P_m(1) = 1$, what we have to exhibit is a sequence $\{c_m\}$, subject to the condition (6.4), for which the series

$$\sum_{m=0}^{\infty} c_m P_m(t), \tag{6.9}$$

is a rational function, but for which

$$\sum_{m=0}^{\infty} c_m w^m \tag{6.10}$$

is not. For this we invoke the classical formula of Heine (c.f. [WW], p. 322)

$$\frac{1}{t-z} = \sum_{m=0}^{\infty} (2m+1) Q_m(t) P_m(z), \tag{6.11}$$

where $Q_m(t)$ denotes the mth order Legendre function of the second kind. For fixed complex t not on the segment $[-1,1]$, the numbers $c_m := (2m+1)Q_m(t)$ satisfy (6.4), and the series on the right in (6.11) converges uniformly for z in any compact subset of the domain bounded by the ellipse with foci at -1 and 1 and passing through t. The Nehari transform of the series in (6.11), i.e., the result of replacing $P_m(z)$ by w^m, is

$$F_t(w) = \sum_{m=0}^{\infty} (2m+1) Q_m(t) w^m \tag{6.12}$$

and we claim that for fixed complex t not in the segment $[-1,1]$, the function $F_t(w)$, given by (6.12), is not a rational function of w. Indeed, from the identity ([WW], p. 321, Example 20)

$$\sum_{m=0}^{\infty} Q_m(t) w^m = \frac{1}{(1-2tw+w^2)^{1/2}} \cosh^{-1}\left(\frac{w-t}{(t^2-1)^{1/2}}\right), \tag{6.13}$$

we can easily derive a closed expression for $F_t(w)$. If we denote the function in (6.13) by $G_t(w)$, then we have

$$F_t(w^2) = \frac{d}{dw}(wG_t(w)). \tag{6.14}$$

Thus, if F_t were rational, then $d/dw(wG_t(w^2)$ would be as well. But, it is not; indeed, it is not even algebraic, and the same is therefore true of $F_t(w)$. We do not insist on all details, because the same conclusion follows from our earlier results: We have shown (cf. the proof of Theorem (2.1)) that the solution to the Dirichlet problem (1.1) in $\mathbb{R}^3$ with data $f(x) = 1/(2 - x_1)$ (and the same could be easily be shown with data $f_t(x) = 1/(t - x_1)$ and t not in the segment $[-1, 1]$) restricted to the x_1-axis is not algebraic. By using the Heine expansion (6.11) and solving the Dirichlet problem as in the proof of Theorem 6.1, we conclude that the Nehari transform $F_t(w)$ (given by (6.12)) is not algebraic. The argument sketched in this section yields an alternative proof, and interpretation, of Theorem 2.1, and we have given it here to show the link with ultraspherical polynomials and the Nehari transform. From a purely technical point of view, it appears that our earlier method, based on the Poisson integral, is simpler and also does not rely on the knowledge of Heine's expansion and (6.13).

One could also, as a tour de force, solve the axially symmetric Dirichlet problem (1.1) in $\mathbb{R}^n$, for all odd n greater than 1, with the special data

$$\frac{1}{(t - x_1)^{(n-1)/2}}$$

by an explicit series of Legendre functions as follows. By differentiating (6.11) r times with respect to z, we obtain an expansion of the function

$$\frac{1}{(t - z)^{r+1}}$$

in terms of rth order derivatives of Legendre polynomials P_n, which in view of the recursion formula (6.2) are ultraspherical polynomials with parameter $\lambda = r + 1/2$ and that correspond to zonal harmonics in the ball in dimension $2r + 3$. Observe that for odd n larger than 3, the exponent $(n - 1)/2$ is smaller than that, $2k - 1 = n - 2$, appearing as the exponent in Theorem 2.1. Thus, for odd $n \geq 5$, it may seem at first glance that we do not get an alternative approach to Theorem 2.1 using the expansion (6.11). But, in fact, we can explicitly obtain the expansion of $1/(t - z)^s$ for every integer $s \geq (n - 2)/2$ in terms of ultraspherical polynomials with parameter $\lambda = (n - 2)/2$ if we differentiate (6.11) $(n - 3)/2$ times with respect to z and $s - (n - 3)/2$ times with respect to t. Presumably this could lead to an alternative proof of Theorem (2.1) and also further generalizations, but we shall not pursue that here.

References

[E] P. Ebenfelt, *Singularities encountered by the analytic continuation of solutions to Dirichlet's problem*, Complex Variables Theory Appl. **20** (1992), 75–92.

[EKS] P. Ebenfelt, D. Khavinson and H.S. Shapiro, *Analytic continuation of Jacobi polynomial expansions*, Indag. Math. **8** (1997), 19–31.

[F] M.N. Ferrers, *On the potentials of ellipsoids and ellipsoidal shells*, Quarterly J. Pure and Applied Math. **14** (1877), 1–22.

[GS] P. Gorkin and J. Smith, *Dirichlet: his life, his principle, and his problem*, Math. Magazine, (to appear).

[K] D. Khavinson, *Holomorphic partial differential equations and classical potential theory.* Universidad de La Laguna, Departamento de Análisis Matemático, La Laguna, 1996. ii+123 pp. ISBN: 84-600-9323-9

[KS] D. Khavinson and H.S. Shapiro, *Dirichlet's problem when the data is an entire function*, Bull. London Math. Soc. **24** (1992), 456–468.

[N] Z. Nehari, *On the singularities of Legendre expansions*, Arch. Rational Mech. Anal. **5** (1956), 987–992.

[Sh] H.S. Shapiro, *An algebraic theorem of E. Fischer, and the holomorphic Goursat problem*, Bull. London Math. Soc. **21** (1989), 513–537.

[SW] E.M. Stein and G. Weiss, *Introduction to Fourier Analysis on Euclidian Spaces*, Princeton University Press, Princeton, 1971.

[S] G. Szegö, *Orthogonal Polynomials*, Amer. Math. Soc. Colloquium Publ., *23*, New York, 1959.

[WW] E.T. Whittaker, G.N. Watson, *A Course of Modern Analysis*, 4th Ed., Cambridge, 1958.

P. Ebenfelt
Department of Mathematics
University of California, San Diego
La Jolla, CA 92093–0112, USA
e-mail: pebenfel@math.ucsd.edu

D. Khavinson
Department of Mathematics
University of Arkansas
Fayetteville, AR 72701, USA
e-mail: dmitry@comp.uark.edu

H.S. Shapiro
Department of Mathematics
Royal Institute of Technology
S-100 44 Stockholm, Sweden
e-mail: shapiro@math.kth.se

Operator Theory:
Advances and Applications, Vol. 156, 173–194

Linear Analysis of Quadrature Domains. IV

Björn Gustafsson and Mihai Putinar

Dedicated to Harold S. Shapiro on the occasion of his seventy-fifth birthday

Abstract. The positive definiteness of the exponential transform of a planar domain is proved by elementary means. This direct approach avoids the heavy machinery of the theory of hyponormal operators and leads to a better understanding of the linear data associated in previous works to a quadrature domain.

Mathematics Subject Classification (2000). Primary 65D32; Secondary 47B20, 31A10.

Keywords. Exponential transform, positive definite kernel, quadrature domain, hyponormal operator.

1. The exponential transform

Let Ω be a bounded open subset of the complex plane and let dA stand for the Lebesgue planar measure. The *exponential transform* of the set Ω is the function

$$E_\Omega(z,w) = \exp\left[-\frac{1}{\pi}\int_\Omega \frac{dA(\zeta)}{(\zeta - z)(\overline{\zeta} - \overline{w})}\right]. \tag{1.1}$$

The integral is convergent for all values of $z, w \in \mathbf{C}$ avoiding the diagonal

$$\Delta = \{(z,w);\ z = w \in \overline{\Omega}\}.$$

In case $(z,w) \in \Delta$ and the integral is divergent (necessarily to infinity) we adopt the convention $\exp(-\infty) = 0$. Thus $E_\Omega(z,w)$ is defined everywhere on $\mathbf{C}^2$ and one proves that the resulting function is uniformly bounded and separately continuous in each variable, see [10]. We shall occasionally use the notation (1.1) also when the set Ω is not open.

The above exponential transform has appeared in operator theory as a determining function for a class of hyponormal operators ([18], [20], [2], [3], [4]). Later

Paper supported by the Swedish Research Council and the National Science Foundation grant DMS 0100367.

it was analyzed in purely function theoretic terms and was used in proving the regularity of certain free boundaries ([10]) or in image reconstruction ([8]). More generally, the exponential transform was regarded as a renormalized Riesz potential, and was instrumental in converting the measure $\chi_\Omega dA$ from its moments, see also [13] for a multivariable generalization. In this process, an exact reconstruction algorithm corresponding to the special class of quadrature domains Ω was discovered ([21], [11], [22]). A key positivity property of the exponential transform remained however available only from its operator theoretic origins; and these involved the highly sophisticated theory of the principal function of a semi-normal operator.

The aim of the present note is to make a short cut by proving the basic positivity property of the exponential transform by elementary arguments, accessible to function theorists. We mention that this positivity is a specific phenomenon to two real dimensions, [13].

We recall first some identities satisfied by the transform $E_\Omega(z,w)$. Their simple proofs can be found in [10]. Since we keep the set Ω fixed, we sometimes denote $E = E_\Omega$. Also, to simplify notation we write $\overline{\partial}_z = \partial_{\overline{z}} = \frac{\partial}{\partial \overline{z}}$ and $\partial_w = \frac{\partial}{\partial w}$.

Remark that $E(z,w) = \overline{E(w,z)}$ for all values of $z, w \in \mathbf{C}$ and that $E(z,w)$ is analytic in $z \in \overline{\Omega}^c$ and antianalytic in $w \in \overline{\Omega}^c$. The Taylor expansion at infinity starts with the terms

$$E(z,w) = 1 - \frac{1}{\pi}\int_\Omega \frac{dA(\zeta)}{(\zeta - z)(\overline{\zeta} - \overline{w})} + O(z^{-2}, \overline{w}^{-2}). \tag{1.2}$$

The following identities hold in the sense of distributions in $\mathbf{C}^2$:

$$\overline{\partial}_z E(z,w) = E(z,w)\frac{\chi_\Omega(z)}{\overline{z} - \overline{w}},$$

$$\partial_w E(z,w) = -E(z,w)\frac{\chi_\Omega(w)}{z - w}.$$

Note that the right hand members are given by locally integrable functions in $\mathbf{C}^2$. Moreover,

$$\overline{\partial}_z \partial_w E(z,w) = -E(z,w)\frac{\chi_\Omega(z)\chi_\Omega(w)}{|z-w|^2}, \tag{1.3}$$

again as distributions, at least in iterated integrals sense, see formula (2.20) and the related comments in [10].

We define the *interior exponential transform* by

$$H_\Omega(z,w) = \frac{E_\Omega(z,w)}{|z-w|^2}, \quad z, w \in \Omega, \tag{1.4}$$

so that

$$H(z,w) = -\overline{\partial}_z \partial_w E(z,w), \quad z, w \in \Omega. \tag{1.5}$$

It turns out by elementary computations that $H(z,w)$ is an analytic function in $z \in \Omega$ and antianalytic in $w \in \Omega$. By applying Cauchy's formula twice to the function $1 - E(z, w)$ (which vanishes at infinity in each variable) we obtain the integral representation:

$$1 - E(z, w) = \frac{1}{\pi^2} \int_\Omega \int_\Omega H(u, v) \frac{dA(u)}{u - z} \frac{dA(v)}{\overline{v} - \overline{w}}, \quad z, w \in \mathbf{C}. \tag{1.6}$$

The right member should be interpreted as an iterated convolution in the distribution sense.

A second remarkable feature of the interior transform H is the following complementarity relation, valid for a pair of *disjoint* sets Ω_1 and Ω_2:

$$H_{\Omega_1 \cup \Omega_2}(z, w) = H_{\Omega_1}(z, w) E_{\Omega_2}(z, w), \quad z, w \in \Omega_1. \tag{1.7}$$

To prove it one just notices from (1.1) that $E_{\Omega_1 \cup \Omega_2}(z, w) = E_{\Omega_1}(z, w) E_{\Omega_2}(z, w)$ holds everywhere. Applying $\overline{\partial}_z \partial_w$ for $z, w \in \Omega_1$ to both members gives (1.7) (in view of (1.5)).

It is not necessary that Ω_2 is open in (1.7), but Ω_1 and $\Omega_1 \cup \Omega_2$ should be. If Ω_1 and Ω_2 are both open then $\Omega_1 \cup \Omega_2$ is disconnected, and it is interesting to notice that the restriction of $H_{\Omega_1 \cup \Omega_2}$ to Ω_1 does not agree with H_{Ω_1}; the other part Ω_2 influences via the factor E_{Ω_2} in (1.7). Thus although H_Ω has some similarity with classical domain functions, like the Szegö kernel, it has drastically different behaviour in some respects. Another example of this is that there seems to be very little of conformal invariance properties for E_Ω and H_Ω (see [10] for behaviour under Möbius transformations).

Example 1. The case of the unit disk $\Omega = \mathbf{D}$ is relevant for the rest of the article. One finds by direct computation:

$$E_{\mathbf{D}}(z, w) = \begin{cases} 1 - \frac{1}{z\overline{w}} & |z| \geq 1, |w| \geq 1, \\ 1 - \frac{z}{\overline{w}} & |z| < 1, |w| \geq 1, \\ 1 - \frac{\overline{w}}{z} & |z| \geq 1, |w| < 1, \\ \frac{|z-w|^2}{1 - z\overline{w}} & |z|, |w| < 1. \end{cases} \tag{1.8}$$

Thus the interior transform is:

$$H_{\mathbf{D}}(z, w) = \frac{1}{1 - z\overline{w}} = \sum_{n=0}^{\infty} z^n \overline{w}^n, \quad |z|, |w| < 1. \tag{1.9}$$

We note that $H_{\mathbf{D}}$ agrees with the Szegö kernel in this case.

A kernel function $K : I \times I \longrightarrow \mathbf{C}$ is called *positive semidefinite* if for any function $\lambda : I \longrightarrow \mathbf{C}$ of finite support one has:

$$\sum_{i,j \in I} K(i, j) \lambda(i) \overline{\lambda(j)} \geq 0.$$

The kernel is said to be *positive definite* if the equality sign occurs only for the identically equal to zero function λ.

It is obvious that any sum (even an infinite one) of positive semidefinite kernels is positive semidefinite. If at least one of the terms is positive definite then the whole sum is definite. We recall also Schur's theorem [15] saying that the pointwise product of two (or more) positive semidefinite kernels is again positive semidefinite.

Any kernel $K(z,w)$ which can be written on the form

$$K(z,w)=\sum_{n=0}^{\infty} f_n(z)\overline{f_n(w)} \tag{1.10}$$

(with absolute convergence for each z and w) for some functions f_n is obviously positive semidefinite. Thus, by (1.9), $H_{\mathbf{D}}(z,w)$ is positive semidefinite. It is even positive definite since it agrees with the Szegö kernel, which is known to be positive definite. The same is true for

$$\frac{1}{E_{\mathbf{D}}(z,w)}=\frac{1}{1-\frac{1}{z\overline{w}}}=\sum_{n=0}^{\infty}\frac{1}{z\overline{w}}, \quad z,w\notin\overline{\mathbf{D}}.$$

We finally notice that $1-E_{\mathbf{D}}(z,w)=1/z\overline{w}$, for $z,w\notin\overline{\mathbf{D}}$, is positive semidefinite but not definite.

The main result of the note is the following theorem.

Theorem 1.1. *Let Ω be a bounded open planar set. The kernels*

$$\frac{1}{E_\Omega(z,w)}, \quad z,w\in\overline{\Omega}^c,$$

$$H_\Omega(z,w), \quad z,w\in\Omega,$$

are positive definite, and

$$1-E_\Omega(z,w), \quad z,w\in\mathbf{C},$$

is positive semidefinite.

Remark 1. Even though $H(z,w)$ is positive definite in the above linear algebra sense there may still be functions $h\neq 0$ such that

$$\int_\Omega\int_\Omega H(z,w)h(z)\overline{h(w)}\,dA(z)dA(w)=0.$$

For example, if $\Omega=\mathbf{D}$ then $h(z)=z$ is such a function. See Proposition 3.3 for a general statement in this respect, and also for a refinement of the statement concerning $1-E_\Omega$.

Proof. By Definition (1.1):

$$\frac{1}{E_\Omega(z,w)}=\exp\Big[\frac{1}{\pi}\int_\Omega\frac{dA(\zeta)}{(\zeta-z)(\overline{\zeta}-\overline{w})}\Big]=\sum_{n=0}^{\infty}\frac{1}{\pi^n n!}\Big[\int_\Omega\frac{dA(\zeta)}{(\zeta-z)(\overline{\zeta}-\overline{w})}\Big]^n.$$

Clearly the integral kernel here is positive definite:

$$\sum_{i,j=1}^{n}\int_\Omega \frac{dA(\zeta)}{(\zeta-z_i)(\overline{\zeta}-\overline{z_j})}\lambda_i\overline{\lambda_j}=\int_\Omega|\sum_{j=1}^{n}\frac{\lambda_j}{\zeta-z_j}|^2 dA(\zeta)\geq 0$$

with strict inequality only when $\sum_{j=1}^{n}\frac{\lambda_j}{\zeta-z_j}=0$ (identically), i.e., only when all the λ_j are zero.

By Schur's theorem the powers $[\int_\Omega \frac{dA(\zeta)}{(\zeta-z)(\overline{\zeta}-\overline{w})}]^n$ are then also positive semidefinite, and the positivity is preserved under the summation and limit processes. Thus $1/E_\Omega(z,w)$ is positive semidefinite for $z,w\in\overline{\Omega}^c$, and it is even positive definite since the term with $n=1$ is so.

We note from the proof so far that $1/E_\Omega$ will be positive semidefinite even if the set Ω is not open. This will be needed below.

To prove that $H_\Omega(z,w)$ is positive definite, choose a disc $D=D(0,R)$ with $\overline{\Omega}\subset D$. By (1.7) we have

$$H_\Omega(z,w)=H_D(z,w)\cdot\frac{1}{E_{D\setminus\Omega}(z,w)}$$

for $z,w\in\Omega$. Here both factors on the right are positive definite and it follows that the product is positive semidefinite.

Moreover, expanding $1/E_{D\setminus\Omega}$ as in the beginning of the proof and H_D as in (1.9) and multiplying these expansions we get a series of positive semidefinite kernels having at least one term which is positive definite (namely the term coming from the linear term in $1/E_{D\setminus\Omega}$ times the constant term in H_D). Thus H_Ω is positive definite.

Finally, having proved that H_Ω is positive definite the positive semidefiniteness of $1-E_\Omega(z,w)$ follows from the representation (1.6). □

In the next section we shall need the following consequence of Theorem 1.1.

Corollary 1.2. *For R sufficiently large, the kernel $(R^2-z\overline{w})H_\Omega(z,w)$ is positive definite.*

Indeed, with R chosen so that $\overline{\Omega}\subset D(0,R)$ we have

$$(R^2-z\overline{w})H_\Omega(z,w)=\frac{H_\Omega(z,w)}{H_{D(0,R)}(z,w)}=\frac{1}{E_{D(0,R)\setminus\Omega}(z,w)},$$

which is positive definite by the theorem (or rather its proof).

We wish to point out that there is an even more elementary way, not using Schur's theorem, to prove that $1/E_\Omega$ is positive semidefinite when Ω is open. Just exhaust Ω by mutually disjoint discs $D_n=D(a_n,r_n)$ so that

$$\Omega=(\cup_{n=1}^{\infty}D_n)\cup N$$

where $|N| = 0$. For each finite union $\Delta_n = \cup_{j=1}^n D_j$ we have, outside $\overline{\Omega}$ and using scaled versions of (1.8),

$$\frac{1}{E_{\Delta_n}(z,w)} = \prod_{j=1}^n \frac{1}{E_{D_j}(z,w)} = \prod_{j=1}^n \frac{1}{1 - \frac{r_j^2}{(z-a_j)(\overline{w}-\overline{a}_j)}} = \prod_{j=1}^n \sum_{k=0}^\infty \frac{r_j^{2k}}{(z-a_j)^k(\overline{w}-\overline{a}_j)^k}$$

$$= \sum_{(k_1,\dots,k_n)} \prod_{j=1}^n \frac{r_j^{2k_j}}{(z-a_j)^{k_j}(\overline{w}-\overline{a}_j)^{k_j}} = \sum_{(k_1,\dots,k_n)} \prod_{i=1}^n \frac{r_i^{k_i}}{(z-a_i)^{k_i}} \prod_{j=1}^n \frac{r_j^{k_j}}{(\overline{w}-\overline{a}_j)^{k_j}},$$

where $(k_1, \dots, k_n)$ ranges over all n-tuples of nonnegative integers. Choosing an ordering of this set brings $1/E_{\Delta_n}$ onto the form (1.10), hence it is positive semi-definite. Since $\frac{1}{E_{\Delta_n}(z,w)} \to \frac{1}{E_\Omega(z,w)}$ as $n \to \infty$ for $z, w \in \overline{\Omega}^c$ the statement follows.

Theorem 1.1 implies, via a direct computation or standard arguments familiar to complex geometers (or see [13]), that the function $\log(1 - E_\Omega(z,z))$ is subharmonic on the complement of $\overline{\Omega}$. The diagonal versions $E_\Omega(z,z)$ and $H_\Omega(z,z)$ can naturally be extended to any number of variables. However, the subharmonicity of $\log(1 - E_\Omega(z,z))$ does not hold in higher dimensions, although $1 - E_\Omega(z,z)$ remains subharmonic there [13].

2. A Hilbert space factorization

A celebrated and widely used theorem of Kolmogorov asserts that a positive semi-definite kernel $K(i,j)$, $i,j \in I$, can always be factored as

$$K(i,j) = \langle k_i, k_j \rangle,$$

with k_i belonging to an auxiliary Hilbert space. Many spectral decompositions, interpolation and prediction questions, inverse problems depend on such factorizations, see for instance [7], [24].

The positivity results proved in the preceding section invite to study the Hilbert space factorizations of the kernels H_Ω and $1 - E_\Omega$. There are at least three convergent ways of understanding the fine structure of the factorization of these kernels, cf. [4], [19] and respectively [12]. We briefly recall the construction contained in the latter reference.

Throughout this section we assume that Ω is a bounded open set of $\mathbf{C}$ having smooth boundary. Then the boundary behavior of H_Ω is comparable to that of a disk and implies (see more precisely Appendix, Section 5)

$$\int_\Omega \int_\Omega |H_\Omega(u,v)| dA(u) dA(v) < \infty. \tag{2.1}$$

On the space $L^\infty(\Omega)$ we consider the scalar product:

$$\langle f, g \rangle = \frac{1}{\pi^2} \int_\Omega \int_\Omega H_\Omega(u,v) f(u) \overline{g(v)} dA(u) dA(v),$$

and denote by $\mathcal{H}(\Omega)$ the associated separated Hilbert space completion. Thus the map

$$L^\infty(\Omega) \longrightarrow \mathcal{H}(\Omega),$$

has dense range. Notice that even the image of all test functions $\mathcal{D}(\Omega)$ is dense in $\mathcal{H}(\Omega)$.

As a matter of fact $\mathcal{H}(\Omega)$ "contains" many other elements, for instance images of distributions (or even analytic functionals) $\sigma \in \mathcal{E}'(\mathbf{C})$ such that

$$\langle \sigma, \sigma \rangle = \frac{1}{\pi^2} \int_\Omega \int_\Omega H_\Omega(u, v) \sigma(u) \overline{\sigma(v)} dA(u) dA(v) < \infty.$$

A typical example being Dirac's distributions $\delta_z,\ z \in \Omega$, which produce the factorization of H_Ω:

$$H_\Omega(z, w) = \pi^2 \langle \delta_z, \delta_w \rangle, \quad z, w \in \Omega.$$

The constant function $\mathbf{1}$ belongs to $\mathcal{H}(\Omega)$, as well as all simple rational functions

$$k_z(\zeta) = \frac{1}{\zeta - z}, \quad \zeta \in \Omega, \ z \in \mathbf{C} \setminus \partial\Omega.$$

The regularity assumption on the boundary of Ω implies that the map $z \mapsto k_z \in \mathcal{H}(\Omega)$ extends across $\partial\Omega$ and it is weakly continuous on the entire complex plane.

Using (1.6) we have

$$\langle k_z, k_w \rangle = \frac{1}{\pi^2} \int_\Omega \int_\Omega H(u, v) \frac{dA(u)}{u - z} \frac{dA(v)}{\overline{v} - \overline{w}} = 1 - E_\Omega(z, w), \tag{2.2}$$

for all values $z, w \in \mathbf{C}$. It is worth mentioning at this moment that k_z inherits some regularity from E_Ω. For instance k_z is analytic in $z \in \overline{\Omega}^c$ and doubly analytic for $z \in \Omega$ (the latter means that $\overline{\partial}_z^2 k_z = 0$ as an element of $\mathcal{H}(\Omega)$).

By integrating counterclockwise the relation (1.2) on a large circle we obtain

$$\frac{-1}{\pi} \int_\Omega \frac{dA(u)}{u - z} = \frac{1}{2\pi i} \int_{|w|=R} E_\Omega(z, w) d\overline{w}, \quad z \in \mathbf{C},$$

or, equivalently, via Stokes' theorem

$$\frac{-1}{\pi} \int_\Omega \frac{dA(u)}{u - z} = \frac{-1}{\pi} \int_{\mathbf{C}} \partial_w E_\Omega(z, w) dA(w), \quad z \in \mathbf{C}.$$

By taking a partial derivative with respect to $\overline{z}$ this gives

$$\chi_\Omega(z) = -\frac{1}{\pi} \int_{\mathbf{C}} \overline{\partial}_z \partial_w E_\Omega(z, w) dA(w), \quad z \in \mathbf{C},$$

and hence

$$1 = \frac{1}{\pi} \int_\Omega H_\Omega(z, w) dA(w), \quad z \in \Omega.$$

Thus, for any $h \in L^\infty(\Omega)$ we find

$$\langle h, \mathbf{1} \rangle = \frac{1}{\pi} \int_\Omega h dA, \tag{2.3}$$

and more generally

$$\langle hk_z, \mathbf{1}\rangle = \frac{1}{\pi}\int_\Omega \frac{h(\zeta)dA(\zeta)}{\zeta - z}, \quad z \in \mathbf{C}.$$

As a special case we have the Cauchy transform identity

$$\langle k_z, \mathbf{1}\rangle = \frac{1}{\pi}\int_\Omega \frac{dA(u)}{u - z}, \quad z \in \mathbf{C}.$$

One step further we can consider the multiplication operator

$$(Tf)(z) = zf(z), \quad f \in \mathcal{H}(\Omega).$$

Corollary 1.2 assures that T is a linear bounded operator on $\mathcal{H}(\Omega)$. The adjoint turns out to be an elementary singular integral operator:

$$(T^*f)(z) = \overline{z}f(z) - \frac{1}{\pi}\int_\Omega \frac{f(\zeta)}{\zeta - z}dA(\zeta).$$

Indeed, notice that $\partial_w[(z-w)H_\Omega(z,w)] = -H(z,w)$ and denote the Cauchy transform by

$$\hat{\psi}(z) = \frac{-1}{\pi}\int_\Omega \frac{\psi(\zeta)dA(\zeta)}{\zeta - z},$$

so that $\partial_{\overline{z}}\hat{\psi} = \psi$. For a pair of test functions $\phi, \psi \in \mathcal{D}(\Omega)$ we find by partial integration (Stokes), and using in the last steps (1.4), the boundedness of $E(z,w)$ and the decay of $\hat{\psi}$ at infinity:

$$\begin{aligned}
&\langle z\phi(z), \psi(z)\rangle - \langle \phi(z), \overline{z}\psi(z)\rangle \\
&= \frac{1}{\pi^2}\int_\Omega\int_\Omega H_\Omega(z,w)(z-w)\phi(z)\overline{\psi(w)}dA(z)dA(w) \\
&= \frac{1}{\pi^2}\int_\Omega\int_\Omega H_\Omega(z,w)(z-w)\phi(z)\overline{\partial_{\overline{w}}\hat{\psi}(w)}dA(z)dA(w) \\
&= -\frac{1}{\pi^2}\int_\Omega\int_\Omega \partial_w(H_\Omega(z,w)(z-w))\phi(z)\overline{\hat{\psi}(w)}dA(z)dA(w) \\
&\quad - \frac{1}{2i\pi^2}\int_\Omega\int_{\partial\Omega} \frac{E_\Omega(z,w)}{\overline{z}-\overline{w}}\phi(z)\overline{\hat{\psi}(w)}d\overline{w}dA(z) \\
&= \frac{1}{\pi^2}\int_\Omega\int_\Omega H_\Omega(z,w)\phi(z)\overline{\hat{\psi}(w)}dA(z)dA(w) = \langle \phi, \hat{\psi}\rangle.
\end{aligned}$$

A direct computation using (2.3) now leads to the commutator identity

$$[T, T^*] = \mathbf{1}\otimes\mathbf{1} = \mathbf{1}\langle\cdot, \mathbf{1}\rangle,$$

or, equivalently, on elements:

$$([T,T^*]f)(z) = \frac{1}{\pi}\int_\Omega f dA, \quad f \in \mathcal{H}(\Omega).$$

In particular this shows that $[T,T^*] \geq 0$, that is, T is a cohyponormal operator.

Remark also the simple identity

$$(T - z)k_z(\zeta) = (\zeta - z)\frac{1}{\zeta - z} = \mathbf{1}, \quad z \in \mathbf{C}.$$

Thus we can denote by convention $(T - z)^{-1}\mathbf{1} = k_z$ even for points $z \in \overline{\Omega}$.

By putting together the above computations we can state a partial result.

Proposition 2.1. *Let Ω be a bounded open set with smooth boundary. There exists a canonically associated cohyponormal operator $T \in L(\mathcal{H}(\Omega))$ with rank one self-commutator $[T, T^*] = \mathbf{1} \otimes \mathbf{1}$ and whose localized generalized resolvent factors the exponential transform:*

$$1 - E_\Omega(z, w) = \langle (T - z)^{-1}\mathbf{1}, (T - w)^{-1}\mathbf{1} \rangle, \quad z, w \in \mathbf{C}. \tag{2.4}$$

Originally this decomposition was obtained the other way around, from Hilbert space operators to their functional spectral invariants, see [3, 4, 2, 17]. In the case of an arbitrary bounded open set Ω one can use an exhaustion with smooth domains $\Omega_n \uparrow \Omega$ and prove that the weak operator limits $T_n \to T$, $T_n^* \to T^*$ exist, so that the factorization (2.4) holds for Ω and T. Indeed, since for $\Omega_1 \subset \Omega_2$ the difference $H_{\Omega_1}(z, w) - H_{\Omega_2}(z, w) = (1 - E_{\Omega_2 \setminus \Omega_1}(z, w))H_{\Omega_1}(z, w)$ $(z, w \in \Omega_1)$ is positive semidefinite there is a natural embedding ("extension by zero") $\mathcal{H}(\Omega_1) \to \mathcal{H}(\Omega_2)$ which decreases the norm. This gives good enough monotonicity to pass to the limit for $\Omega_n \uparrow \Omega$.

Alternatively, one can argue as in Section VII.3 of [17]. Namely, for an arbitrary domain Ω, by using the positive semidefiniteness of the kernel $1 - E_\Omega(z, w)$, one introduces the Hermitian form

$$\langle \phi, \psi \rangle = -\int_{\mathbf{C}} \int_{\mathbf{C}} E_\Omega(z, w) \partial_{\overline{z}} \phi(z) \partial_w \overline{\psi(w)} dA(z) dA(w), \quad \phi, \psi \in \mathcal{D}(\mathbf{C}),$$

and considers on the associated Hilbert space the multiplication operator $T = M_z$. Then a formula for the adjoint as before, and the factorization (2.4) will follow. We do not expand here the details of either proof.

3. Quadrature domains

The Hilbert space factorization (2.4) is particularly simple and relevant for the class of quadrature domains. We explore below some constructive aspects of this relationship between quadrature domains and their associated operators T.

A bounded domain $\Omega \subset \mathbf{C}$ is called a *quadrature domain* if there exists a distribution $u \in \mathcal{E}'(\Omega)$ with finite support in Ω satisfying

$$\int_\Omega h \, dA = u(h), \quad h \in AL^1(\Omega, dA),$$

where the latter means the space of all integrable analytic functions in Ω.

For instance a disk is a quadrature domain, due to Gauss mean value property. By abuse of terminology we will accept non-connected open sets Ω carrying such a quadrature identity and still call them quadrature domains. Then a finite disjoint union of disks is also a quadrature domain.

The pioneering work [1] of Aharonov and Shapiro can be considered as the formal birth place of quadrature domains. Since then the study of quadrature domains has achieved maturity; many unexpected ramifications to different fields of pure and applied mathematics were discovered in the last decades. The monograph [26] treats in a unifying format part of these applications of quadrature domains. The reader can also consult [5, 9, 25, 28]. In the sequel, as a direct continuation of the articles [21, 11, 22], we confine ourselves to investigate quadrature domains and the factorization (2.4) of their exponential transform. As shown elsewhere [8, 23] this study can be motivated by image reconstruction problems.

We recall that quadrature domains have real algebraic boundaries, with a limited variety of possible singular points.

To fix ideas we consider a quadrature domain Ω with distinct quadrature nodes $a_1, \dots a_n \in \Omega$ and corresponding weights $c_1, \dots, c_n \in \mathbf{C}$:

$$\int_\Omega h dA = c_1 h(a_1) + \cdots + c_n h(a_n), \quad h \in AL^1(\Omega, dA). \tag{3.1}$$

Formula (1.6) (or (2.2)) becomes

$$1 - E_\Omega(z,w) = \frac{1}{\pi^2} \sum_{i,j=1}^{n} \frac{c_i \overline{c_j} H(a_i,a_j)}{(z-a_i)(\overline{w}-\overline{a_j})}, \quad z,w \in \overline{\Omega}^c. \tag{3.2}$$

According to (2.4), the function $\langle (T-z)^{-1}\mathbf{1}, (T-w)^{-1}\mathbf{1} \rangle$ is then rational for z, w exterior to the closure of Ω. Thus the Hilbert subspace $K = \vee_{j=0}^{\infty} T^j \mathbf{1}$ is finite dimensional and invariant under the operator T. Let $A \in L(K)$ be the restriction of T to this subspace, so that $A^* = P_K T^*|_K$, where P_K denotes the orthogonal projection onto K. In view of these observations we obtain

$$\langle (A-z)^{-1}\mathbf{1}, (A-w)^{-1}\mathbf{1} \rangle = \frac{1}{\pi^2} \sum_{i,j=1}^{n} \frac{c_i \overline{c_j} H(a_i,a_j)}{(z-a_i)(\overline{w}-\overline{a_j})}, \quad z,w \in \mathbf{C}. \tag{3.3}$$

Let

$$P(z) = (z-a_1)\dots(z-a_n)$$

be the monic polynomial of degree n vanishing at the quadrature nodes. It is easy to see from the preceding identities that the matrix A is cyclic, with $\mathbf{1}$ as a cyclic vector, and $P(z)$ is its minimal polynomial. Moreover $\dim K = n$. It turns out that the polynomial

$$\begin{aligned} Q(z,w) &= P(z)\overline{P(w)} E_\Omega(z,w) \\ &= P(z)\overline{P(w)} - P(z)\overline{P(w)} \langle (A-z)^{-1}\mathbf{1}, (A-w)^{-1}\mathbf{1} \rangle, \quad z,w \in \overline{\Omega}^c, \end{aligned} \tag{3.4}$$

has minimal degree among all symmetric polynomials describing Ω as

$$\Omega \equiv \{z \in \mathbf{C};\ Q(z,z) < 0\},$$

where $\equiv$ means equality up to a finite set. For proofs see [11].

In general, a quadrature domain Ω is not determined by its quadrature data $(a_1, \dots, a_n; c_1, \dots, c_n)$. On the other hand, the matrix A and the distinguished cyclic vector $\mathbf{1}$ do determine Ω. On this ground, the correspondence

$$(a_1, \dots, a_n; c_1, \dots, c_n) \mapsto (A, \mathbf{1}),$$

when well defined, is fundamental in understanding this break of uniqueness. Section 4 of the paper is devoted to the constructive aspects of the latter correspondence in the very particular case of a disjoint union of disks. Arguably, based on fluid mechanics interpretations, the disjoint unions of disks generate by a natural expansion process all quadrature domains (having positive weights).

As a preparation for this we consider a consequence of Theorem 1.1 which might be of independent interest.

Lemma 3.1. *Let $D_i = D(a_i, r_i)$, $1 \le i \le n$, be disjoint disks and let*

$$Q(z, w) = \prod_{i=1}^{n} [(z - a_i)(\overline{w} - \overline{a_i}) - r_i^2],$$

be the polarized equation defining their union. Then the matrix $(-Q(a_i, a_j))_{i,j=1}^n$ is positive definite.

Proof. Let $\Omega = \cup_{i=1}^n D(a_i, r_i)$. Since the union is disjoint, Ω is a quadrature domain with nodes at $a_1, a_2, \dots, a_n$. Let $P(z)$ be the monic polynomial vanishing at these points.

For large values of $|z|, |w|$, due to the multiplicativity of the exponential transform we find:

$$E_\Omega(z, w) = \prod_{i=1}^{n} E_{D_i}(z, w) = \prod_{i=1}^{n} [1 - \frac{r_i^2}{(z - a_i)(\overline{w} - \overline{a_i})}].$$

Thus we see directly in this case that

$$Q(z, w) = P(z)\overline{P(w)} E_\Omega(z, w).$$

Using (3.2) gives

$$Q(z, w) = P(z)\overline{P(w)} - \frac{1}{\pi^2} \sum_{i,j=1}^{n} \frac{c_i P(z) \overline{c_j} \overline{P(w)} H_\Omega(a_i, a_j)}{(z - a_i)(\overline{w} - \overline{a_j})}.$$

Hence

$$Q(a_i, a_j) = -\frac{1}{\pi^2} P'(a_i) c_i H_\Omega(a_i, a_j) \overline{c_j} \overline{P'(a_j)},$$

which is negative definite by Theorem 1.1. □

It would be interesting to find an elementary proof for Lemma 3.1. For small values of n it is certainly possible to check everything directly (see Example 2 below for the case $n = 2$), but for general n it becomes messy. Anyhow we notice that the above proof works (using (3.4)) for any quadrature domain as in (3.1). We proceed to prove a more general statement.

Let Ω be any quadrature domain as in (3.1). We keep the previous notation and consider for any $w \in \mathbf{C} \cup \{\infty\}$ the n solutions $z_1, \dots, z_n$ (some of which could coincide) of

$$Q(z_j, w) = 0. \tag{3.5}$$

For $w = \infty$ we have $z_j = a_j$ (up to a permutation) and for any $w \in \overline{\Omega}^c$ it is known that $z_1, \dots, z_n \in \Omega$ (see [26], Theorem 5.2, for example). The following theorem can be viewed as a strengthened form of that fact, and simultaneously as a generalization of Lemma 3.1.

Theorem 3.2. *The matrix*

$$(-Q(z_i, z_j))_{i,j=1}^n$$

is positive definite for any $w \in \overline{\Omega}^c$ (including $w = \infty$).

Proof. The case $w = \infty$ works exactly as in Lemma 3.1, so we assume from now on that $w \in \mathbf{C} \setminus \overline{\Omega}$. By (3.4),

$$Q(z, z) = |P(z)|^2 (1 - \langle (A - z)^{-1}\mathbf{1}, (A - z)^{-1}\mathbf{1} \rangle)$$

for $z, w \in \overline{\Omega}^c$. Being an identity between rational functions (see (3.3)) the relation remains valid everywhere. It follows that the assumption $w \notin \overline{\Omega}$ (i.e., $Q(w, w) > 0$) means that

$$\|(A - w)^{-1}\mathbf{1}\| < 1 \tag{3.6}$$

and that the definition (3.5) of $z_1, \dots, z_n$ can be written

$$\langle (A - z_j)^{-1}\mathbf{1}, (A - w)^{-1}\mathbf{1} \rangle = 1.$$

Thus for any complex numbers $t_1, \dots, t_n$:

$$\langle \sum_{j=1}^n t_j (A - z_j)^{-1}\mathbf{1}, (A - w)^{-1}\mathbf{1} \rangle = \sum_{j=1}^n t_j$$

so that

$$|\sum_{j=1}^n t_j| \le \|\sum_{j=1}^n t_j (A - z_j)^{-1}\mathbf{1}\| \cdot \|(A - w)^{-1}\mathbf{1}\| \le \|\sum_{j=1}^n t_j (A - z_j)^{-1}\mathbf{1}\|.$$

By (3.6) the last inequality is strict unless the right member is zero. For any $\lambda_1, \dots, \lambda_n$ we get, setting $t_i = P(z_i)\lambda_i$,

$$\sum_{i,j=1}^n Q(z_i, z_j)\lambda_i \overline{\lambda_j} = \sum_{i,j=1}^n P(z_i)\lambda_i \overline{P(z_j)\lambda_j} (1 - \langle (A - z_i)^{-1}\mathbf{1}, (A - z_j)^{-1}\mathbf{1} \rangle)$$

$$= |\sum_{i=1}^n t_i|^2 - \|\sum_{i=1}^n t_i (A - z_i)^{-1}\mathbf{1}\|^2 \le 0,$$

proving the positive semidefinitenes of $(-Q(z_i, z_j))_{i,j=1}^n$.

To show that $(-Q(z_i, z_j))_{i,j=1}^n$ is actually definite assume there is equality in the last inequality. In view of the comment after the previous inequality we then have

$$\|\sum_{i=1}^{n} t_i(A - z_i)^{-1}\mathbf{1}\| = 0. \tag{3.7}$$

Since we assumed at the beginning that $w \neq \infty$, the points $\{z_i\}$ are not the nodes $\{a_i\}$ of the quadrature identity. Therefore $P(z_i) \neq 0$, so in order to show that $\lambda_1 = \cdots = \lambda_n = 0$ it is enough to show that $t_1 = \cdots = t_n = 0$.

So assume that the t_j are not all zero. Then (3.7) says that the vectors $(A - z_j)^{-1}\mathbf{1}$, $j = 1, \ldots, n$, are linearly dependent. But it follows from the detailed analysis carried out in Section 4 of [11] that this is not the case. Indeed, it was shown that the map $z \mapsto (A - z)^{-1}\mathbf{1}$, regarded as rational map from $\mathbf{C}$ to $\mathbf{C}^n$ (or between the corresponding projective spaces), is linearly equivalent to the Veronese embedding $z \mapsto (z, z^2, \ldots, z^n)$, for which the corresponding linear independence is well-known (it amounts to the nonvanishing of a Vandermonde determinant). This finishes the proof. □

Remark 2. Let $S(z)$ be the algebraic function associated to $Q(z, w)$, i.e., the function defined by

$$Q(z, \overline{S(z)}) = 0, \ z \in \mathbf{C}.$$

Since $Q(z, z) = 0$ on $\partial\Omega$ one of the branches of $S(z)$ satisfies $S(z) = \bar{z}$ on $\partial\Omega$, hence this branch is the Schwarz function [6], [26] of $\partial\Omega$. The definition (3.5) of z_j in terms of w now says that $\overline{S(z_j)} = w$, i.e., that

$$S^{-1}(\overline{w}) = \{z_1, \ldots, z_n\}.$$

Therefore Theorem 3.2 can be conveniently expressed as saying that

$$-Q(S^{-1}(\overline{w}), S^{-1}(\overline{w})) > 0$$

(positive definite) for every $w \notin \overline{\Omega}$.

Example 2. Consider the union Ω of two discs $D_i = D(a_i, r_i)$ $(i = 1, 2)$. When the discs are disjoint we have, keeping the notation from Lemma 3.1 and thereafter,

$$1 - E_\Omega(z, w) = 1 - E_{D_1}(z, w)E_{D_2}(z, w) = \frac{r_1^2}{(z - a_1)(\overline{w} - \bar{a}_1)}$$

$$+ \frac{r_2^2}{(z - a_2)(\overline{w} - \bar{a}_2)} - \frac{r_1^2 r_2^2}{(z - a_1)(z - a_2)(\overline{w} - \bar{a}_1)(\overline{w} - \bar{a}_2)}$$

for large $|z|$ and $|w|$.

This function is positive semidefinite by Theorem 1.1. It is an interesting fact that it remains positive semidefinite even if the discs overlap a little. Indeed, a straightforward calculation (which we omit) shows that the function $1 - E_{D_1}(z, w)E_{D_2}(z, w)$ is positive semidefinite if and only if

$$r_1^2 + r_2^2 \leq |a_1 - a_2|^2. \tag{3.8}$$

Similarly, with $Q(z,w) = ((z-a_1)(\overline{w}-\overline{a_1})-r_1^2)((z-a_2)(\overline{w}-\overline{a_2})-r_2^2)$ the matrix $(-Q(a_i,a_j))$ is positive semidefinite if and only if (3.8) holds. On the other hand, turning to Theorem 3.2, the matrix $(-Q(z_i,z_j))$ will not be positive semidefinite for all choices of $w \in \overline{\Omega}^c$ if the discs overlap. Indeed, in case $D_1 \cap D_2 \neq \emptyset$ we can choose w so that $z_1 \in D_1 \cap D_2$. Then $Q(z_1,z_1) > 0$ and therefore $\sum_{i,j=1}^{2}(-Q(z_i,z_j))\lambda_i\overline{\lambda}_j < 0$ with $\lambda_1 = 1$, $\lambda_2 = 0$.

It also turns out that the induction process to be performed in Section 4 will be destroyed if overlappings are allowed: if $D_1 \cap D_2 \neq \emptyset$ with $1 - E_{D_1}(z,w)E_{D_2}(z,w)$ positive semidefinite then adding a third disc D_3, disjoint from D_1 and D_2, $1 - E_{D_1}(z,w)E_{D_2}(z,w)E_{D_3}(z,w)$ will not always be positive semidefinite.

Finally in this section we wish to make Theorem 1.1 a little more precise. We shall then use the word quadrature domain in its full sense, i.e., we shall allow in the quadrature identity (3.1) also derivatives of h in the right member. For simplicity we keep the notation (3.1) however, thinking of a repeated occurrence of a node a_i as representing a derivative at a_i.

Proposition 3.3. *Let Ω be a bounded planar open set with $\partial\Omega$ smooth. The following statements are equivalent.*

a) *Ω is a quadrature domain.*
b) *$1 - E_\Omega(z,w)$ is not positive definite outside $\overline{\Omega}$ (only semidefinite).*
c) *There exists a polynomial $p \neq 0$ such that*

$$\int_\Omega \int_\Omega H(z,w)p(z)\overline{p(w)}\,dA(z)dA(w) = 0$$

(i.e., such that $p = 0$ as an element of the Hilbert space $\mathcal{H}(\Omega)$).

Proof. a) $\Rightarrow$ b): This follows easily from the representation (3.2) of $1 - E(z,w)$ as a *finite* sum of the type (1.10) when Ω is a quadrature domain.

b) $\Rightarrow$ c): That $1 - E(z,w)$ is only semidefinite means in view of (2.2) that there exists a rational function

$$R(z) = \sum_{i=1}^{n} \frac{\lambda_i}{z - a_i} = \sum_{i=1}^{n} \lambda_i k_{a_i}(z)$$

($a_i \in \overline{\Omega}^c$), not identically zero, so that

$$\begin{aligned}\langle R, R\rangle &= \frac{1}{\pi^2}\sum_{i,j=1}^{n} \int_\Omega\int_\Omega H(u,v)\frac{\lambda_i}{u-a_i}\frac{\overline{\lambda_j}}{\overline{v}-\overline{a_j}}\,dA(u)dA(v) \\ &= \sum_{i,j=1}^{n}(1 - E(a_i,a_j))\lambda_i\overline{\lambda}_j = 0.\end{aligned}$$

Let $P(z) = \prod_{i=1}^{n}(z - a_i)$. The multiplication operator $h(z) \mapsto P(z)h(z)$ is a bounded linear operator $\mathcal{H}(\Omega) \to \mathcal{H}(\Omega)$ because it is a linear combination of repeated uses of the operator T in Section 2 (indeed, it is $P(T)$). Thus from

$\langle R, R\rangle = 0$ follows $\langle PR, PR\rangle = 0$, and since PR is a polynomial this is exactly the assertion of c).

$c) \Rightarrow a)$: From $\langle p, p\rangle = 0$ (with p a polynomial) it follows that $\langle hp, hp\rangle = 0$ for any function h which is analytic in a neighbourhood of $\overline{\Omega}$. To see this one may, e.g., repeat the calculation of the adjoint of T to obtain

$$\langle h\phi, \psi\rangle - \langle \phi, \overline{h}\psi\rangle = \langle \phi, \overline{h'}\hat{\psi}\rangle$$

for, say, $\phi, \psi \in C^\infty(\overline{\Omega})$. Choosing here $\psi = h\phi$ and using the Cauchy-Schwarz inequality gives

$$\langle h\phi, h\phi\rangle \leq C\langle \phi, \phi\rangle.$$

Now take $\phi = p$.

From $\langle hp, hp\rangle = 0$ we deduce, using (2.3), that

$$\int_\Omega hp\, dA = \langle hp, \mathbf{1}\rangle = 0.$$

Having such a relation holding for all h analytic in a neighbourhood of $\overline{\Omega}$ easily implies an identity (3.1), with the a_i being the zeros of p. □

4. Adding an external disc

We consider the same disjoint union of disks $\Omega_n = \cup_{i=1}^n D(a_i, r_i)$ as in Lemma 3.1, to which we add a new disjoint disk; let $\Omega_{n+1} = \cup_{i=1}^{n+1} D(a_i, r_i)$ be the enlarged set. At each stage we have a finite dimensional Hilbert space K, a cyclic vector $\mathbf{1} \in K$ and an operator $A \in L(K)$ as after (3.2). In terms of matricial representations we write, at stage k:

$$E_{\Omega_k}(z, w) = 1 - \langle (A_k - z)^{-1}\xi_k, (A_k - w)^{-1}\xi_k\rangle, \ |z|, |w| \gg 1,$$

where $A_k \in L(K_k)$ has cyclic vector ξ_k and $\dim K_k = k,\ k = n, n+1$.

Our aim is to understand the structure of the matrix A_{n+1} and its cyclic vector ξ_{n+1} as functions of the previous data (A_n, ξ_n) and the new disk $D(a_{n+1}, r_{n+1})$. Henceforth we assume that the closed disks $\overline{D(a_i, r_i)}$ are still disjoint. In order to simplify notation we suppress for a while the index $n+1$, e.g., $a = a_{n+1},\ r = r_{n+1}$ etc. The following computations are based on standard realization techniques in linear systems theory, see for instance [7].

Due to the multiplicativity of the external exponential transform for disjoint domains we find:

$$[1 - \langle (A_n - z)^{-1}\xi_n, (A_n - w)^{-1}\xi_n\rangle][1 - \frac{r^2}{(z-a)(\overline{w} - \overline{a})}]$$

$$= 1 - \langle (A - z)^{-1}\xi, (A - w)^{-1}\xi\rangle.$$

Equivalently,

$$\langle (A_n - z)^{-1}\xi_n, (A_n - w)^{-1}\xi_n\rangle + \frac{r^2}{(z-a)(\overline{w}-\overline{a})}$$
$$= \langle \frac{r}{z-a}(A_n - z)^{-1}\xi_n, \frac{r}{w-a}(A_n - w)^{-1}\xi_n\rangle + \langle (A-z)^{-1}\xi, (A-w)^{-1}\xi\rangle.$$

Thus, for each z avoiding the poles, the norm of the vector

$$f(z) = \begin{pmatrix} (A_n - z)^{-1}\xi_n \\ \frac{r}{z-a} \end{pmatrix} \in K_n \oplus \mathbf{C}$$

equals that of the vector

$$g(z) = \begin{pmatrix} \frac{r}{z-a}(A_n - z)^{-1}\xi_n \\ (A-z)^{-1}\xi \end{pmatrix} \in K_n \oplus K.$$

And moreover, the same is true for any linear combination

$$\|\lambda_1 f(z_1) + \cdots + \lambda_r f(z_r)\| = \|\lambda_1 g(z_1) + \cdots + \lambda_r g(z_r)\|.$$

Because the span of $f(z)$, $z \in \mathbf{C}$, is the whole space $K_n \oplus \mathbf{C}$, there exists a unique isometric linear operator $V : K_n \oplus \mathbf{C} \longrightarrow K_n \oplus K$ mapping $f(z)$ to $g(z)$. We write, corresponding to the two direct sum decompositions

$$V = \begin{pmatrix} B & \beta \\ C & \gamma \end{pmatrix},$$

where $B : K_n \longrightarrow K_n$, $\beta \in K_n$, $C : K_n \longrightarrow K_{n+1}$, $\gamma \in K$. Since $Vf(z) = g(z)$ for all z, we find by coefficient identification:

$$B = r(A_n - a)^{-1}, \quad \beta = (A_n - a)^{-1}\xi_n.$$

The isometry condition $V^*V = I$ written at the level of the above 2×2 matrix yields the identities

$$\begin{cases} r^2(A_n^* - \overline{a})^{-1}(A_n - a)^{-1} + C^*C = I, \\ r(A_n^* - \overline{a})^{-1}(A_n - a)^{-1}\xi_n + C^*\gamma = 0, \\ \|(A_n - a)^{-1}\xi_n\|^2 + \|\gamma\|^2 = 1. \end{cases} \tag{4.1}$$

In particular we deduce that $(A_n^* - \overline{a})^{-1}(A_n - a)^{-1} \leq r^{-2}$ and since this operator inequality is valid for every radius which makes the disks disjoint, we can enlarge slightly r and still have the same inequality. Thus, the defect operator

$$\Delta = [I - r^2(A_n^* - \overline{a})^{-1}(A_n - a)^{-1}]^{1/2} : K_n \longrightarrow K_n \tag{4.2}$$

is strictly positive.

The identity $C^*C = \Delta^2$ shows that the polar decomposition of the matrix $C = U\Delta$ defines without ambiguity an isometric operator $U : K_n \longrightarrow K$. Since $\dim K = \dim K_n + 1$ we will identify $K = K_n \oplus \mathbf{C}$, so that the map U becomes the natural embedding of K_n into the first factor. Thus the second line of the isometry V becomes

$$(C \ \ \gamma) = \begin{pmatrix} \Delta & d \\ 0 & \delta \end{pmatrix} : K_n \oplus \mathbf{C} \longrightarrow K_n \oplus \mathbf{C} = K,$$

where $d \in K_n,\ \delta \in \mathbf{C}$. We still have the freedom of a rotation of the last factor, and can assume $\delta \geq 0$. One more time, equations (4.1) imply

$$\begin{cases} d = \frac{1}{r}(\Delta\xi_n - \Delta^{-1}\xi_n), \\ \delta = [1 - \|(A_n - a)^{-1}\xi_n\|^2 - \|d\|^2]^{1/2}. \end{cases} \tag{4.3}$$

From relation $Vf(z) = g(z)$ we deduce

$$\begin{pmatrix} \Delta & d \\ 0 & \delta \end{pmatrix} \begin{pmatrix} (A_n - z)^{-1}\xi_n \\ \frac{r}{z-a} \end{pmatrix} = (A - z)^{-1}\xi.$$

This shows that $\delta > 0$ because the operator A has the point a in its spectrum.

At this point straightforward matrix computations lead to the following exact description of the pair $(A, \xi) = (A_{n+1}, \xi_{n+1})$ (by restoring the indices):

$$A_{n+1} = \begin{pmatrix} \Delta A_n \Delta^{-1} & -\delta^{-1}\Delta(A_n - a_{n+1})\Delta^{-1}d \\ 0 & a_{n+1} \end{pmatrix}, \quad \xi = \begin{pmatrix} \Delta^{-1}\xi_n \\ -\delta r_{n+1} \end{pmatrix}. \tag{4.4}$$

It is sufficient to verify these formulas, that is

$$\begin{pmatrix} \Delta(A_n - z)\Delta^{-1} & -\delta^{-1}\Delta(A_n - a)\Delta^{-1}d \\ 0 & a - z \end{pmatrix} \begin{pmatrix} \Delta & d \\ 0 & \delta \end{pmatrix} \begin{pmatrix} (A_n - z)^{-1}\xi_n \\ \frac{r}{z-a} \end{pmatrix}$$

$$= \begin{pmatrix} \Delta^{-1}\xi_n \\ -\delta r \end{pmatrix}.$$

And this is done by direct multiplication:

$$\Delta\xi_n + \Delta(A_n - z)\Delta^{-1}\frac{rd}{z-a} - \Delta(A_n - a)\Delta^{-1}\frac{rd}{z-a} = \Delta^{-1}\xi_n,$$

which is equivalent to the known relation $dr = \Delta\xi_n - \Delta^{-1}\xi_n$.

Summing up, we can formulate the transition laws of the linear data of a disjoint union of disks.

Proposition 4.1. *Let $\overline{D(a_i, r_i)}, 1 \leq i \leq n+1$, be a disjoint family of closed disks, and let $\Omega_k = \cup_{i=1}^k D(a_i, r_i),\ 1 \leq k \leq n+1$.*

The linear data (A_k, ξ_k) of the quadrature domain Ω_k can be inductively obtained by the formula (4.4), with the aid of the definitions (4.2), (4.3).

Remark that letting $r = r_{n+1} \to 0$ we obtain $\Delta \to I$ and $d \to 0$, which is consistent with the fact that $\Omega_{n+1} \to \Omega$, in measure, where Ω is a bounded domain. Moreover, in this case the vectors ξ_n will converge to a ξ and A_n will converge in the weak operator topology to a bounded operator A, namely the one factoring $1 - E_\Omega$:

$$E_\Omega(z, w) = 1 - \langle (A - z)^{-1}\xi, (A - w)^{-1}\xi \rangle, \ |z|, |w| \gg 1.$$

5. Appendix on integrability of the interior transform

The representation formula (1.6) (or (2.2)) is crucial for the whole theory. It depends on the distributional identity (1.3) together with the definition (1.4) of $H(z,w)$. It is desirable that $H(z,w)$ is integrable over $\Omega \times \Omega$ because then the right member of (1.3) makes immediate sense as a distribution and there is no question about the meaning of (1.6) (the right member will be a convolution between distributions).

Thus setting

$$\|H\|_p = \left(\int_\Omega \int_\Omega |H(z,w)|^p \, dA(z) dA(w)\right)^{1/p} \tag{5.1}$$

for $0 < p < \infty$ we would like to have at least that $\|H\|_1 < \infty$. We do not know whether this is always the case but here are at least some partial results.

Lemma 5.1. *Let Ω be a bounded planar open set. Then*

a) $\|H\|_p < \infty$ *for all* $p < 1$.

b) *If $\partial\Omega$ is Lipschitz then $\|H\|_p < \infty$ for all $p < 3/2$.*

c) *If $\partial\Omega$ is smooth real analytic (or if Ω is a quadrature domain) then $\|H\|_p < \infty$ for all $p < 3$ (but not for $p = 3$).*

Proof. Let for $t \geq 0$

$$m(t) = |\{(z,w) \in \Omega \times \Omega : |H(z,w)| > t\}|$$

be the distribution function of H ($|\dots|$ here denotes Lebesgue measure in $\mathbf{C} \times \mathbf{C}$). Then

$$\|H\|_p^p = -\int_0^\infty t^p \, dm(t).$$

The integral from zero to one is certainly finite since Ω is bounded, so $\|H\|_p$ will be finite if and only if $-\int_1^\infty t^p \, dm(t) < \infty$.

For $z \in \Omega$, let $d(z)$ denote the distance from z to Ω^c, and for $\delta > 0$ let

$$f(\delta) = |\{(z \in \Omega : d(z) < \delta\}|.$$

Thus $f(\delta)$ is the area of a δ-neighbourhood of $\partial\Omega$. It was shown in [10] (Lemma 2.4 and 2.5) that

$$|E(z,w)| \leq 2 \ \ (z, w \in \mathbf{C})$$

and

$$|H(z,w)| \leq 2 \min\{\frac{1}{d(z)^2}, \frac{1}{d(w)^2}\}.$$

Combining these estimates gives (with (1.4))

$$|H(z,w)| \leq 2 \min\{\frac{1}{d(z)^2}, \frac{1}{d(w)^2}, \frac{1}{|z-w|^2}\}. \tag{5.2}$$

By (5.2)

$$m(t) \leq M(t),$$

where

$$M(t) = |\{(z,w) \in \Omega \times \Omega : d(z) < \sqrt{\frac{2}{t}},\ |z-w| < \sqrt{\frac{2}{t}}\}|.$$

Clearly

$$M(t) \leq f(\sqrt{\frac{2}{t}}) \cdot \pi(\sqrt{\frac{2}{t}})^2 = \frac{2\pi}{t} f(\sqrt{\frac{2}{t}}).$$

Since t^p is an increasing function of t the above inequalities imply

$$-\int_1^\infty t^p\, dm(t) \leq -\int_1^\infty t^p\, dM(t) \leq -2\pi \int_1^\infty t^p\, d(\frac{f(\sqrt{\frac{2}{t}})}{t}).$$

We now turn to the particular assertions of the lemma. Since $f(\delta) \leq C < \infty$ for all δ we have, for $0 < p < 1$,

$$-\int_1^\infty t^p\, d(\frac{f(\sqrt{\frac{2}{t}})}{t}) \leq -C \int_1^\infty t^p\, d(\frac{1}{t})) = C \int_1^\infty t^{p-2}\, dt < \infty,$$

proving a).

If $\partial\Omega$ is Lipschitz we have $f(\delta) \leq C\delta$, which gives

$$-\int_1^\infty t^p\, d(\frac{f(\sqrt{\frac{2}{t}})}{t}) \leq -C \int_1^\infty t^p\, d(\frac{1}{t^{3/2}})) = C \int_1^\infty t^{p-5/2}\, dt < \infty$$

for all $p < 3/2$, proving b).

To prove c) we need a better estimate of $H(z,w)$. What we have when $\partial\Omega$ is analytic is essentially (5.2) without the squares, namely:

$$|H(z,w)| \leq C \min\{\frac{1}{d(z)}, \frac{1}{d(w)}, \frac{1}{|z-w|}\}. \tag{5.3}$$

Assuming this for a moment and inserting it in the estimate of $M(t)$ above gives

$$M(t) \leq \frac{\pi C^2}{t^2} f(\frac{C}{t}).$$

Thus, still using $f(\delta) \leq C\delta$,

$$-\int_1^\infty t^p\, dm(t) \leq -C \int_1^\infty t^p\, d(\frac{f(\frac{C}{t})}{t^2}) \leq C \int_1 t^{p-4}\, dt < \infty$$

when $p < 3$. For $p = 3$ it is easy to check that $\|H\|_p = +\infty$ even for the unit disc.

It remains to prove (5.3) when $\partial\Omega$ is analytic. Let $S(z)$ be the Schwarz function of $\partial\Omega$, so that $S(z)$ is analytic in a neighbourhood of $\partial\Omega$ and satisfies $S(z) = \overline{z}$ on $\partial\Omega$. When $\partial\Omega$ is smooth real analytic the exponential transform has an analytic/antianalytic continuation from the exterior of Ω across $\partial\Omega$, see [10]. This means that there exists a function $F(z,w)$ analytic/antianalytic in a neighbourhood of $\partial\Omega \times \partial\Omega$ such that $F(z,w) = E(z,w)$ for $z, w \in \Omega^c$.

Inside Ω (but close to $\partial\Omega$) we have

$$F(z,w) = (z - \overline{S(w)})(S(z) - \overline{w})H(z,w)$$

(see [10]). We shall use this to estimate $H(z,w)$. We immediately get

$$H(z,w) = \frac{F(z,w)}{(z-\overline{S(w)})(S(z)-\overline{w})}. \tag{5.4}$$

As $E(z,z)$ vanishes on $\partial\Omega$ so does $F(z,z)$:

$$F(z,z) = 0, \ \ z \in \partial\Omega.$$

Since also $S(z)-\overline{z}=0$ on $\partial\Omega$ it follows that $F(z,w)$ contains $S(z)-\overline{w}$ as a factor. Hence one of the factors in the denominator of (5.4) cancels and we obtain

$$H(z,w) = \frac{G(z,w)}{z-\overline{S(w)}}, \tag{5.5}$$

where also $G(z,w)$ is analytic/antianalytic in a neighbourhood of $\partial\Omega \times \partial\Omega$. In particular we get the estimate

$$|H(z,w)| \leq \frac{C}{|z-\overline{S(w)}|} \tag{5.6}$$

for $z,w \in \Omega$ close to $\partial\Omega$.

It remains to notice that the estimate (5.6) is equivalent to (5.3). Indeed, $\overline{S(w)}$ is the conformally reflected point of $w \in \Omega$, so $d(w)$ is comparable to $\frac{1}{2}|w-\overline{S(w)}|$, and in addition $|z-w| \leq C|z-\overline{S(w)}|$ (for points $z,w \in \Omega$ close enough to $\partial\Omega$). Now one can pass between (5.6) and (5.3) by using triangle inequalities.

In case Ω is a quadrature domain $\partial\Omega$ is analytic but there may be singular points. However it turns out that these singularities go the right way so that $H(z,w)$ will actually be less singular than at smooth points.

To be a little more precise, when Ω is a quadrature domain $F(z,w)$ is the rational function

$$F(z,w) = \frac{Q(z,w)}{P(z)\overline{P(w)}}$$

(in the notation of Section 3), the Schwarz function $S(z)$ in (5.4) is meromorphic in all Ω and it is one of the branches of the algebraic function defined by the polynomial $Q(z,w)$ (i.e., $Q(z,\overline{S(z)})=0$ identically). Hence $F(z,w)$ still contains $S(z)-\overline{w}$ as a factor and one obtains (5.5).

What happens at a singular point is roughly speaking that $F(z,w)$ contains one more factor $S(z)-\overline{w}$ (for a different branch of $S(z)$). Just think of the simplest example: the touching point of two touching discs. More precisely we have that $G(z_0,z_0)=0$ when $z_0 \in \partial\Omega$ is singular, and this really improves the behaviour of $H(z,w)$. Therefore (5.3) remains valid. □

References

[1] Aharonov, D., Shapiro, H.S., *Domains on which analytic functions satisfy quadrature identities*, J. Analyse Math. **30** (1976), 39–73.

[2] Carey, R.W. and Pincus, J.D., *An exponential formula for determining functions,* Indiana Univ. Math.J. **23** (1974), 1031–1042.

[3] Clancey, K., *A kernel for operators with one-dimensional self-commutator*, Integral Eq. Operator Theory **7** (1984), 441–458.

[4] Clancey, K., *Hilbert space operators with one-dimensional self-commutator*, J. Operator Theory **13** (1985), 265–289.

[5] Crowdy, D. and Marshall, J., *Constructing multiply-connected quadrature domains*, SIAM J. Appl. Math. **64** (2004), 1334–1359.

[6] Ph.J. Davis, *The Schwarz function and its applications*, Carus Math. Mono. vol. 17, Math. Assoc. Amer., 1974.

[7] Foiaş, C. and Frazho, A.E., *The commutant lifting approach to interpolation problems*, Birkhäuser Verlag, Basel, 1990.

[8] Golub, G., Gustafsson, B., Milanfar, P., Putinar, M. and Varah, J., *Shape reconstruction from moments: theory, algorithms, and applications*, Signal Processing and Image Engineering, SPIE Proceedings vol. **4116** (2000), Advanced Signal Processing, Algorithms, Architecture, and Implementations X (Franklin T. Luk, ed.), pp. 406–416.

[9] Gustafsson, B., *Quadrature identities and the Schottky double*, Acta Appl. Math. **1** (1983), 209–240.

[10] Gustafsson, B. and Putinar, M., *An exponential transform and regularity of free boundaries in two dimensions*, Ann. Sc. Norm. Sup. Pisa, **26** (1998), 507–543.

[11] Gustafsson, B. and Putinar, M., *Linear analysis of quadrature domains. II*, Israel J. Math. **119** (2000), 187–216.

[12] Gustafsson, B. and Putinar, M., *Analytic continuation of Cauchy and exponential transforms*, in S.Saitoh et al. (eds.), Analytic extension Formulas and their Applications, pp. 47–57, Kluwer Academic Publishers, Netherlands, 2001.

[13] Gustafsson, B. and Putinar, M., *The exponential transform: a renormalized Riesz potential at critical exponent*, Indiana Univ. Math. J. **52** (2003), 527–568.

[14] Helton, J.W. and Howe, R., *Traces of commutators of integral operators*, Acta Math. **135** (1975), 271–305.

[15] Horn, R.A. and Johnson, C.R., *Matrix Analysis*, Cambridge Univ. Press, Cambridge, 1985.

[16] Krein, M.G., *On a trace formula in perturbation theory* (in Russian), Mat. Sbornik **33** (1953), 597–626.

[17] Martin, M. and Putinar, M., *Lectures on Hyponormal Operators*, Birkhäuser, Basel, 1989.

[18] Pincus, J.D., *Commutators and systems of singular integral equations. I*, Acta Math. **121** (1968), 219–249.

[19] Pincus, J.D., Xia, J. and Xia, D., *The analytic model of a hyponormal operator with rank-one self-commutator*, Integral Eq. Operator Theory **7** (1984), 516–535.

[20] Pincus, J.D. and Rovnyak, J., *A representation for determining functions*, Proc. Amer. Math. Soc. **22** (1969), 498–502.

[21] Putinar, M., *Linear analysis of quadrature domains*, Ark. Mat. **33** (1995), 357–376.

[22] Putinar, M., *Linear analysis of quadrature domains. III*, J. Math. Analysis Appl. **239** (1999), 101–117.

[23] Putinar, M., *A renormalized Riesz potential and applications*, Advances in Constructive Approximation XXX, (M. Neamtu, E. Saff, eds.), Nashboro Press, Brentwood, TN, 433–465.

[24] Riesz, F. and Sz.-Nagy, B., *Functional analysis*, Dover Publ., New York, 1990.

[25] Sakai, M., *Quadrature Domains*, Lect. Notes Math. **934**, Springer-Verlag, Berlin-Heidelberg 1982.

[26] Shapiro, H.S., *The Schwarz function and its generalization to higher dimensions*, Univ. of Arkansas Lect. Notes Math. Vol. 9, Wiley, New York, 1992.

[27] Xia, D., *Spectral Theory of Hyponormal Operators*, Birkhäuser, Basel, 1983.

[28] Xia, D., *Hyponormal operators with finite rank self-commutator and quadrature domains*, J. Math. Anal. Appl. **203** (1996), 540–559.

Björn Gustafsson
Department of Mathematics
The Royal Institute of Technology
S-10044 Stockholm, Sweden
e-mail: `gbjorn@math.kth.se`

Mihai Putinar
Mathematics Department
University of California
Santa Barbara, CA 93106, USA
e-mail: `mputinar@math.ucsb.edu`

Operator Theory:
Advances and Applications, Vol. 156, 195–205

Restriction, Localization and Microlocalization

Makoto Sakai

Abstract. We discuss methods for studying quadrature domains for subharmonic functions, using three modifications of measures, which we call restriction, localization, and microlocalization. Applying these methods, we discuss the shape of the blob in the Hele-Shaw flow free-boundary problem.

Mathematics Subject Classification (2000). Primary 31A15; Secondary 30C85.

Keywords. Restricted quadrature domain, Localization, Microlocalization.

1. Introduction

In this expository paper we discuss methods for studying quadrature domains for subharmonic functions. Detailed proofs and applications will be published elsewhere.

First, we introduce quadrature domains for subharmonic functions. Next, we discuss *restricted quadrature domains and measures*. Applying the notion of restricted quadrature domains and measures, we construct a new measure ν. We call the process of constructing the measure *localization*. Finally, we construct a more complicated measure. We call the process of constructing this measure *microlocalization*. In all, we discuss three modifications of measures. These methods have many applications. A typical application is seen in the Hele-Shaw flow free-boundary problem. For a given initial domain $\Omega(0)$, we discuss the shape of domain $\Omega(t)$ after a small interval of time t.

This paper is organized as follows: In Section 2 we explain why we introduce quadrature domains for subharmonic functions, but not for harmonic functions. In Section 3 we introduce restricted quadrature domains and measures. Localization and microlocalization are discussed in Sections 4 and 5, respectively. In the final section, Section 6, we comment briefly on applications.

2. Quadrature domains for subharmonic functions

Let μ be a finite positive Borel measure on the complex plane $\mathbb{C}$ with compact support. A bounded open set Ω in $\mathbb{C}$ is called a *quadrature domain of μ for harmonic functions* if

(i) $\mu(\mathbb{C}\setminus\Omega)=0$;

(ii) for every harmonic and integrable function h in Ω, the integral $\int h d\mu$ is definite and
$$\int h d\mu = \iint_\Omega h(z)dxdy,$$
where $z=x+iy$.

A simple example is a quadrature domain of the Dirac measure δ_a at point a. It consists of the disk $B(a,1/\sqrt{\pi})$ with center a and radius $1/\sqrt{\pi}$. Davis gives another example [1]:
$$\int_{-1}^{1} h(x)dx = \iint_\Omega h(z)dxdy$$
for every harmonic and integrable function h in $\Omega = f(B(0,1))$, where
$$z=f(w)=\frac{1}{\pi}\log\Big(\frac{1+\alpha w}{1-\alpha w}\Big)$$
with $\log 1 = 0$ and $\alpha = \sqrt{(e^\pi-1)/(e^\pi+1)} = 0.957\cdots$.

Let $\rho(x)=2(1-\sqrt{|x|})$ for $x\in[-1,1]$. A more complicated example is given in [3]:
$$\int_{-1}^{1} h(x)\rho(x)dx = \iint_\Omega h(z)dxdy$$
for every harmonic and integrable function h in $\Omega=\{z=x+iy : -1<x<1, |y|<(1-x^2)/2\}$. For other examples and the fundamental properties of quadrature domains, see [3] and [7].

In the above examples, the quadrature domain is uniquely determined, but this does not hold for a general measure. Let σ be a uniform measure on the unit circle $\partial B(0,1)$ satisfying $d\sigma = d\theta/2\pi$ and set $\mu = t\sigma$ for $t>0$.

If $0<t\le\pi$, then the quadrature domain of μ is uniquely determined and it is a ring domain $\{z\in\mathbb{C} : \sqrt{\alpha/\pi}<|z|<\sqrt{(\alpha+t)/\pi}\}$, where $\alpha=\alpha(t)$ is the uniquely determined solution of the equation
$$\int_{\sqrt{\alpha/\pi}}^{\sqrt{(\alpha+t)/\pi}} (\log r) r dr = 0.$$
If $\pi<t\le e\pi$, then there are two quadrature domains. One is the ring domain determined above and the other is the disk $B(0,\sqrt{t/\pi})$. If $t>e\pi$, then the quadrature domain is uniquely determined and it is the disk $B(0,\sqrt{t/\pi})$.

Therefore, for t with $\pi<t\le e\pi$, we have two different quadrature domains for harmonic functions. To distinguish these two domains, we introduce the notion of quadrature domains for subharmonic functions.

A bounded open set Ω is called a *quadrature domain of μ for subharmonic functions* if

(i) $\mu(\mathbb{C} \setminus \Omega) = 0$; and

(ii$'$) for every subharmonic and integrable function s in Ω,

$$\int s d\mu \leq \iint_\Omega s(z) dx dy.$$

Note that the integral $\int s d\mu$ for a subharmonic integrable function s in Ω is definite, because $s^+ = \max\{s, 0\}$ is subharmonic, integrable, and $\int s^+ d\mu \leq \iint_\Omega s^+(z) dx dy < \infty$. Since both h and $-h$ are subharmonic if h is harmonic, a quadrature domain for subharmonic functions is a quadrature domain for harmonic functions.

In the above examples, all quadrature domains are also quadrature domains for subharmonic functions except for the case $\mu = t\sigma$ with $\pi < t \leq e\pi$. We have two quadrature domains for harmonic functions in this case. The ring domain $\{z \in \mathbb{C} : \sqrt{\alpha/\pi} < |z| < \sqrt{(\alpha + t)/\pi}\}$ is a quadrature domain for subharmonic functions, but the disk $B(0, \sqrt{t/\pi})$ is not a quadrature domain for subharmonic functions. Hence, the quadrature domain of $\mu = t\sigma$ is uniquely determined for every $t > 0$, if we restrict ourselves to quadrature domains for subharmonic functions.

The important feature of quadrature domains for subharmonic functions is their uniqueness: the quadrature domain for subharmonic functions is uniquely determined in the sense that the characteristic functions χ_{Ω_1} and χ_{Ω_2} coincide a.e. for quadrature domains Ω_1 and Ω_2 of μ (see [4], Lemma 6.1). Furthermore, in many cases, there is a smallest quadrature domain of μ. In this case we denote it as $\Omega(\mu)$. In what follows, we use λ to denote the two-dimensional Lebesgue measure and write $\lambda|D$ for the restriction of λ onto a bounded open set D. If η is a finite positive measure on D with $\eta(D) > 0$, then there exists a smallest quadrature domain $\Omega(\lambda|D + \eta)$ of $\lambda|D + \eta$. We denote it as $\Omega(D + \eta)$.

The advantage of introducing quadrature domains for subharmonic functions is their uniqueness in the above sense. We discuss below quadrature domains that are described as $\Omega(D + \eta)$. For such cases, we can develop methods of modifying measures, which we also discuss here.

The solution $\Omega(t)$ in the Hele-Shaw flow free-boundary problem can be expressed using quadrature domains for subharmonic functions. The problem is the following: given a domain $\Omega(0)$ of the initial blob and point z_0 in $\Omega(0)$ at which further fluid is injected, find the domain $\Omega(t)$ of the blob after time t. We can express the solution $\Omega(t)$ as

$$\Omega(\Omega(0) + t\delta_{z_0}),$$

where δ_{z_0} denotes the Dirac measure at z_0.

A fundamental question of quadrature domains is to find the shape of the quadrature domain or to find the basic property of the shape. For example, in the Hele-Shaw flow free-boundary problem, what can we say about the shape of $\Omega(t)$ for small $t > 0$ if the initial domain $\Omega(0)$ has a corner or a cusp on the boundary?

3. Restriction

Let R be a domain in $\mathbb{C}$ with a piecewise smooth boundary that may not be bounded. We call R a *restriction domain.* Let μ be a finite positive Borel measure on R with compact support: $\mu(\mathbb{C}\setminus R)=0$. The bounded open subset Ω_R of R and the measure ν_R on $(\partial\Omega_R)\setminus(R\cap\partial\Omega_R)\subset\partial R$ are called a *restricted quadrature domain and measure in R of μ* if

(i) $\mu(R\setminus\Omega_R)=0$; and

(ii″) for every integrable and subharmonic function s on $\overline{\Omega_R}\setminus(R\cap\partial\Omega_R)$,

$$\int s d\mu \le \iint_{\Omega_R} s(z)dxdy + \int s d\nu_R,$$

where $\partial\Omega_R$ denotes the boundary of Ω_R and $\overline{\Omega_R}$ denotes the closure of Ω_R. Here we interpret ν_R as 0 if $(\partial\Omega_R)\setminus(R\cap\partial\Omega_R)$ is an empty set and we say that s is subharmonic on $\overline{\Omega_R}\setminus(R\cap\partial\Omega_R)$ if s is subharmonic in some open set containing $\overline{\Omega_R}\setminus(R\cap\partial\Omega_R)$.

The restricted quadrature domain and measure may not exist. If they exist, then the domain Ω_R is determined, except for a set of Lebesgue measure zero, and the measure ν_R is uniquely determined. The measure ν_R is singular with respect to the two-dimensional Lebesgue measure λ and is absolutely continuous with respect to the arc length.

If there exists a smallest domain Ω_R, we denote Ω_R and ν_R by (Ω_R,ν_R). A fundamental relationship between $\Omega(\mu)$ and (Ω_R,ν_R) is the following:

$$\Omega(\Omega_R+\nu_R)=\Omega(\mu).$$

It is very difficult to calculate ν_R explicitly. We can obtain good estimates of ν_R by using balayage measures. Let G be a bounded domain with a smooth boundary. Let μ be a measure on $\overline{G}$. If a measure $\beta=\beta(\mu,G)$ on ∂G satisfies

$$\int h d\mu = \int h d\beta$$

for every function h continuous on $\overline{G}$ and harmonic in G, we call it the *balayage measure of μ on ∂G.* It exists and is uniquely determined. If G does not have a smooth boundary, we interpret $\beta(\mu,G)$ appropriately.

Theorem 3.1. *Let R be a domain with a smooth boundary. Let D be a bounded open subset of R and let η be a finite positive measure on D. Let (Ω_R,ν_R) be the restricted quadrature domain and measure in R of $\lambda|D+\eta$; and, for s with $0<s\le 1$, let $(\Omega_R(s),\nu_R(s))$ be the restricted quadrature domain and measure in R of $\lambda|D+s\eta$. Then*

$$\nu_R=\nu_R(1)=\int_0^1 \beta(\eta,\Omega_R(s))|\partial R\, ds \equiv \lim_{n\to\infty}\sum_{j=1}^{n}\beta\Big(\frac{\eta}{n},\Omega_R\big(\frac{j}{n}\big)\Big)|\partial R,$$

where the convergence is the strong convergence.

Corollary 3.2. *It follows that*

$$\beta(\eta, D)|\partial R \leq \nu_R \leq \beta(\eta, \Omega_R)|\partial R.$$

For the fundamental properties and related results, see [2].

4. Localization

In this section we present the following localization theorem.

Theorem 4.1. *Let R be a restriction domain. Let Ω^0 be a bounded open set and let D be a subdomain of Ω^0 such that $\Omega^0 \setminus R \subset D$. Let η be a finite positive measure on R with compact support. Assume that there exists a smallest restricted quadrature domain and measure in R of $\lambda|\Omega^0 \cap R + \eta$, then there exists a measure ν on ∂R such that*

$$\Omega(D + \nu) \subset \Omega(\Omega^0 + \eta)$$

and

$$\Omega(D + \nu) \setminus R = \Omega(\Omega^0 + \eta) \setminus R.$$

Why do we call Theorem 4.1 the localization theorem? To see why, we consider R, the exterior $\mathbb{C} \setminus \overline{B(a, \rho)}$ of disk $B(a, \rho)$. Then $\Omega(D+\nu) \setminus R = \Omega(\Omega^0 + \eta) \setminus R$ implies

$$\Omega(D + \nu) \cap \overline{B(a, \rho)} = \Omega(\Omega^0 + \eta) \cap \overline{B(a, \rho)}.$$

This means that $\Omega(D+\nu)$ equals to $\Omega(\Omega^0+\eta)$ inside disk $B(a, \rho)$. Given Ω^0 and η, we choose a suitable D and find ν. When we study the shape of $\Omega(\Omega^0 + \eta)$ around a, we only need to know the mass of ν. If we modify Ω^0 somewhere, even at a place far from a, then the shape of $\Omega(\Omega^0 + \eta)$ around a will change, so it is very difficult to describe the shape. Theorem 4.1 enables us to understand the shape by estimating the measure ν.

Another application of the localization theorem is the following: let R be the lower half-plane $\{x + iy : y < 0\}$. Assume that $\Omega^0 \subset R$ and take $D = \emptyset$. Our localization theorem asserts that there is a measure ν on the real axis ∂R such that

$$\Omega(\nu) \subset \Omega(\Omega^0 + \eta)$$

and

$$\Omega(\nu) \setminus R = \Omega(\Omega^0 + \eta) \setminus R.$$

The quadrature domain $\Omega(\nu)$ of measure ν on the real axis can be expressed as

$$\Omega(\nu) = \{x + iy : x \in I, |y| < g(x)\},$$

where I denotes the union of open intervals on the real axis and g denotes a positive real-analytic function on I. This implies that the reflection of $\Omega(\Omega^0 + \eta) \cap \{x + iy : y > 0\}$ with respect to the real axis is contained in $\Omega(\Omega^0 + \eta) \cap \{x + iy : y < 0\}$.

Next, we explain how we construct the measure ν. Set $D_0 = D$ and let (Ω_1, ν_1) be the restricted quadrature domain and measure in R of $\lambda|\Omega^0 \cap R + \eta$. Let $D_1 = \Omega(D_0 + \nu_1)$ and, for $j \geq 1$, $(\Omega_{j+1}, \nu_{j+1})$ be the restricted quadrature domain and measure in R of $\lambda|(\Omega_j \cup D_j) \cap R + \lambda|\Omega_j \cap (D_j \setminus D_{j-1})$ and set $D_{j+1} = \Omega(D_j + \nu_{j+1}) = \Omega(D_0 + \nu_1 + \cdots + \nu_{j+1})$. Then $\{\Omega_j \cup D_j\}_{j=1}^{\infty}$ is an increasing sequence of domains and satisfies

$$\Omega(\Omega^0 + \eta) = \bigcup(\Omega_j \cup D_j).$$

The sequence $\{D_j\}_{j=1}^{\infty}$ of domains is also increasing and satisfies

$$\Omega(D_0 + \sum \nu_j) = \bigcup D_j.$$

We set $\nu = \sum \nu_j$. Then $\Omega(D_0 + \nu) = \bigcup D_j$. Since $D_j \subset \Omega_j \cup D_j$, we obtain $\Omega(D_0 + \nu) \subset \Omega(\Omega^0 + \eta)$. Since $\Omega_j \subset R$, we obtain $D_j \setminus R = (\Omega_j \cup D_j) \setminus R$. Hence

$$\begin{aligned}\Omega(D_0 + \nu) \setminus R &= \left(\bigcup D_j\right) \setminus R \\ &= \bigcup(D_j \setminus R) \\ &= \bigcup((\Omega_j \cup D_j) \setminus R) \\ &= \left(\bigcup(\Omega_j \cup D_j)\right) \setminus R \\ &= \Omega(\Omega^0 + \eta) \setminus R.\end{aligned}$$

Therefore, the measure ν is constructed by applying an infinite process that involves making restricted quadrature domains and measures. It is very complicated, but we can estimate ν by applying the following lemma.

Lemma 4.2. *Let R be a restriction domain. Let η^0 be a finite positive measure on R with compact support. Let $\{D_j\}_{j=0}^{\infty}$ be an increasing sequence of bounded open sets such that $\lambda|D_0 \cap R + \eta^0$ defines the restricted quadrature domain and measure (Ω_1, ν_1) in R. For $j \geq 1$, let $(\Omega_{j+1}, \nu_{j+1})$ be the restricted quadrature domain and measure in R of $\lambda|(\Omega_j \cup D_j) \cap R + \lambda|\Omega_j \cap (D_j \setminus D_{j-1})$. Then $(\Omega_{j+1}, \nu_1 + \nu_2 + \cdots + \nu_{j+1})$ is the restricted quadrature domain and measure in R of $\lambda|D_j \cap R + \eta^0$ and $(\bigcup \Omega_j, \sum \nu_j)$ is the restricted quadrature domain and measure in R of $\lambda|(\bigcup D_j) \cap R + \eta^0$.*

Proposition 4.3. *Let R, Ω^0, D, η and ν be as in Theorem 4.1. Set $\Omega = \Omega(D + \nu)$. Then $(\Omega(\Omega^0 + \eta) \cap R, \nu)$ is the restricted quadrature domain and measure in R of $\lambda|\Omega \cap R + \lambda|(\Omega^0 \setminus D) \cap R + \eta = \lambda|(\Omega \cup \Omega^0) \cap R + \lambda|(\Omega \setminus D) \cap \Omega^0 \cap R + \eta$. In particular, setting $\hat{\eta} = \eta|(\Omega \cup \Omega^0) \cap R$, we obtain*

$$\begin{aligned}&\beta(\lambda|(\Omega \setminus D) \cap \Omega^0 \cap R + \hat{\eta}, (\Omega \cup \Omega^0) \cap R)|\partial R \\ &\leq \nu \\ &\leq \beta(\lambda|(\Omega \setminus D) \cap \Omega^0 \cap R + \eta, \Omega(\Omega^0 + \eta) \cap R)|\partial R.\end{aligned}$$

5. Microlocalization

To explain microlocalization, we need two restriction domains. We write one of them as R, as above, and, for the sake of simplicity, we let $R = \mathbb{C} \setminus \overline{B(0, \rho)}$ and take the left or right half-plane as the other restriction domain. We write the left half-plane $\{x+iy : x<0\}$ as $H = H^-$ and use H^+ to denote the right half-plane $\{x+iy : x>0\}$. We set

$$R^- = R \cap H^-, \quad R^+ = R \cap H^+$$

and

$$(\partial R)^- = (\partial R) \cap H^-, \quad (\partial R)^+ = (\partial R) \cap H^+.$$

For a measure μ, we set

$$\mu^- = \mu|H^-, \quad \mu^+ = \mu|H^+.$$

We define $S = (\partial H) \cap \overline{R}$.

Theorem 5.1. *Let Ω^0 be a bounded domain and let η be a measure on $R = \mathbb{C}\setminus\overline{B(0,\rho)}$ with compact support such that $\lambda|\Omega^0 \cap R^- + \eta^-$ defines the restricted quadrature domain and measure in R^- and $\lambda|\Omega^0 \cap R^+ + \eta^+$ defines the restricted quadrature domain and measure in R^+. Set*

$$\hat{\eta}^- = \eta^-|\Omega(\Omega^0 + \eta|S) \cap R^-, \quad \hat{\eta}^+ = \eta^+|\Omega(\Omega^0 + \eta|S) \cap R^+.$$

Then there exists a measure ν on $(\partial R)^- \cup S \cup (\partial R)^+$ such that

(1) $\Omega(\Omega^0 + \nu) \subset \Omega(\Omega^0 + \eta)$;
(2) $\Omega(\Omega^0 + \nu) \setminus R = \Omega(\Omega^0 + \eta) \setminus R$;
(3) $\beta(\hat{\eta}^-, \Omega(\Omega^0 + \eta|S) \cap R^-)|(\partial R)^- \le \nu|(\partial R)^- \le \beta(\eta^-, \Omega(\Omega^0 + \eta) \cap R^-)|(\partial R)^-$;
(4) $\beta(\hat{\eta}^+, \Omega(\Omega^0 + \eta|S) \cap R^+)|(\partial R)^+ \le \nu|(\partial R)^+ \le \beta(\eta^+, \Omega(\Omega^0 + \eta) \cap R^+)|(\partial R)^+$;

and

$$\begin{aligned} (5) \quad & \eta|S + \beta(\hat{\eta}^-, \Omega(\Omega^0 + \eta|S) \cap R^-)|S + \beta(\hat{\eta}^+, \Omega(\Omega^0 + \eta|S) \cap R^+)|S \\ & \le \nu|S \\ & \le \eta|S + \beta(\eta^-, \Omega(\Omega^0 + \eta) \cap R^-)|S + \beta(\eta^+, \Omega(\Omega^0 + \eta) \cap R^+)|S. \end{aligned}$$

We call Theorem 5.1 the microlocalization theorem. If we apply the localization theorem to the case $R = \mathbb{C} \setminus \overline{B(0,\rho)}$ and $D = \Omega^0$, then, from Proposition 4.3, we see that there is a measure ν on ∂R such that

(1) $\Omega(\Omega^0 + \nu) \subset \Omega(\Omega^0 + \eta)$;
(2) $\Omega(\Omega^0 + \nu) \setminus R = \Omega(\Omega^0 + \eta) \setminus R$; and
(3) $\beta(\hat{\eta}, \Omega(\Omega^0 + \nu) \cap R)|\partial R \le \nu \le \beta(\eta, \Omega(\Omega^0 + \eta) \cap R)|\partial R$,

where $\hat{\eta} = \eta|\Omega(\Omega^0 + \nu) \cap R$.

The balayage measure $\beta(\eta, \Omega(\Omega^0 + \eta) \cap R)|\partial R$ may be large and we have no information on the left ν^- or right ν^+ parts of measure ν. Our microlocalization theorem is more accurate and we can compare ν^- and ν^+. The main part of ν is situated on S.

6. Applications

We apply our methods to the Hele-Shaw flow free-boundary problem, which we introduced at the end of Section 2. We discuss the shape of $\Omega(t)$ for small $t > 0$ for the case in which the initial domain $\Omega(0)$ has a corner or a cusp on the boundary.

Let c be a point on the boundary $\partial\Omega(0)$ of the initial domain $\Omega(0)$. If $c \in \partial\Omega(t)$ for some $t > 0$, we call c a *stationary point.* If c is not a stationary point, then $c \in \Omega(t)$ for every $t > 0$. For a more concrete example, we treat a corner with interior angle φ. Assume that $(\partial\Omega(0)) \cap B$ is a continuous simple arc passing through c for a small disk B with center c. Further, assume that $B \setminus (\partial\Omega(0))$ consists of two connected components and $\Omega(0) \cap B$ is one of them. We express $(\partial\Omega(0)) \cap B$ as the union of two continuous simple arcs $\Gamma_1(0)$ and $\Gamma_2(0)$; $(\partial\Omega(0)) \cap B = \Gamma_1(0) \cup \Gamma_2(0)$ and $\Gamma_1(0) \cap \Gamma_2(0) = \{c\}$, and assume that both $\Gamma_1(0)$ and $\Gamma_2(0)$ are of class C^1 and regular up to endpoint c. Then the intersection of $\Omega(0)$ and the circle with center c and small radius makes a circular arc. We say that c is a *corner with interior angle* φ if the ratio of the length of the circular arc to the radius tends to φ as the radius tends to 0. It follows that $0 \leq \varphi \leq 2\pi$. If $\varphi = \pi$, we interpret c as a smooth boundary point of $\Omega(0)$. If $\varphi = 2\pi$, we call c a *cusp* of $\partial\Omega(0)$ rather than a corner with interior angle 2π.

If c is a corner with interior angle φ, then the discussion can involve more than simply whether it is a stationary point. We introduce the following notion: Assume that c is a corner with interior angle φ and it is a stationary point. Further, assume that there is a small disk B with center c such that $(\Omega(t)) \cap B$ is a continuous simple arc for every small $t > 0$ and $(\Omega(t)) \cap B$ can be expressed as the union of two continuous simple arcs $\Gamma_1(t)$ and $\Gamma_2(t)$; $(\partial\Omega(t)) \cap B = \Gamma_1(t) \cup \Gamma_2(t)$ and $\Gamma_1(t) \cap \Gamma_2(t) = \{c\}$, and both $\Gamma_1(t)$ and $\Gamma_2(t)$ are of class C^1 and regular up to endpoint c, and are real-analytic except for c. If c is a corner of $\partial\Omega(t)$ with interior angle φ, and if φ does not depend on small $t > 0$, we call c a *laminar-flow stationary corner with interior angle* φ.

Next, assume that it is not a stationary point. If there exists a small disk B with center c such that $(\partial\Omega(t)) \cap B$ is a regular real-analytic simple arc for every small $t > 0$, we call c a *laminar-flow point.* If $(\partial\Omega(t)) \cap B$ is a regular real-analytic simple arc for small $t > 0$, except for a countable number of t at most, and $(\partial\Omega(t)) \cap B$ is a real-analytic simple arc with a finite number of cusps for the exceptional value t, we call c a *quasi-laminar-flow point.* If, for every small disk B with center c, there is $t > 0$ such that $(\partial\Omega(t)) \cap B$ has at least two connected components, one of which is an arc and the other is a closed curve, we call c a *turbulent-flow point.*

Fundamental tools used to study these corners and cusps are the notion of restricted quadrature domains and measures, and the localization theorem. In addition, we need to introduce the notion of a continuous reflection property: Let H be an open half-plane. We denote the exterior of H using H^e, which is also an open half-plane. Let H_0 be an open half-plane and let Ω be an open set in $\mathbb{C}$. We say that Ω has a *continuous reflection property with respect to* H_0, if for every open half-plane H which contains $\overline{H_0}$ and has the boundary ∂H parallel to ∂H_0, the reflection of $\Omega \cap H^e$ with respect to ∂H is contained in Ω. This notion is very useful, because we easily obtain the following lemma.

Lemma 6.1. *Let $H_0 = \{x + iy : y < 0\}$ and let Ω be an open set with a continuous reflection property with respect to H_0. Let G be a non-empty bounded connected component of $\Omega \cap H_0^e$. Then there is an open interval $I \subset \partial H_0$ and a positive lower semicontinuous function g defined in I such that*

$$\{x + iy : x \in I,\ -g(x) < y \le 0\} \subset \Omega,$$

and G can be expressed as

$$G = \{x + iy : x \in I,\ 0 < y < g(x)\}.$$

Conversely, if g is a positive lower semicontinuous function defined on an interval $I \subset \partial H_0$, then

$$\{x + iy : x \in I,\ -g(x) < y < g(x)\}$$

has a continuous reflection property with respect to H_0.

Applying the localization theorem and the above lemma, we obtain the following theorem.

Theorem 6.2. *Let c be a corner with interior angle φ on $\partial\Omega(0)$.*

(1) *If $0 \le \varphi < \pi/2$, then c is a laminar-flow stationary corner with interior angle φ.*
(2) *If $\varphi = \pi/2$, then c is a laminar-flow stationary corner with interior angle $\pi/2$ or a laminar-flow point.*
(3) *If $\pi/2 < \varphi < 2\pi$, then c is a laminar-flow point.*
(4) *If $\varphi = 2\pi$, i.e., if c is a cusp, then c is a laminar-flow, a quasi-laminar-flow, or turbulent-flow point.*

In Theorem 6.2, critical values of interior angle φ are $\pi/2$ and 2π. Applying again the localization theorem and Lemma 6.1, we obtain a sufficient condition for a corner with interior angle $\pi/2$ to be a laminar-flow point. To give a sufficient condition for a cusp to be a laminar-flow point, we need to consider this in more detail. We apply the microlocalization theorem and a more detailed reflection theorem, which we call the *local-reflection theorem.*

To derive the local-reflection theorem, we introduce some additional notation. As above, we set $H = H^- = \{x + iy : x < 0\}$, $H^+ = \{x + iy : x > 0\}$, $R = \{re^{i\theta} : r > \rho\}$, and $S = \{re^{i\theta} : r \ge \rho,\ \theta = \pm\pi/2\}$. Let G be a subset of H^+. We use $G^{\partial H}$ to denote the reflection of G with respect to the imaginary axis ∂H. Let D be a set in $\mathbb{C}$. If $(D^+)^{\partial H} = (D \cap H^+)^{\partial H}$ is contained in $D^- = D \cap H^-$, we say that D^+ is *reflexible in* D^-. Furthermore, we set

$$E_{(1)} = \{re^{i\theta} : r = \rho,\ -7\pi/8 \le \theta \le -5\pi/8\},$$

and

$$E_{(2)} = \{re^{i\theta} : r = \rho,\ 5\pi/8 \le \theta \le 7\pi/8\}.$$

Let g be a continuous function defined on $]a, b[$ with $0 \le a < b$ and assume that the graph $\{x+ig(x) : a < x < b\}$ intersects exactly one point on $\partial B(0, r)$ for every r in interval I in the positive real axis. That is to say, we can write the graph as $\{re^{i\theta} : \theta = \theta(r), r \in I\}$ for some continuous function θ defined on interval I. We denote this using $\theta[g]$. If g is defined on $]-b, -a[$, then $g(-x)$ is a function defined on $]a, b[$. We denote this using $(g \circ -)(x)$. In what follows, we treat functions g defined on $[-4\rho, 4\rho]$. We write $\theta[g]$ for $\theta[g|]0, 4\rho]]$ and $\theta[g\circ-]$ for $\theta[(g|]-4\rho, 0[)\circ-]$.

Our local-reflection theorem is the following theorem.

Theorem 6.3. *Let $0 < \alpha < \beta < 2^{-31}$ and let $0 < \rho < 1$. Let D be a bounded open set such that*

(i) *D^+ is reflexible in D^-;*

(ii) *D^+ consists of two connected components $D_{(1)}$ and $D_{(2)}$ satisfying*

$$(D_{(1)})^{\partial H} \supset E_{(1)} \quad \text{and} \quad (D_{(2)})^{\partial H} \supset E_{(2)};\ \text{and}$$

(iii) *$D \cap B(0, 4\rho)$ can be expressed as*

$$D \cap B(0, 4\rho) = \{x + iy : -4\rho < x < 4\rho,\ y < g_1(x) \text{ or } y > g_2(x)\} \cap B(0, 4\rho).$$

Here g_k denote functions of class C^1 defined on $[-4\rho, 4\rho]$ such that

(iv) *$g_1(0) < 0 < g_2(0)$ and $g_1(x) < g_2(x)$ on $[-4\rho, 4\rho]$;*

(v) *$g_2(0) - g_1(0) + 2(\tan\beta)\rho \le 2^{-25} \cdot \rho/45$;*

(vi) *$|g_1'(x)| \le \tan\beta$ on $[-4\rho, 4\rho]$;*

(vii) *$|g_2'(x)| \le \tan\beta$ on $[-4\rho, 4\rho]$;*

(viii) *$((g_1 \circ -) - g_1)'(x) \ge 0$ on $[10\rho/45, 36\rho/45]$ and $\theta[g_1 \circ -](r) - \theta[g_1](r) \ge 2\alpha$ on $[11\rho/45, 33\rho/45]$; and*

(ix) *$(g_2 - (g_2 \circ -))'(x) \ge 0$ on $[10\rho/45, 36\rho/45]$ and $\theta[g_2](r) - \theta[g_2 \circ -](r) \ge 2\alpha$ on $[11\rho/45, 33\rho/45]$.*

Let μ be a measure on $(\partial R)^- \cup S \cup (\partial R)^+$ with compact support such that $\Omega(D+\mu)^+$ consists of two connected components $\Omega_{(1)}$ and $\Omega_{(2)}$ satisfying $\Omega_{(1)} \supset D_{(1)}$ and $\Omega_{(2)} \supset D_{(2)}$. Furthermore, assume that

(x) *$\mu(\Omega_{(k)} \cap (\partial R)^+) \le C_+\rho^2||\mu||$ and $\mu(E_{(k)}) \ge C_-\rho^2||\mu||$ for $k = 1$ and 2,*

where $C_+ = C_+(\rho)$ and $C_- = C_-(\rho)$ are positive constants that depend only on ρ, and $||\mu||$ denotes the total variation of μ. The constants satisfy

(xi)
$$C_+ \le CC_-,$$

where C denotes an absolute constant; and

(xii)
$$||\mu|| \le \frac{1}{10^9} \frac{\alpha^{1+2\psi}}{C_+},$$

where $\psi = 250\sqrt{(g_2(0) - g_1(0))/\rho + 2\beta}$.

Then $\Omega(D + \mu)^+ \cap B(0, \rho/45)$ is reflexible in $\Omega(D + \mu)^- \cap B(0, \rho/45)$.

The proof of the local-reflection theorem is very complicated and long.

Finally, we define a conformal cusp. Let

$$f(w) = c + a_1 w + a_2 w^2 + a_3 w^3 + \cdots$$

be a holomorphic function in the unit disk $\{w \in \mathbb{C} : |w| < 1\}$ such that $a_1 = 0$ and f is univalent on the closure $\overline{U}$ of a half disk $U = \{w = u + iv \in \mathbb{C} : |w| < 1, u < 0\}$. If $\Gamma_1(0) \cup \Gamma_2(0) \subset f(\{w = u + iv \in \mathbb{C} : u = 0, -1 < v < 1\})$ and $f(\overline{U}) \subset \overline{\Omega(0)}$, we call c a *conformal cusp*. We note that the cusp that appeared in the Regularity Theorem on a boundary having a Schwarz function is a conformal cusp(see [5]). We divide the conformal cusps into two types: cusps with index $1/2$ or with index $-1/2$(see [6]). The cusps that appeared in the Hele-Shaw flow free-boundary problem are conformal cusps with index $-1/2$.

Applying the microlocalization and local-reflection theorems, we obtain the following theorem.

Theorem 6.4. *Let c be a conformal cusp. Then it is a laminar-flow point.*

We can apply our methods to more general cusps, such as a power cusp. The proof is long, even for the simplest case, the conformal cusp.

References

[1] Davis, Philip J., *The Schwarz function and its applications*, The Carus Mathematical Monographs, No. 17. The Mathematical Association of America, Buffalo, N. Y., 1974.

[2] Gustafsson, Björn and Sakai, Makoto, *Properties of some balayage operators, with applications to quadrature domains and moving boundary problems*, Nonlinear Anal. **22** (1994), 1221–1245.

[3] Sakai, Makoto, *Quadrature domains*, Lecture Notes in Mathematics Vol. 934, Springer-Verlag, Berlin-New York, 1982.

[4] Sakai, Makoto, *Solutions to the obstacle problem as Green potentials*, J. Analyse Math. **44** (1985), 97–116.

[5] Sakai, Makoto, *Regularity of a boundary having a Schwarz function*, Acta Math. **166** (1991), 263–297.

[6] Sakai, Makoto, *Regularity of free boundaries in two dimensions*, Ann. Scuola Norm. Sup. Pisa Cl. Sci. (4) **20** (1993), 323–339.

[7] Shapiro, Harold S., *The Schwarz function and its generalization to higher dimensions*, University of Arkansas Lecture Notes in the Mathematical Sciences. Vol. 9, A Wiley-Interscience Publication. John Wiley & Sons, Inc., New York, 1992.

Makoto Sakai
Department of Mathematics
Tokyo Metropolitan University
Minami-Ohsawa 1-1, Hachioji-shi
Tokyo 192-0397, Japan
e-mail: `sakai@comp.metro-u.ac.jp`

Operator Theory:
Advances and Applications, Vol. 156, 207–215

Quadrature Domains and Brownian Motion (A Heuristic Approach)

Henrik Shahgholian

To Harold S. Shapiro on the occasion of his 75th birthday

Abstract. In this note we will make an attempt to link the theory of the so-called quadrature domains (QD) to stochastic analysis. We show that a QD, with the underlying measure μ, can be represented as the set of points x, for which the expectation value (average reward)

$$E^x\left(-\theta + \int_0^\theta \mu(X_t)\right),$$

is positive for some (bounded) stopping time θ. Here X_t denotes the Brownian motion starting at the point x, and E^x denotes the expectation with respect to the underlying probability measure P^x.

Mathematics Subject Classification (2000). Primary 35R35, 60J65, 60J45.

Keywords. Brownian motion, quadrature domains, variational inequalities.

1. Setting and backgrounds

Our objective in this note is to find a stochastic interpretation of the so-called quadrature domains. To fix the idea, let D be a bounded domain in $\mathbf{R}^n$ $(n \geq 2)$, and

$$\mu = M\chi_D, \qquad (M > 1) \tag{1.1}$$

where χ_D is the characteristic function. One can consider a more general class of functions, or measures μ. However, for simplicity and clarity we stick to the case of multiples of characteristic functions.

Supported in part by the Swedish Research Council.

A quadrature domain (QD) with respect to the function μ, and a class of functions H, is a bounded domain Ω with properties:

$$D \subset \Omega, \qquad \int_\Omega h dx \geq \int h d\mu \qquad \forall\, h \in H. \tag{1.2}$$

Actually the more restrictive condition $\overline{D} \subset \Omega$ is prefered, but this in general is much harder to achieve.

We adopt the notation

$$\Omega \in QD(\mu, H)$$

to denote that Ω is a QD w.r.t. μ and the class H. Throughout this paper we mainly consider the class of integrable subharmonic functions over Ω, denoted by $SL^1(\Omega)$.

A PDE point of view of QD is the consideration of the function (also called the modified Schwarz potential (MSP))

$$u(x) := c_n \int_\Omega |y-x|^{2-n} (dy - d\mu_y) \qquad x \in \Omega. \tag{1.3}$$

Here we have assumed $n \geq 3$, and for $n = 2$, we replace $|y-x|^{2-n}$ with $-\log|y-x|$. Also, the constant c_n is a normalization factor.

Elementary PDE then tells us that u, the MSP, satisfies

$$\Delta u = \chi_\Omega - \mu, \qquad u \geq 0, \quad \text{in } \mathbf{R}^n, \qquad u = 0 \quad \text{in } \mathbf{R}^n \setminus \Omega, \tag{1.4}$$

where the Laplacian is taken in the sense of distributions.

Conversely if (1.4) holds for a triple (u, μ, Ω) then we can show, using Green's identity, that $\Omega \in QD(\mu, SL^1)$.

So we have

$$\Omega \in QD(\mu, SL^1) \qquad \Longleftrightarrow \qquad (u, \Omega, \mu) \quad \text{solves (1.4)}. \tag{1.5}$$

We refer the reader to the papers [Sak83], [Gus90], and [GS], for background and further results.

2. Brownian Motion and stopping times

In this section we will recall some definitions and facts about Brownian motion. Let $W_t = (W_t^1, \ldots, W_t^n)$ be a standard Brownian motion in $\mathbf{R}^n$, i.e., for each $j = 1, \ldots, n$, W_t^j is a real-valued, continuous stochastic process with independent and stationary increments. Being standard means

$$W_0 = 0, \qquad E(W_t^j) = 0, \qquad E((W_t^j)^2) = t, \qquad j = 1, \ldots, n.$$

Let us now fix a point in $\mathbf{R}^n$, and consider a Brownian motion X_t starting at x, i.e., $X_t = W_t + x$.

Let P^x denote the underlying probability measure, and E^x, the mathematical expectation w.r.t. P^x. Let also $\mathcal{F}_t$ denote the natural filtration of increasing family of σ-algebras generated by W_t.

A stopping time τ is a random variable with values in $\mathbf{R}^+ \cup \infty$, and such that

$$\{\tau \leq t\} \in \mathcal{F}_t.$$

The first exit time of a bounded domain D,

$$\tau_D := \inf\{t > 0, \ X_t \notin D\},$$

is a stopping time.

Now for a given bounded function μ and a (finite valued) stopping time θ, we consider the expected reward of stopping the process X_t at θ,

$$U_\theta(x) = U_{\theta,\mu}(x) := E^x \left(-\theta + \int_0^\theta \mu(X_t) \, dt \right). \tag{2.1}$$

Observe that if $\theta \equiv 0$, then $U_\theta(x) = 0$.

Remark 2.1. For a game theoretic interpretation see the last section.

Our prime goal will be to show that if we take the supremum value of U_θ over all finite stopping times θ,

$$\sup_\theta U_\theta(x), \tag{2.2}$$

then the resulting function is (a multiple of) the MSP and the (interior of the) support of this function is a QD for the measure μ.

However, before moving on into the next section, and finding out about relations between Brownian motion and QD, we need to recall a couple of facts in stochastic PDE.

We start with the infinitesimal generator of the Brownian motion, and recall from [Dyn] (see also [Oks]) that the operator

$$\frac{1}{2}\Delta$$

is the infinitesimal generator of the n-dimensional Brownian motion B_t. In other words,

$$\frac{1}{2}\Delta f(x) = \lim_{t\to 0} \frac{E^x(f(X_t)) - f(x)}{t} \qquad x \in \mathbf{R}^n,$$

which (by integration and using that $X_0 = x$) implies the Ito/Dynkin formula

$$E^x\left(\frac{1}{2}\int_0^\tau \Delta f(X_s) ds\right) = E^x(f(X_\tau)) - f(x), \tag{2.3}$$

for all functions f, with bounded Laplacian, and all bounded stopping times τ, with $E^x(\tau) < \infty$. A good source of reference to this formula is [Dyn], Chapter 5.

3. Connection between MSP and the expected reward

Before we start to show connections between QDs and expected reward, we should mention that there is a vast literature on the topic of variational inequalities and stochastic differential equation. A quite fresh reference is the book of B. Øksendal [Oks], and also nice (but unpublished) lecture notes by L.C. Evans [Ev]. We also bring the reader's attention to the (by now classical) book of A. Bensoussan and J.L. Lions [BL].

The variational formulation (in complementary form) of a QD, as we deduced above (1.4), is very well studied. The author of this note does NOT claim solving such problems for the first time. However, we believe that the (only) novelty of this paper is indeed the simple observation of the connection between QD and the Brownian motion, from a variational point of view.

Let us start posing a couple of questions, before formulating any results. After answering these questions, we will formulate a main theorem just for the sake of future references.

Problem 1. For each bounded stopping time θ set

$$\Omega_\theta := \{x \in \mathbf{R}^n : \ U_\theta(x) > 0\}, \qquad \Omega := \bigcup_\theta \Omega_\theta,$$

where U_θ is as in (2.1). Show that Ω is bounded.

Heuristically this seems obvious. First, one observes that for $U_\theta(x)$ to be positive, we need that the Brownian path enters the support of μ. Now if a point x is far from the support of μ, in this case $\overline{D}$, then the Brownian path will need much longer time to reach the set D. In other words, when x is far away from D, then the probability to reach D decreases. Moreover we need θ to be large to reach the set D. Since

$$U_\theta(x) = -E^x(\theta) + ME^x\left(\int_0^\theta \chi_D(X_t)dt\right),$$

intuitively, the negative part should dominate, for θ large. And one expects this to be negative if θ is large. See below for the rigorous proof.

Problem 2. Prove that $D \subset \Omega$.

Let μ be as in (1.1). Then for $x \in D$, and $\theta = \tau_D$ (first exit time from D) we have

$$U_{\tau_D}(x) \geq (M-1)E^x(\tau_D) > 0. \tag{3.1}$$

Hence $D \subset \Omega$.

Actually, if D is somewhat smooth then one can show that $\overline{D} \subset \Omega$. Also if M is large enough then one may get the same result. These statements follow from the work of M. Sakai [Sak83]. The probabilistic way of seeing this is that if D is smooth, then the Brownian path, starting at $x \in \partial D$, has good chances of entering into D, for short time intervals. Hence the contribution of the term $M\int_0^\theta \chi_D$ can be significant in relation to the the negative term $-\theta$.

Problem 3. Define
$$u(x) := \frac{1}{2}\sup_\theta U_\theta(x),$$
where supremum is taken over all bounded stopping times θ. Show that u is a solution to (1.4). This, in view of (1.5), will show that $\Omega \in QD(\mu, SL^1)$.

Observe also that $u(x) \geq 0$, since we can always choose $\theta \equiv 0$.

Problem 4. Conversely, show that if $\Omega \in QD(\mu, SL^1)$ then u, the MSP of Ω as defined in (1.3), has the representation
$$u(x) = \frac{1}{2}\sup_\theta U_\theta(x).$$

Problem 5. Is there any stopping time θ, for which the supremum value, $\sup U_\theta$, is attained?

All the above problems are answered by the following theorem.

Theorem 3.1. *Let $\mu = M\chi_D$, with $M > 1$, and D a bounded set in $\mathbf{R}^n$. Then the maximal expected reward*
$$\frac{1}{2}\sup_\theta U_{\theta,\mu}$$
is the unique solution to the complementary problem (1.4)*. Moreover, the domain $\Omega := \{x : \sup_\theta U_{\theta,\mu} > 0\}$ is a (bounded) QD w.r.t. μ, and the supremum above is attained for the first exit time from Ω.*

Proof. Let us start with small restriction of the class of stopping times. We first consider a fixed ball $B_R = B(0, R)$ with R large enough, and such that $D \subset B_R$. Then we consider all stopping times of the form
$$\theta_R = \min(\theta, \tau_R),$$
where τ_R is the first exit time from the ball B_R. Now we set
$$u_R(x) := \frac{1}{2}\sup_{\theta_R} U_{\theta_R}(x), \qquad \Omega_R = \{u_R > 0\}.$$

It is apparent from the definition of u_R, that if we define
$$2v_R := \sup_{\theta_R} E^x\left(-\theta_R + M\int_0^{\theta_R}\chi_{B_r}(X_t)dt\right),$$
with $D \subset B_r \subset B_R$, then $u_R \leq v_R$. Hence to show that Ω_R is uniformly bounded (independent of R) it suffices to do so for $\{v_R > 0\}$.

Define now $\tilde v_R$ according to
$$\tilde v_R = \frac{1}{2n}|x|^2 - \frac{M}{2n}|x|^2 + \frac{r^2}{2(n-2)}\left(M - M^{2/n}\right) \qquad \text{for } |x| \leq r$$
$$\tilde v_R = \frac{1}{2n}|x|^2 + \frac{Mr^2}{n(n-2)}|x|^{2-n} - \frac{r^2}{2(n-2)}M^{2/n} \qquad \text{for } r < |x| \leq \rho := rM^{1/n},$$
$$\tilde v_R = 0 \quad \text{in } B_R \setminus B_\rho,$$

where we have assumed $R > \rho$. Observe that also $\tilde{v}_R > 0$ in B_ρ, and $\tilde{v}_R$ is independent of R. Moreover, we have

$$\Delta \tilde{v}_R = \chi_{B_\rho} - M\chi_{B_r}. \tag{3.2}$$

According to standard theory of stochastic PDE one can show

$$v_R = \tilde{v}_R. \tag{3.3}$$

From here it follows that u_R ($\leq v_R$) has compact support, independent of R.

To prove (3.3), we see that by Ito's formula (2.3), and (3.2) we have (observe that $\theta_R = \tau_\rho$, the exit time from B_ρ)

$$\begin{aligned} E^x\left(-\tau_\rho + M\int_0^{\tau_\rho} \chi_{B_r}(X_t)dt\right) &= E^x\left(\int_0^{\tau_\rho} -\Delta\tilde{v}_R(X_s)ds\right) \\ &= 2E^x(-\tilde{v}_R(X_{\tau_\rho})) + 2\tilde{v}_R(x). \end{aligned}$$

Since $X_{\tau_\rho} \in \partial B_\rho$, and $\tilde{v}_R = 0$ on ∂B_ρ we are left with

$$E^x\left(-\tau_\rho + M\int_0^{\tau_\rho} \chi_{B_r}(X_t)dt\right) = 2\tilde{v}_R(x). \tag{3.4}$$

Hence $v_R \geq \tilde{v}_R$. Next by Ito's formula, (3.2), and $\tilde{v}_R \geq 0$, we have

$$\begin{aligned} 2\tilde{v}_R(x) &= E^x\left(\int_0^{\theta_R} -\Delta\tilde{v}_R(X_s)ds\right) + 2E^x(\tilde{v}_R(X_{\theta_R})) \\ &\geq E^x\left(\int_0^{\theta_R} \left(\chi_{B_\rho} - M\chi_{B_r}\right)(X_t)dt\right), \end{aligned}$$

for all θ_R. This gives the desired result.

From now on one can consider the problem in B_ρ, and we may replace θ_R by θ_ρ.

Next, we see that by (3.1) we have $D \subset \Omega$.

Finally, in order to prove that u solves the complementary problem (1.4), we need to show that there exists a unique solution $\tilde{u}$ to the complementary problem and that it can be represented by

$$u(x) = \frac{1}{2}E^x\left(-\tau_G + \int_0^{\tau_G} \mu(X_t)\,dt\right) = \frac{1}{2}\sup_{\theta_\rho} U_{\theta_\rho}(x),$$

with τ_G being the first exit time from $G := \{\tilde{u} > 0\}$.

The existence and uniqueness of a solution to (1.4) has been shown earlier by [Sak83]. Cf. also [Gus90], [GS]. In general one can consider either a penalized version or a variational form of the problem at hand.

Now, having a unique solution to (1.4), we can make a similar analysis as above to obtain $u_\rho = \tilde{u}$. □

Remark 3.2. We give a direct heuristic prove of the statement that u solves (1.4). Let $\delta > 0$ and suppose that the system runs at least until time δ. Then the new state at time δ is X_δ, and the optimal cost after this time is $u(X_\delta)$. Hence

$$\begin{aligned} 2u(x) &= \sup_{\theta_R} E^x \left(-\theta_R + M \int_0^{\theta_R} \chi_D(X_s)ds \right) \\ &\geq E^x \left(-\delta + M \int_0^{\delta} \chi_D(X_s)ds \right) + 2u(X_\delta). \end{aligned}$$

Now using Ito's formula for $u(x)$, we obtain

$$\begin{aligned} E^x \left(\int_0^{\delta} -\Delta u(X_s)ds \right) + 2E^x(u(X_\delta)) &= 2u(x) \\ &\geq E^x \left(-\delta + M \int_0^{\delta} \chi_D(X_s)ds + 2u(X_\delta) \right), \end{aligned}$$

i.e.,

$$E^x \left(\int_0^{\delta} -\Delta u(X_s)ds \right) \geq E^x \left(-\delta + M \int_0^{\delta} \chi_D(X_s)ds \right).$$

Dividing by δ and letting δ tend to zero we end up with

$$E^x \left(-\Delta u(X_0) \right) \geq E^x \left(-1 + M\chi_D(X_0) \right),$$

i.e.,

$$-\Delta u(x) \geq -1 + M\chi_D(x).$$

Observe that the pointwise Laplacian $\Delta u(x)$ may not exist, and the above argument needs to be carried out in the weak sense.

Remark 3.3. It is also noteworthy that if ∂D satisfies an interior sphere condition, then the explicit form of v_R can be used to show that $\overline{D} \subset \Omega$. To see this, let $B(y,s) \subset D$ be any ball touching ∂D at some point(s). Then $u_R \geq w_R$, where

$$2w_R := \sup_{\theta_R} E^x \left(-\theta_R + \int_0^{\theta_R} \chi_{B(y,s)}(X_t)dt \right).$$

Moreover, w_R is given (as above) by

$$w_R = \frac{s^2}{r^2} \tilde{v}_R\left(\frac{r(x-y)}{s}\right).$$

Hence $B(y, M^{1/n}s) \subset \Omega$.

4. Game theoretic interpretation

In this section we try to make a game theoretic interpretation of (2.2). However, at this time, we have no particular game in mind, played and run, by the rules U_θ. But the possibility always exists.

Suppose there is a game offered (nowadays on the internet) and the rules for playing it are governed by (2.2). Suppose further that the state of affairs offered by the game are $X_t \in \mathbf{R}^n$. This can mean almost anything, depending on the game.

Now a gambler enters into the game (at time $t = 0$) with state of affairs $X_0 = x$ (the player bets on this state). Now if the gambler chooses not to start the process, and pulls out, there will be neither loss nor win; $U_0(x) = 0$. If the gambler lets the game start and continues for a certain time length $t = \delta$, then the amount the gambler gains or looses (depending on the sign of U_δ) is $|U_\delta(x)|$.

Several questions arise:

Q1) *Can the gambler ever win?*

Q2) *What is the best time, for the gambler, to quit the game?*

Q3) *What state of affairs x should the gambler choose, for the best outcome?*

Q4) *Is it possible that the gambler never wins?*

These questions are important to both the gambler and those who offer the game. E.g., if the gambler chooses a state $x \in D = \text{Interior}(\ \text{supp}\mu)$ then obviously he wins for short time intervals $t = \delta$, provided

$$X_t \in D \qquad \forall\, 0 < t < \delta.$$

Indeed, $U_\delta(x) = (-1+M)\delta$. Hence he can always choose to stop the process before the exit time from D, τ_D.

On the other hand, if the gambler starts at points far from D then $U_\delta(x) = -\delta$, as long as X_t has not entered into D, for $t \le \delta$.

A third situation is that if the gambler starts at ∂D, then the outcome is not so clear anymore.

Naturally the company offering the game has to make sure that no arbitrage, i.e., risk-free profit, takes place. So the state of affairs $x \in D$ should be out of question. The same should go for affairs $x \in \overline{D}^c$, since there is an immediate loss for the gambler, and no guarantee of future gains.

There can be given many variants of combinations of state of affairs such that the loss/gain of short time intervals is not obvious to see. E.g. if the state $x \in \partial D$ as suggested above then it is not apparent what will happen after the game starts. One can also consider a strict rule from the company that for any chosen affair $x \in D$ one should chose a second affair y a certain distance d_x, away from D.

Let us now see what kind of strategy should the gambler choose, once he has started at a point $X_0 = x$. According to what we know from last section, if $x \notin \Omega$, then the gambler will never win. So the best is that he does not enter the game at such states. If on the other hand $x \in \Omega$ (observe that the location of Ω is not known in advance) then the best he can do is to continue until $t = \tau_\Omega$, a value unknown to the gambler. At this time the reward is the highest possible.

The set Ω (here called a QD) is called the continuation region and the complement of it the stopping region. Also at time t the gambler has access to all information on the past and present state but not on the future state. Mathematically, this is defined in terms of the σ-algebra $\mathcal{F}_t$. Also the decision/control variable θ is then a stopping time.

The problem of finding the value τ_Ω is naturally of great importance to both parties in the game.

References

[BL] A. Bensoussan, J.L. Lions, *Applications of variational inequalities in stochastic control.* Translated from the French. Studies in Mathematics and its Applications, 12. North-Holland Publishing Co., Amsterdam-New York, 1982.

[Dyn] E.B. Dynkin, *Markov processes. Vols. I, II.* Translated with the authorization and assistance of the author by J. Fabius, V. Greenberg, A. Maitra, G. Majone. Die Grundlehren der Mathematischen Wissenschaften, Bnde 121, 122 Academic Press Inc., Publishers, New York; Springer-Verlag, Berlin-Göttingen-Heidelberg 1965 Vol. I: xii+365 pp.; Vol. II: viii+274 pp.

[Ev] L.C. Evans *An introduction to stochastic differential equations.* Lecture notes (Version 1.2) for an advanced undergraduate course.
Can be found at http://math.berkeley.edu/~evans/SDE.course.pdf

[Gus90] B. Gustafsson *On quadrature domains and an inverse problem in potential theory.* J. d'Analyse Math. **55** (1990), 172–216.

[GS] B. Gustafsson, H. Shahgholian *Existence and geometric properties of solutions of a free boundary problem in potential theory.* J. Reine Angew. Math. **473** (1996), 137–179.

[Oks] B. Øksendal, *Stochastic differential equations. An introduction with applications.* Fifth edition. Universitext. Springer-Verlag, Berlin, 1998.

[Sak83] M. Sakai *Application of variational inequalities to the existence theorem on quadrature domains.* Trans. Amer. Math. Soc. 1983, vol. 276, pp. 267–279.

Henrik Shahgholian
Department of Mathematics
Royal Institute of Technology
S-100 44 Stockholm, Sweden
e-mail: henriksh@math.kth.se

Operator Theory:
Advances and Applications, Vol. 156, 217–237

Weighted Composition Operators Associated with Conformal Mappings

Serguei Shimorin

Dedicated to Harold Shapiro

Abstract. For conformal self-maps φ of the unit disk, we study weighted composition operators $f \mapsto f \circ \varphi \cdot (\varphi')^b$. We are interested in their boundedness, compactness and contractivity properties as operators acting in weighted Djrbashian-Bergman spaces.

Mathematics Subject Classification (2000). Primary 47B33; Secondary 30C35.

Keywords. Composition operators, conformal mappings.

1. Introduction

Let φ be an analytic function defined in the unit disk $\mathbb{D}$ of the complex plane $\mathbb{C}$ and mapping $\mathbb{D}$ into itself (an analytic self-map of $\mathbb{D}$). The classical composition operator C_φ associated with φ is defined as a mapping $f \mapsto f \circ \varphi$. If, in addition, φ is locally univalent (i.e., $\varphi' \neq 0$ throughout $\mathbb{D}$), then for each $b \in \mathbb{R}$ one can define a weighted composition operator C_φ^b as a mapping

$$C_\varphi^b : f \mapsto f \circ \varphi \cdot (\varphi')^b \tag{1.1}$$

(with an appropriate choice of the branch of $(\varphi')^b$). The present paper is devoted to the study of operators C_φ^b for univalent φ, i.e., conformal self-maps of the unit disk. There are at least two reasons for such a study. First, the operators C_φ^b are as well natural as the classical composition operators. In particular, they satisfy a similar semigroup identity

$$C_\varphi^b \cdot C_\psi^b = C_{\psi \circ \varphi}^b. \tag{1.2}$$

In fact, for positive integers b, the operators C_φ^b correspond to usual composition operators applied to differentials (of order b) $f(z)(dz)^b$ instead of functions. Next, as we shall see later, the properties of C_φ^b are related with the problem of estimating the integral means of derivatives of univalent functions. In particular, famous

Brennan's conjecture for conformal mappings can be reformulated as a fact that the operators C^b_φ with $b \in (-1,0)$ are bounded in the classical Bergman space A^2 (see Theorem 2.9 below and the discussion after it).

We shall study the operators C^b_φ as acting in standard weighted Djrbashian-Bergman spaces. For $\alpha > -1$ and $p \in [1,+\infty)$, let A^p_α be the Banach space of functions f analytic in $\mathbb{D}$ and having a finite norm

$$\|f\|_{\alpha,p} := \left(\int_{\mathbb{D}} |f(z|^p(\alpha+1)(1-|z|^2)^\alpha\, dA(z)\right)^{1/p},$$

where dA is the normalized area measure in $\mathbb{D}$. Spaces A^p_0 will be denoted simply by A^p and the Hilbert space norms $\|\cdot\|_{\alpha,2}$ will be denoted simply by $\|\cdot\|_\alpha$. The first basic question is when C^b_φ are bounded operators in A^p_α. We shall see below that this is true (for fixed p and b) if α is sufficiently big but not necessarily true for all α and p. In particular, for $p = 2$ (and the general case can be reduced to this), we introduce two following functions:

$$\alpha_\varphi(t) := \inf\{\beta > 0 \,:\, C^{t/2}_\varphi \text{ is bounded in } A^2_{\beta-1}\} \tag{1.3}$$

and

$$\mathcal{A}(t) := \sup_\varphi \alpha_\varphi(t) \tag{1.4}$$

(the supremum is taken over all conformal self-maps φ of $\mathbb{D}$). In some sense, these functions correspond to the integral means spectrum function $\beta_\varphi(t)$ and the universal integral means spectrum $B(t)$. For a function φ univalent in $\mathbb{D}$ and $t \in \mathbb{R}$, the function $\beta_\varphi(t)$ is usually defined as the infimum of those $\beta > 0$ that there exists a positive constant $C(\beta, f)$ such that for any $r \in (0,1)$

$$\int_{\mathbb{T}} |\varphi'(r\zeta)|^t\, dm(\zeta) \leqslant \frac{C(\beta,\varphi)}{(1-r)^\beta}$$

(where $\mathbb{T} = \partial\mathbb{D}$ is the unit circle and dm is the normalized arc measure on $\mathbb{T}$). The (bounded) universal integral means spectrum $B(t)$ is then the supremum of $\beta_\varphi(t)$ taken over all bounded univalent φ. It is easy to see that if $(\varphi')^{t/2} \in A^2_\alpha$, then $\beta_\varphi(t) \leqslant \alpha+1$ and if $\beta_\varphi(t) \leqslant \beta_0$, then $(\varphi')^{t/2} \in A^2_\alpha$ for any $\alpha > \beta_0 - 1$. Hence,

$$\beta_\varphi(t) = \inf\{\beta > 0 \,:\, (\varphi')^{t/2} \in A^2_{\beta-1}\}.$$

We see immediately that $\beta_\varphi(t) \leqslant \alpha_\varphi(t)$ (it is enough to apply the operator $C^{t/2}_\varphi$ to the constant function 1). Hence, $\mathcal{A}(t) \geqslant B(t)$. We conjecture that in fact the equality holds (in other words, the only obstacle for the operators C^b_φ to be bounded is that the power of the derivative $(\varphi')^b$ is outside the space in question). As a support for this conjecture, in Section 2 we find a number of estimates for $\mathcal{A}(t)$ which correspond to known estimates for the function $B(t)$. We prove also that Brennan's conjecture (which can be formulated as the property that $(\varphi')^b \in A^2$ for any $b \in (-1,0)$ and conformal mappings φ of $\mathbb{D}$ or as $B(-2) = 1$) is equivalent to the equality $\mathcal{A}(t) = |t| - 1$ for $t \leqslant -2$.

Section 3 of the paper is devoted to the compactness properties of the operators C_φ^b. The situation here is similar to that of the classical composition operators C_φ. We shall see that the condition of nonexistence of the angular derivative at any point of the unit circle $\mathbb{T}$ is a necessary and "almost sufficient" (in the sense precised below, see Theorem 3.3) for compactness of C_φ^b.

Finally, in Section 4 we discuss the weighted composition operators as contractions. Under the assumption $\varphi(0) = 0$, we prove that for $b \geqslant 0$ the operators C_φ^b are contractions in A_α^2 for any $\alpha \geqslant 2\max(1, b) - 2$, and for $b < 0$, the modified operators

$$C_\varphi^{b,\gamma} := f \mapsto (f \circ \varphi)(z) \cdot (\varphi'(z))^b \cdot \left(\frac{\varphi(z)}{z}\right)^\gamma$$

with sufficiently big positive γ are contractions in A_α^2 for $\alpha \geqslant 2|b|$. We discuss also contractivity of $C_\varphi^{b,\gamma}$ with respect to Hilbert space norms of the form

$$\|f\|^2 = \sum_{n \geqslant 0} |\hat{f}(n)|^2 w_n.$$

Our proofs are based on the study of certain natural reproducing kernels associated with $C_\varphi^{b,\gamma}$ and the techniques of Löwner chains.

Throughout the paper, φ denotes a conformal self-map of $\mathbb{D}$. We often assume the normalization $\varphi(0) = 0$ and $\varphi'(0) > 0$. The general case can be easily reduced to this by a suitable Möbius change of variables.

2. Boundedness

A first easy observation related to the operators C_φ^b is that their boundedness in the spaces A_α^p is equivalent to certain Carleson measure condition. For the classical composition operators, this fact was noted first in [7] and [8] (in the context of compact composition operators). A finite positive measure μ in $\mathbb{D}$ is called an A_α^p-Carleson measure if the space A_α^p is continuously embedded in $L^p(\mathbb{D}, d\mu)$, or, in other words, the inequality

$$\int_{\mathbb{D}} |f(z)|^p \, d\mu(z) \leqslant C \int_{\mathbb{D}} |f(z)|^p (1 - |z|^2)^\alpha \, dA(z)$$

holds for any $f \in A_\alpha^p$.

Lemma 2.1. *Let φ be a conformal self-map of $\mathbb{D}$. Then C_φ^b is bounded in A_α^p if and only if the measure $d\mu$ defined as*

$$\mu(E) := \int_{\varphi^{-1}(E)} |\varphi'(z)|^{pb} (1 - |z|^2)^\alpha \, dA(z) \tag{2.1}$$

is an A_α^p-Carleson measure.

The proof follows at once by the change of variables in the condition (definition of boundedness of C_φ^b)

$$\int_{\mathbb{D}} |f(\varphi(z))|^p \, |\varphi'(z)|^{pb} \, (1-|z|^2)^\alpha \, dA(z) \leqslant C \int_{\mathbb{D}} |f(z)|^p \, (1-|z|^2)^\alpha \, dA(z).$$

The description of A_α^p-Carleson measures is well-known (see, e.g., [15], Chapter 6). A measure μ is an A_α^p-Carleson measure if and only if

$$\mu\big(D(\lambda, \frac{1}{2}(1-|\lambda|))\big) \leqslant C(\mu)(1-|\lambda|)^{\alpha+2} \tag{2.2}$$

for any $\lambda \in \mathbb{D}$, where $D(z,r)$ denotes the disk of radius r centered at z. A further equivalent condition is

$$\int_{\mathbb{D}} \frac{d\mu(z)}{|1-\bar{\lambda}z|^\gamma} \leqslant \frac{C(\mu,\gamma)}{(1-|\lambda|^2)^{\gamma-\alpha-2}} \tag{2.3}$$

for some $\gamma > \alpha+2$ and any $\lambda \in \mathbb{D}$. In particular, the property of a measure μ to be an A_α^p-Carleson measure depends only on α but not on p. Together with Lemma 2.1, this implies that the property of C_φ^b to be bounded in A_α^p depends only on α and the product pb. Hence, it is enough to study only the Hilbert space case $p=2$. In what follows we shall consider only that case.

Lemma 2.2. *Assume that φ is a conformal self-map of $\mathbb{D}$, and C_φ^b is bounded in $A_{\alpha_0}^2$ for certain $\alpha_0 > 1$. Then, for any $\alpha > \alpha_0$, C_φ^b is bounded in A_α^2.*

Proof. For each $\alpha > -1$, let

$$\mu_\alpha(E) := \int_{\varphi^{-1}(E)} |\varphi'(z)|^{2b}(1-|z|^2)^\alpha \, dA(z),$$

and for each $\lambda \in \mathbb{D}$ let $D_\lambda := D(\lambda, (1-|\lambda|)/2)$. Without loss of generality, we assume that $\varphi(0) = 0$, which implies $|\varphi(z)| \leqslant |z|$. We have then for each $z \in \varphi^{-1}(D_\lambda)$ the inequality $1-|z| \leqslant 1-|\varphi(z)| \leqslant \frac{3}{2}(1-|\lambda|)$. Hence, we have for $\alpha > \alpha_0$ and any $\lambda \in \mathbb{D}$

$$\begin{aligned} \mu_\alpha(D_\lambda) &\leqslant \int_{\varphi^{-1}(D_\lambda)} |\varphi'(z)|^{2b} \, (1-|z|^2)^{\alpha_0} \cdot (3(1-|\lambda|))^{\alpha-\alpha_0} dA(z) \\ &\leqslant (3(1-|\lambda|))^{\alpha-\alpha_0} \mu_{\alpha_0}(D_\lambda). \end{aligned} \tag{2.4}$$

Since μ_{α_0} is an $A_{\alpha_0}^2$-Carleson measure, we have $\mu_{\alpha_0}(D_\lambda) \leqslant C(\mu_{\alpha_0})(1-|\lambda|)^{\alpha_0+2}$, which implies that $\mu_\alpha(D_\lambda)$ is an A_α^2-Carleson measure. □

The lemma implies that the function $\alpha_\varphi(t)$ defined in the introduction is the critical value such that the operator $C_\varphi^{t/2}$ is bounded in $A_{\beta-1}^2$ for any $\beta > \alpha_\varphi(t)$ and unbounded in $A_{\beta-1}^2$ for any $\beta < \alpha_\varphi(t)$.

In what follows we shall need the following criterion of boundedness of C_φ^b in A_α^2 obtained by the application of the Carleson measure condition (2.3) to the measure μ defined by (2.1):

Lemma 2.3. *The operator C_φ^b is bounded in A^2_α if and only if for some $\gamma > \alpha + 2$*

$$\int_{\mathbb{D}} \frac{|\varphi'(u)|^{2b}}{|1-\bar{\lambda}\varphi(u)|^\gamma}(1-|u|^2)^\alpha\, dA(u) \leqslant \frac{C(\varphi,\gamma)}{(1-|\lambda|^2)^{\gamma-\alpha-2}}, \qquad \forall \lambda \in \mathbb{D}. \tag{2.5}$$

In other words, it is enough to test C_φ^b only on functions of the form $(1-\bar{\lambda}z)^{-\gamma/2}$ with $\gamma > \alpha+2$. Moreover, if the inequality (2.5) holds for some $\gamma > \alpha+2$, then it holds also for all $\gamma > \alpha+2$. The function $(1-\bar{\lambda}z)^{-\alpha-2}$ is the reproducing kernel for the space A^2_α. Therefore, condition (2.5) with $\gamma = 4\alpha+4$ becomes a variant of the following folklore "Reproducing Kernel Thesis" (see, e.g., [9], p. 131): to test boundedness of a certain operator defined on a reproducing kernel Hilbert space one should check it on reproducing kernels.

The next Proposition establishes some basic properties of the functions $\alpha_\varphi(t)$.

Proposition 2.4. *Let φ be a conformal self-map of $\mathbb{D}$. Then the function $\alpha_\varphi(t)$ is convex and satisfies*

$$|\alpha_\varphi(t_1) - \alpha_\varphi(t_2)| \leqslant |t_1 - t_2|. \tag{2.6}$$

Proof. Let α_1 and α_2 be such that $\alpha_\varphi(t_1) < \alpha_1$ and $\alpha_\varphi(t_2) < \alpha_2$. Then $C_\varphi^{t_k/2}$ is bounded in $A^2_{\alpha_k-1}$, $k=1,2$ and hence by Lemma 2.3

$$\int_{\mathbb{D}} \frac{|\varphi'(u)|^{t_k}}{|1-\bar{\lambda}\varphi(u)|^\gamma}\left(1-|u|^2\right)^{\alpha_k-1} dA(u) \leqslant C_k(1-|\lambda|^2)^{\alpha_k+1-\gamma}, \quad \lambda\in\mathbb{D}, \quad k=1,2.$$

If now $t=(1-s)t_1+st_2$ and $\alpha=(1-s)\alpha_1+s\alpha_2$ with $s\in(0,1)$, then applying the Hölder inequality with $1/p=(1-s)$, $1/q=s$, one obtains

$$\begin{aligned}
&\int_{\mathbb{D}} \frac{|\varphi'(u)|^{t}}{|1-\bar{\lambda}\varphi(u)|^\gamma}\left(1-|u|^2\right)^{\alpha-1} dA(u)\\
&\leqslant \left(\int_{\mathbb{D}} \frac{|\varphi'(u)|^{t_1}}{|1-\bar{\lambda}\varphi(u)|^\gamma}\left(1-|u|^2\right)^{\alpha_1-1} dA(u)\right)^{1-s}\\
&\quad\cdot\left(\int_{\mathbb{D}} \frac{|\varphi'(u)|^{t_2}}{|1-\bar{\lambda}\varphi(u)|^\gamma}\left(1-|u|^2\right)^{\alpha_2-1} dA(u)\right)^{s}\\
&\leqslant C_1^{1-s}C_2^s(1-|\lambda|^2)^{\alpha+1-\gamma},
\end{aligned}$$

which implies boundedness of $C_\varphi^{t/2}$ in A^2_α and shows convexity of $\alpha_\varphi(t)$.

The proof of inequality (2.6) is based on the following Lemma:

Lemma 2.5. *If φ is a conformal self-map of $\mathbb{D}$ satisfying $\varphi(0)=0$, then*

$$\begin{aligned}
&\text{(i)} \quad \frac{|\varphi'(z)|}{1-|\varphi(z)|^2} \leqslant \frac{1}{1-|z|^2}, \qquad z\in\mathbb{D}\\
&\textit{and } \text{(ii)} \quad \frac{|\varphi'(z)|^{-1}\left|\dfrac{\varphi(z)}{z}\right|^2}{1-|\varphi(z)|^2} \leqslant \frac{1}{1-|z|^2}, \qquad z\in\mathbb{D}.
\end{aligned} \tag{2.7}$$

The inequality (2.7(i)) is well-known (see, e.g., [1], p.3), it simply reflects the fact that φ is a contraction in hyperbolic metric. The proof of the inequality (2.7(ii)) is postponed until Section 4.

Now, the inequality (2.6) is equivalent to two estimates

$$\alpha_\varphi(t+r) \leqslant \alpha_\varphi(t) + r \tag{2.8}$$

and

$$\alpha_\varphi(t-r) \leqslant \alpha_\varphi(t) + r \tag{2.9}$$

valid for $r \geqslant 0$. Assuming that $\varphi(0) = 0$, we shall prove (2.9). Let α be such that $\alpha_\varphi(t) < \alpha$ and $\gamma > \alpha + r + 1$. Then we have by (2.7(ii)) and obvious inequalities $|\varphi(z)| \geqslant c|z|$ and $|1 - \bar{\lambda}\varphi(u)| \geqslant (1 - |\varphi(u)|^2)/2$

$$\int_{\mathbb{D}} \frac{|\varphi'(u)|^{t-r}}{|1-\bar{\lambda}\varphi(u)|^{\gamma}} (1-|u|^2)^{\alpha+r-1}\, dA(u)$$

$$\leqslant \left(\frac{2}{c^2}\right)^r \int_{\mathbb{D}} \frac{|\varphi'(u)|^t}{|1-\bar{\lambda}\varphi(u)|^{\gamma-r}} \cdot \frac{|\varphi'(u)|^{-r}\left|\dfrac{\varphi(u)}{u}\right|^{2r}}{(1-|\varphi(u)|^2)^r} (1-|u|^2)^r \cdot (1-|u|^2)^{\alpha-1}\, dA(u)$$

$$\leqslant \left(\frac{2}{c^2}\right)^r \int_{\mathbb{D}} \frac{|\varphi'(u)|^t}{|1-\bar{\lambda}\varphi(u)|^{\gamma-r}} (1-|u|^2)^{\alpha-1}\, dA(u) \leqslant \frac{C(\varphi,\gamma)}{(1-|\lambda|^2)^{\gamma-r-\alpha-1}},$$

which shows that $C_\varphi^{(t-r)/2}$ is bounded in $A^2_{\alpha+r-1}$ and hence $\alpha_\varphi(t-r) \leqslant \alpha + r$. The proof of (2.8) is similar (with the only difference that one has to use (2.7(i)) instead of (2.7(ii))). □

Corollary 2.6. *(a trivial estimate of $\alpha_\varphi(t)$). For any φ,*

$$\alpha_\varphi(t) \leqslant |t|. \tag{2.10}$$

Proof. This is a special case of (2.6) with $t_2 = 0$. □

Corollary 2.7. *The universal function $\mathcal{A}(t)$ is convex and satisfies*

$$|\mathcal{A}(t_1) - \mathcal{A}(t_2)| \leqslant |t_1 - t_2|. \tag{2.11}$$

Remark. The corresponding property of the integral means spectral function $B(t)$ is well-known (see, e.g., [12]).

For $t \geqslant 2$, the universal function $\mathcal{A}(t)$ can be determined explicitly.

Proposition 2.8. *For $t \geqslant 2$,*

$$\mathcal{A}(t) = t - 1.$$

Proof. Boundedness (in fact, contractivity) of C_φ^1 in the unweighted Bergman space $A^2 = A_0^2$ is easily obtained by a standard change of variables. Hence, we have $\mathcal{A}(2) \leqslant 1$ and by (2.6) we have $\mathcal{A}(t) \leqslant t-1$ for $t \geqslant 2$. The inequality $\mathcal{A}(t) \geqslant t-1$ follows from the fact that $\mathcal{A}(t) \geqslant B(t)$ and $B(t) = t-1$ for $t \geqslant 2$ (see [12], in fact, the inequality $B(t) \geqslant t-1$ for $t \geqslant 1$ can be obtained by considering conformal maps $\varphi(z) = (1-z)^\delta$ with small positive δ). □

The exact values of $\mathcal{A}(t)$ can be found also for big negative t. We recall first that Carleson and Makarov [4] proved that $B(t) = |t| - 1$ for $t \leqslant t_0$ and $B(t) > |t| - 1$ for $t > t_0$, where $t_0 \in (-\infty, -2]$ is some still unknown critical value. It turns out that $\mathcal{A}(t)$ has the same property.

Theorem 2.9. *Let t_0 be above critical value in the Carleson-Makarov theorem. Then*

$$\mathcal{A}(t) = |t| - 1 \quad \textit{for} \quad t \leqslant t_0$$

and

$$\mathcal{A}(t) > |t| - 1 \quad \textit{for} \quad t > t_0.$$

Proof. Obviously, it is enough to prove that $\mathcal{A}(t) \leqslant |t| - 1$ for $t \leqslant t_0$. Our proof is based on the following strong result obtained by D. Bertilsson in his thesis (Theorem 3.7(e) of [3]).

Theorem 2.10. *(Bertilsson). For $t \leqslant t_0$, there exists a constant $C = C(t)$ such that for any $f \in S$*

$$\int_{\mathbb{T}} \left| r^2 \frac{f'(r\zeta)}{f^2(r\zeta)} \right|^t dm(\zeta) \leqslant \frac{C}{(1-r)^{|t|-1}}. \tag{2.12}$$

Here, S is the usual class of functions univalent in $\mathbb{D}$ and normalized by $f(0) = 0$, $f'(0) = 1$.

Remark. The corresponding result is formulated in [3] for functions g of the class Σ. We used a standard relationship between functions from classes S and Σ.

Now, let φ be a conformal self-map of $\mathbb{D}$ such that $\varphi(0) = 0$. Fix some $\lambda \in \mathbb{D}$. The function

$$f(z) = \frac{\varphi'(0)^{-1}\varphi(z)}{(1 - \bar{\lambda}\varphi(z))^2}$$

is in the class S and we have

$$\frac{z^2 f'(z)}{f^2(z)} = \varphi'(0) \left(\frac{z}{\varphi(z)} \right)^2 \varphi'(z)(1 + \bar{\lambda}\varphi(z))(1 - \bar{\lambda}\varphi(z)).$$

Then, for $t \leqslant t_0$, we have by (2.12),

$$\begin{aligned} \int_{\mathbb{T}} \frac{|\varphi'(r\zeta)|^t}{|1 - \bar{\lambda}\varphi(r\zeta)|^{|t|}} dm(\zeta) &\leqslant C(\varphi, t) \int_{\mathbb{T}} \left| r^2 \frac{f'(r\zeta)}{f^2(r\zeta)} \right|^t dm(\zeta) \\ &\leqslant \frac{C_1(\varphi, t)}{(1-r)^{|t|-1}}. \end{aligned} \tag{2.13}$$

We pick now some $\varepsilon \in (0,1)$ and $\gamma > |t| + \varepsilon + 1$ and we estimate

$$\begin{aligned}
&\int_{\mathbb{D}} \frac{|\varphi'(u)|^t}{|1-\bar{\lambda}\varphi(u)|^\gamma}(1-|u|^2)^{|t|-2+\varepsilon}dA(u)\\
&= \int_0^1 \int_{\mathbb{T}} \frac{|\varphi'(r\zeta)|^t}{|1-\bar{\lambda}\varphi(r\zeta)|^{|t|}} \cdot \frac{(1-r^2)^{|t|-2+\varepsilon}}{|1-\bar{\lambda}\varphi(r\zeta)|^{\gamma-|t|}}\, dm(\zeta)\, 2r\, dr\\
&\leqslant C_2 \int_0^1 \int_{\mathbb{T}} \frac{|\varphi'(r\zeta)|^t}{|1-\bar{\lambda}\varphi(r\zeta)|^{|t|}} \cdot \frac{(1-r^2)^{|t|-2+\varepsilon}}{(1-r|\lambda|)^{\gamma-|t|}}\, dm(\zeta)\, dr\\
&\leqslant C_3 \int_0^1 \frac{(1-r)^{\varepsilon-1}}{(1-r|\lambda|)^{\gamma-|t|}}\, dt\\
&= C_3 \left(\int_0^{|\lambda|} + \int_{|\lambda|}^1\right) \frac{(1-r)^{\varepsilon-1}}{(1-r|\lambda|)^{\gamma-|t|}}\, dt = C_3(I_1+I_2).
\end{aligned}$$

Further,

$$I_1 \leqslant (1-|\lambda|)^{\varepsilon-1} \int_0^{|\lambda|} \frac{dr}{(1-r|\lambda|)^{\gamma-|t|}} \leqslant \frac{C_4}{(1-|\lambda|)^{\gamma-|t|-\varepsilon}}$$

and

$$I_2 \leqslant \frac{1}{(1-|\lambda|)^{\gamma-|t|}} \int_{|\lambda|}^1 (1-r)^{\varepsilon-1} dr = \frac{C_5}{(1-|\lambda|)^{\gamma-|t|-\varepsilon}},$$

and we arrive at the estimate

$$\int_{\mathbb{D}} \frac{|\varphi'(u)|^t}{|1-\bar{\lambda}\varphi(u)|^\gamma}(1-|u|^2)^{|t|-2+\varepsilon}dA(u) \leqslant \frac{C_6}{(1-|\lambda|^2)^{\gamma-|t|-\varepsilon}}.$$

The constant C_6 here does not depend on λ and by Lemma 2.3, the operator $C_\varphi^{t/2}$ is bounded in $A^2_{|t|-2+\varepsilon}$, which shows that $\alpha_\varphi(t) \leqslant |t| - 1 + \varepsilon$. Since $\varepsilon \in (0,1)$ was arbitrary, we get the desired conclusion. □

Famous Brennan's conjecture is equivalent to the equality $B(-2) = 1$ or $t_0 = -2$ (see, e.g., [3]). Therefore, Theorem 2.9 implies also that it is equivalent to $\mathcal{A}(-2) = 1$ or to the property that all weighted composition operators C_φ^b with $b \in (-1, 0)$ are bounded in the unweighted Bergman space A^2.

We turn now to estimates of the function $\mathcal{A}(t)$ for those t where the exact values are not known. These estimates are based on the differentiation techniques developed in papers [14] and [6]. In fact, all estimates of the universal integral means spectrum obtained in [6] are valid also for the universal function $\mathcal{A}(t)$ and proofs are only a slight modification of those given in [6]. Without going into details, we outline here briefly the main steps of the differentiation techniques as it applies to operators C_φ^b.

The following Proposition proved in [6] is crucial.

Proposition 2.11. *There exists a constant $C(\alpha)$ such that for any $g \in A^2_\alpha$*

$$0 \leqslant (\alpha+2)(\alpha+3)\|g\|^2_\alpha - \|g'\|^2_{\alpha+2} \leqslant C(\alpha)\|g\|^2_{\alpha+1}.$$

Now, let φ be a conformal self-map of $\mathbb{D}$ such that $\varphi(0) = 0$ and φ is sufficiently smooth. The operator C_φ^b then is automatically bounded in A_α^2. Assume that for appropriate $\alpha = \alpha(b)$ one can estimate the norm of C_φ^b in A_α^2 by a constant depending only on $\varphi'(0)$. Then, by approximation arguments, the operators C_φ^b with all (not necessarily smooth) univalent φ will be bounded in A_α^2 and one obtains the bound

$$\mathcal{A}(2b) \leqslant \alpha + 1.$$

The desired uniform estimate of the norm of C_φ^b in A_α^2 can be obtained as follows. One has for $g \in A_\alpha^2$ and $s \in [0,1]$, $\delta > 0$

$$\begin{aligned}
\|C_\varphi^b g\|_\alpha^2 &= \frac{1}{(\alpha+2)(\alpha+3)}\|\partial_z\left[C_\varphi^b g\right]\|_{\alpha+2}^2 + O(\|C_\varphi^b g\|_{\alpha+1}^2)\\
&= \frac{1}{(\alpha+2)(\alpha+3)}\|(1-s)b\frac{\varphi''}{\varphi'}\cdot C_\varphi^b g + s\partial_z\left[C_\varphi^b g\right] + (1-s)C_\varphi^{b+1}g'\|_{\alpha+2}^2 + O(\|C_\varphi^b g\|_{\alpha+1}^2)\\
&\leqslant \frac{1+\delta}{(\alpha+2)(\alpha+3)}\|(1-s)b\frac{\varphi''}{\varphi'}C_\varphi^b g + s\partial_z\left[C_\varphi^b g\right]\|_{\alpha+2}^2\\
&\quad + \frac{(1+\delta^{-1})(1-s)^2}{(\alpha+2)(\alpha+3)}\|C_\varphi^{b+1}g'\|_{\alpha+2}^2 + O(\|C_\varphi^b g\|_{\alpha+1}^2). \qquad (2.14)
\end{aligned}$$

Two last terms in (2.14) are automatically bounded (with the constants depending only on $|\varphi'(0)|$) by $O(\|g\|_\alpha^2)$ in all interesting cases (and interesting cases are $\alpha > 2b-2$ for $b > 0$ and $\alpha > -2b-2$ for $b < 0$). This can be easily seen by applying arguments leading to the trivial bounds $\mathcal{A}(t) \leqslant |t|$ for $t \leqslant 0$, $\mathcal{A}(t) \leqslant t/2$ for $t \in (0,2)$ and $\mathcal{A}(t) = t-1$ for $t \geqslant 2$. Therefore, if one can show that for appropriate s the inequality

$$\frac{1}{(\alpha+2)(\alpha+3)}\|(1-s)b\frac{\varphi''}{\varphi'}\cdot C_\varphi^b g + s\partial_z\left[C_\varphi^b g\right]\|_{\alpha+2}^2 \leqslant D\|C_\varphi^b g\|_\alpha^2 + O(\|g\|_\alpha^2) \quad (2.15)$$

holds with some positive constant $D < 1$ which does not depend on φ, then (2.14) (with sufficiently small δ) implies the desired uniform estimate $\|C_\varphi^b g\|_\alpha^2 = O(\|g\|_\alpha^2)$.

Inequalities of type (2.15) were obtained in [6] by combination of area-type estimates for univalent functions with the techniques of Bergman spaces in the bidisk. For example, the following inequality holds (see [6], formula (6.3)) for $h \in A_\alpha^2$ and $\theta \in (0,1)$:

$$\frac{(1-\theta)^2}{\sigma(\alpha+2\theta,-\theta)}\|\frac{1}{2}\frac{\varphi''}{\varphi'}h - \frac{1}{\alpha+2}h'\|_{\alpha+2}^2 \leqslant K(\alpha+1,\theta)\|h\|_\alpha^2 + O(\|h\|_{\alpha+\theta}^2),$$

where constants $\sigma(\alpha+2\theta,-\theta)$ and $K(\alpha+1,\theta)$ are given by certain explicit (rather long) expressions. Analysis of these expressions gives the inequality (2.15) for appropriate α depending on b.

In fact, a similar analysis can be performed for second order and higher order derivatives of $C_\varphi^b g$. In the case $g = 1$ and second order derivatives, it was done in [6]. Again, only slight modifications are needed to obtain estimates of the form $\|C_\varphi^b g\|_\alpha^2 = O(\|g\|_\alpha^2)$.

Finally, we present a table which contains bounds $\mathcal{A}^*(t)$ obtained by differentiation techniques, for the universal function $\mathcal{A}(t)$.

t	$\mathcal{A}^*(t)$	$\|t\|-1$
−20.000	19.028	19.000
−10.000	9.040	9.000
−8.000	7.049	7.000
−6.000	5.067	5.000
−5.000	4.082	4.000
−4.000	3.105	3.000
−3.000	2.144	2.000
−2.400	1.582	1.400
−2.200	1.398	1.200
−2.000	1.218	1.000
−1.800	1.042	0.800
−1.600	0.871	0.600
−1.400	0.706	0.400

t	$\mathcal{A}^*(t)$
−1.200	0.549
−1.000	0.403
−0.800	0.272
−0.600	0.159
−0.400	0.072
−0.200	0.0179
−0.100	0.00443
−0.050	0.00110
0.050	0.00141
0.100	0.0065
0.200	0.031
0.300	0.101
0.400	0.314

Table 1.

3. Compactness

For the classical composition operators $C_\varphi : f \mapsto f \circ \varphi$, the characterization of compactness in the weighted Djrbashian-Bergman spaces A^p_α is well-known (it was obtained in [8]): C_φ is compact in A^p_α if and only if

$$\lim_{|\lambda|\to 1-0} \frac{1-|\varphi(\lambda)|^2}{1-|\lambda|^2} = +\infty. \tag{3.1}$$

By the classical Julia-Carathéodory theorem (see, e.g., [1]), this last condition is equivalent to nonexistence of the angular derivative of φ at any point $\zeta \in \mathbb{T}$. For spaces A^p_α, the condition (3.1) is necessary and sufficient for all holomorphic (not necessarily univalent) self-maps φ of $\mathbb{D}$. In the case of Hardy spaces H^p, this condition is necessary for all holomorphic φ and sufficient for univalent φ.

It turns out that the same condition (3.1) is necessary and "almost" sufficient for compactness of weighted operators C^b_φ with univalent φ in spaces A^p_α.

As in [8], we observe first that compactness of C^b_φ corresponds to a certain vanishing Carleson measure condition. We recall first that a finite positive measure μ in $\mathbb{D}$ is called a vanishing A^p_α-Carleson measure if the space A^p_α is compactly embedded into $L^p(\mathbb{D}, d\mu)$. A characterization of vanishing Carleson measures is as follows (see [15], Chapter 6):

$$\mu\Big(D(\lambda, \frac{1}{2}(1-|\lambda|))\Big) = o((1-|\lambda|)^{\alpha+2}) \quad \text{as} \quad |\lambda| \to 1-0$$

or equivalently

$$\int_{\mathbb{D}} \frac{d\mu(z)}{|1-\bar{\lambda}z|^{\gamma}} = o((1-|\lambda|)^{\alpha+2-\gamma}) \quad \text{as} \quad |\lambda| \to 1-0$$

for some $\gamma > \alpha + 2$.

The following Lemma is a compactness counterpart of Lemmas 2.1 and 2.3 in Section 2.

Lemma 3.1. *For a conformal self-map φ of $\mathbb{D}$ the following conditions are equivalent:*

(i) *The operator C_φ^b is compact in A_α^p;*
(ii) *The measure μ defined by* (2.1) *is a vanishing A_α^p-Carleson measure;*
(iii) *For some $\gamma > \alpha + 2$,*

$$\int_{\mathbb{D}} \frac{|\varphi'(u)|^{2b}}{|1-\bar{\lambda}\varphi(u)|^{\gamma}} (1-|u|^2)^{\alpha} dA(u) = o((1-|\lambda|)^{\alpha+2-\gamma}) \quad \textit{as} \quad |\lambda| \to 1-0.$$

The proof is completely analogous to those of Lemmas 2.1 and 2.3. In particular, we obtain that compactness of C_φ^b in A_α^p depends only on α and the product pb, which means that it is enough to consider only the Hilbert space case $p = 2$.

Theorem 3.2. *If the operator C_φ^b is compact in A_α^2, then φ satisfies* (3.1).

Proof. We use a standard method of testing the adjoint operator on reproducing kernels (see, e.g., [13]). Let $k_\lambda^\alpha(z) = (1-\bar{\lambda}z)^{-\alpha-2}$ be the reproducing kernel for the space A_α^2 at the point λ. Then an easy computation shows that

$$(C_\varphi^b)^* k_\lambda^\alpha = \overline{(\varphi'(\lambda))^b} k_{\varphi(\lambda)}^\alpha. \tag{3.2}$$

Functions $k_\lambda^\alpha / \|k_\lambda^\alpha\|_\alpha$ converge weakly in A_α^2 to zero as $|\lambda| \to 1-0$. Hence, compactness of C_φ^b implies that

$$\lim_{|\lambda|\to 1-0} \frac{\|(C_\varphi^b)^* k_\lambda^\alpha\|_\alpha^2}{\|k_\lambda^\alpha\|_\alpha^2} = 0$$

which gives

$$\lim_{|\lambda|\to 1-0} \frac{(1-|\lambda|^2)^{\alpha+2}}{(1-|\varphi(\lambda)|^2)^{\alpha+2}} |\varphi'(\lambda)|^{2b} = 0.$$

It remains to apply the inequality (2.7ii)

$$|\varphi'(\lambda)| \geqslant \left|\frac{\varphi(\lambda)}{\lambda}\right|^2 \frac{1-|\lambda|^2}{1-|\varphi(\lambda)|^2}$$

in the case $b > 0$ and the inequality (2.7i)

$$|\varphi'(\lambda)|^{-1} \geqslant \frac{1-|\lambda|^2}{1-|\varphi(\lambda)|^2}$$

in the case $b < 0$. □

We observe now that if C_φ^b is compact in some space $A^2_{\alpha_0}$, then it is compact also in any space A^2_α with $\alpha > \alpha_0$. The proof of this fact is completely similar to that of Lemma 2.2. It turns out that for each φ satisfying (3.1), the limiting value

$$\alpha_\varphi^{\mathrm{comp}}(t) = \inf\{\beta : C_\varphi^{t/2} \text{ is compact in } A^2_{\beta-1}\}$$

is the same as boundedness limiting value $\alpha_\varphi(t)$.

Theorem 3.3. *If φ satisfies (3.1) and C_φ^b is bounded in some space $A^2_{\alpha_0}$, then C_φ^b is compact in any space A^2_α with $\alpha > \alpha_0$.*

Proof. We use the same notations as in the proof of Lemma 2.2. Let

$$C(\lambda) := \sup_{\varphi(z)\in D_\lambda} \left(\frac{1-|z|^2}{1-|\varphi(z)|^2}\right)$$

(if $\varphi^{-1}(D_\lambda)$ is empty, we set $C(\lambda) = 0$). Clearly, (3.1) implies that $\lim\limits_{|\lambda|\to 1-0} C(\lambda) = 0$. Now, we have

$$\begin{aligned}\mu_\alpha(D_\lambda) &\leqslant \int_{\varphi^{-1}(D_\lambda)} |\varphi'(z)|^{2b}(1-|z|^2)^{\alpha_0}\cdot C(\lambda)^{\alpha-\alpha_0}(1-|\varphi(z)|^2)^{\alpha-\alpha_0}\,dA(z)\\ &\leqslant C(\lambda)^{\alpha-\alpha_0}\big(3(1-|\lambda|)\big)^{\alpha-\alpha_0}\mu_{\alpha_0}(D_\lambda).\end{aligned}$$

Since μ_{α_0} is an $A^2_{\alpha_0}$-Carleson measure, this implies that μ_α is a vanishing A^2_α-Carleson measure. □

Remark. The last theorem is an analog for weighted composition operators of Theorem 5.3(b) in [8].

4. Contractivity

The classical Littlewood's subordination principle implies that the composition operators C_φ are contractions in Hardy spaces H^p for all holomorphic self-maps φ of the unit disk $\mathbb{D}$ satisfying $\varphi(0) = 0$. The same is true for all spaces A^p_α. If, in addition, φ is univalent in $\mathbb{D}$, then C_φ is contractive also in the Dirichlet space D and in many other Hilbert spaces defined by Taylor coefficients.

In this section, we discuss contractivity of the weighted composition operators C_φ^b. Throughout the section, φ is a conformal self-map of $\mathbb{D}$ satisfying $\varphi(0) = 0$ and $\varphi'(0) > 0$. The family of all such φ will be denoted by $\mathcal{S}_1$.

Theorem 4.1.

(i) *If $b \geqslant 1$, then C_φ^b is a contraction in any space A^2_α with $\alpha \geqslant 2b-2$.*
(ii) *If $0 < b < 1$, then C_φ^b is a contraction in the unweighted Bergman space $A^2 = A^2_0$.*

Proof. Let first $b \geqslant 1$. For any $f \in A^2_\alpha$ with $\alpha \geqslant 2b-2$, we have

$$\begin{aligned}
\|C^b_\varphi f\|^2_\alpha &= (\alpha+1)\int_{\mathbb{D}} |f(\varphi(z))|^2|\varphi'(z)|^{2b}(1-|z|^2)^\alpha dA(z) \\
&\leqslant (\alpha+1)\int_{\mathbb{D}} |f(\varphi(z))|^2|\varphi'(z)|^2\left(\frac{1-|\varphi(z)|^2}{1-|z|^2}\right)^{2b-2}(1-|z|^2)^\alpha\, dA(z) \\
&\leqslant (\alpha+1)\int_{\mathbb{D}} |f(\varphi(z))|^2|\varphi'(z)|^2(1-|\varphi(z)|^2)^\alpha dA(z) \\
&= (\alpha+1)\int_{\varphi(\mathbb{D})} |f(u)|^2(1-|u|^2)^\alpha dA(u) \leqslant \|f\|^2_\alpha,
\end{aligned}$$

which proves the assertion.

In the case $0 < b < 1$, we argue as follows:

$$\begin{aligned}
\|C^b_\varphi f\|^2_0 &= \int_{\mathbb{D}} |f(\varphi(z))|^2|\varphi'(z)|^{2b}dA(z) \\
&\leqslant \left(\int_{\mathbb{D}} |f(\varphi(z))|^2|\varphi'(z)|^2 dA(z)\right)^b \left(\int_{\mathbb{D}} |f(\varphi(z))|^2 dA(z)\right)^{1-b} \\
&= \|C^1_\varphi f\|^{2b}_0 \|C_\varphi f\|^{2(1-b)}_0 \leqslant \|f\|^2_0
\end{aligned}$$

for any $f \in A^2$. □

In the case $b < 0$, we cannot expect contractivity of C^b_φ in spaces A^2_α. For example, it is enough to apply C^b_φ with $\varphi(z) = rz$, $r \in (0,1)$ to the constant function 1. On the other hand, for slightly modified operators, it is possible to obtain contractivity property even for negative b.

For each real γ, we define an operator $C^{b,\gamma}_\varphi$ as

$$\left[C^{b,\gamma}_\varphi f\right](z) := f(\varphi(z)) \cdot (\varphi'(z))^b \cdot \left(\frac{\varphi(z)}{z}\right)^\gamma. \tag{4.1}$$

An important feature of these modified operators is that they satisfy the semigroup identity similar to (1.2):

$$C^{b,\gamma}_\varphi \cdot C^{b,\gamma}_\psi = C^{b,\gamma}_{\psi\circ\varphi}.$$

Theorem 4.2. *If $b < 0$, $\alpha \geqslant 2|b|$ and $\gamma \geqslant 2+\alpha+2|b|$, then $C^{b,\gamma}_\varphi$ is a contraction in A^2_α.*

This theorem is not so trivial as all preceding results. Its proof is based on the techniques of Löwner chains. We recall briefly that a conformal self-map φ of the disk $\mathbb{D}$ satisfying $\varphi(0) = 0$ and $\varphi'(0) > 0$ can be included in a one-parametric family $\{\varphi_t\}_{0\leqslant t\leqslant T}$ of such conformal self-maps of $\mathbb{D}$ that

$$\varphi_0(z) = z; \qquad \varphi_T(z) = \varphi(z),$$

and

$$\frac{\partial}{\partial t}\varphi_t(z) = -\varphi_t(z)p_t(\varphi_t(z)), \tag{4.2}$$

where p_t are functions analytic in $\mathbb{D}$ and satisfying $\operatorname{Re} p_t(\zeta) > 0$ in $\mathbb{D}$. In the case where φ is a slit mapping, i.e., it maps $\mathbb{D}$ into $\mathbb{D}\setminus\Gamma$, where Γ is some smooth arc in $\mathbb{D}\setminus\{0\}$ starting at some point of $\mathbb{T}$, the Löwner differential equation (4.2) takes the form

$$\frac{\partial}{\partial t}\varphi_t(z) = -\varphi_t(z)\cdot\frac{1+k_t\varphi_t(z)}{1-k_t\varphi_t(z)}, \tag{4.3}$$

where $|k_t| = 1$ and k_t depends continuously on t. The details of the Löwner method can be found, e.g., in books [10] or [5].

We may assume (for simplicity of further presentation) that φ is a slit mapping and hence it is included in the Löwner chain satisfying (4.3). There is no loss of generality in this assumption since an arbitrary conformal map can be approximated by slit mappings.

We prove first inequality (2.7ii). It is enough to check that

$$\frac{\partial}{\partial t}\log\left[\frac{|\varphi_t'(z)|^{-2}\left|\dfrac{\varphi_t(z)}{z}\right|^4}{(1-|\varphi_t(z)|^2)^2}\right] \leqslant 0.$$

But an explicit computation based on (4.3) shows that the left-hand side of this inequality is

$$-\frac{2}{|1-k_t\varphi_t(z)|^4}\Big(2(1+|\varphi_t(z)|^2)|1-k_t\varphi_t(z)|^2-(1-|\varphi_t(z)|^2)^2\Big)$$

which is negative since

$$2(1+|u|^2)|1-u|^2-(1-|u|^2)^2 \geqslant 0$$

for any $u\in\mathbb{D}$.

We turn now to the proof of Theorem 4.2. It follows from the assumption $\alpha \geqslant 2|b|$ and results of Section 2 that $C_\varphi^{b,\gamma}$ is bounded in A_α^2. Then to prove the Theorem, it is enough to check that

$$I - C_\varphi^{b,\gamma}(C_\varphi^{b,\gamma})^* \geqslant 0. \tag{4.4}$$

Let $k_\lambda^\alpha(z) = (1-\bar\lambda z)^{-\alpha-2}$ be the reproducing kernel for A_α^2. Then we have

$$\Big([I - C_\varphi^{b,\gamma}(C_\varphi^{b,\gamma})^*]k_\lambda^\alpha, k_z^\alpha\Big) = \frac{1}{(1-\bar\lambda z)^{\alpha+2}} - K_\varphi(z,\lambda),$$

where

$$K_\varphi(z,\lambda) = \Big(\varphi'(z)\overline{\varphi'(\lambda)}\Big)^b\left(\frac{\varphi(z)}{z}\frac{\overline{\varphi(\lambda)}}{\lambda}\right)^\gamma\frac{1}{\Big(1-\overline{\varphi(\lambda)}\varphi(z)\Big)^{\alpha+2}}.$$

Since

$$\frac{1}{(1-\bar\lambda z)^{\alpha+2}} = K_{\varphi_0}(z,\lambda),$$

(where $\varphi_0(z) = z$), we realize that (4.4) will follow if for all $t \in [0, T]$

$$-\frac{\partial}{\partial t} K_{\varphi_t}(z, \lambda) \gg 0 \tag{4.5}$$

(where $\gg 0$ means positive definiteness). In fact, it is easy to see that this condition is also necessary for contractivity of all operators $C_{\varphi_t}^{b,\gamma}$, $t \in [0, T]$ in A_α^2.

Then, an explicit calculation shows that

$$-\frac{\partial}{\partial t} K_{\varphi_t}(z, \lambda) = \frac{2\left(\varphi_t'(z)\overline{\varphi_t'(\lambda)}\right)^b \left(\frac{\varphi_t(z)}{z}\frac{\overline{\varphi_t(\lambda)}}{\lambda}\right)^\gamma}{(1 - k_t\varphi_t(z))^2\,\overline{(1 - k_t\varphi_t(\lambda))^2}} \mathbf{X}(u, v), \tag{4.6}$$

where

$$\mathbf{X}(u, v) = \frac{(\alpha + 2 - 2b)(1 - u)\overline{(1 - v)} + b(1 - \bar{v}u)^2}{(1 - \bar{v}u)^{\alpha+2}} + \frac{(\gamma - \alpha - 2 + 2b)(1 - u)\overline{(1 - v)}}{(1 - \bar{v}u)^{\alpha+1}} \tag{4.7}$$

and $u = k_t\varphi_t(z)$, $v = k_t\varphi_t(\lambda)$. We obtain, therefore, that positive definiteness of this kernel $\mathbf{X}(u, v)$ in $\mathbb{D}$ is a necessary and sufficient condition for contractivity of all operators $C_{\varphi_t}^{b,\gamma}$ in A_α^2, and hence to contractivity of all operators $C_\varphi^{b,\gamma}$ with $\varphi \in \mathcal{S}_1$.

The second addend in (4.7) is positive definite by the assumption $\gamma \geqslant \alpha + 2 + 2|b|$. Positive definiteness of the first addend is given by the following

Proposition 4.3. *For each $\beta > 0$ and $\alpha \geqslant 2\beta$, the kernel*

$$L(u, v) := \frac{(\alpha + 2 + 2\beta)(1 - u)\overline{(1 - v)} - \beta(1 - \bar{v}u)^2}{(1 - \bar{v}u)^{\alpha+2}} \tag{4.8}$$

is positive definite in $\mathbb{D}$.

This Proposition, in turn, is a special case of more general fact. For each positive sequence $(w_n)_{n\geqslant 0}$, we define $l^2(w_n)$ to be the Hilbert space of functions f analytic in $\mathbb{D}$ and satisfying

$$\|f\|_{w_n}^2 := \sum_{n\geqslant 0} |\hat{f}(n)|^2 w_n < +\infty.$$

Here, it is natural to identify a function f with the sequence $\left(\hat{f}(n)\right)_{n\geqslant 0}$, and the same notation $l^2(w_n)$ will be used for the corresponding space of sequences.

Proposition 4.4. *Let $(d_n)_{n\geqslant 0}$ and $(e_n)_{n\geqslant 0}$ be positive sequences. Then two following conditions are equivalent:*

(i) *The kernel*

$$(1 - u)\overline{(1 - v)} \cdot \sum_{n\geqslant 0} d_n(u\bar{v})^n - \sum_{n\geqslant 0} e_n(u\bar{v})^n$$

is positive definite;

(ii) *The operator of "tail summation"*

$$T : (x_n)_{n\geqslant 0} \mapsto (y_m)_{m\geqslant 0}; \qquad y_m = \sum_{n\geqslant m} x_n$$

is a contraction from $l^2(d_n)$ *to* $l^2(e_n)$.

Moreover, the following condition is sufficient for both (i) *and* (ii)*:*

(iii) *There exists a strictly decreasing sequence* $(\psi_m)_{m\geqslant 0}$ *such that*

$$\lim_{m\to+\infty} \psi_m = 0$$

and for any $n \geqslant 0$

$$\sum_{m=0}^{n} \psi_m e_m \leqslant (\psi_n - \psi_{n+1}) d_n.$$

Proof. The kernel

$$K_1(u,v) = \sum_{n\geqslant 0} d_n (u\bar{v})^n$$

is the reproducing kernel for the space $l^2\left(\frac{1}{d_n}\right)$. The kernel

$$K_2(u,v) = \frac{1}{(1-u)\overline{(1-v)}} \sum_{n\geqslant 0} e_n (u\bar{v})^n$$

is the reproducing kernel for the space $\frac{1}{1-u} \cdot l^2\left(\frac{1}{e_n}\right)$ supplied with the range-norm

$$\left\| \frac{f(u)}{1-u} \right\|^2 = \|f\|^2_{\frac{1}{e_n}}.$$

Therefore, by the classical theory of reproducing kernels ([2]), $K_1(u,v) - K_2(u,v)$ is positive definite if and only if the multiplication operator

$$M_{\frac{1}{1-u}} : f(u) \mapsto \frac{f(u)}{1-u} =: g(u) \tag{4.9}$$

is a contraction from $l^2\left(\frac{1}{e_n}\right)$ to $l^2\left(\frac{1}{d_n}\right)$. If we interpret now $M_{\frac{1}{1-u}}$ as acting on sequences of Taylor coefficients

$$M_{\frac{1}{1-u}} : \left(\hat{f}(n)\right)_{n\geqslant 0} \mapsto \left(\hat{g}(n)\right)_{n\geqslant 0}, \qquad \hat{g}(n) = \sum_{k=0}^{n} \hat{f}(k),$$

and then go to the adjoint operator from $l^2(d_n)$ to $l^2(e_n)$, we obtain condition (ii).

To prove that (iii) implies (ii), we use the following chain of inequalities (which is, in fact, a version of the classical Schur's test of boundedness of integral operators). We have for any sequence $(x_n)_{n\geqslant 0}$

$$\left|\sum_{n=m}^{\infty} x_n\right|^2 \leqslant \left(\sum_{n=m}^{\infty} \frac{|x_n|^2}{\psi_n - \psi_{n+1}}\right) \cdot \left(\sum_{n=m}^{\infty} (\psi_n - \psi_{n+1})\right) = \psi_m \cdot \sum_{n=m}^{\infty} \frac{|x_n|^2}{\psi_n - \psi_{n+1}}$$

and hence

$$\begin{aligned}\sum_{m=0}^{\infty}\left|\sum_{n=m}^{\infty} x_n\right|^2 e_m &\leqslant \sum_{m=0}^{+\infty} e_m\psi_m \cdot \sum_{n=m}^{\infty} \frac{|x_n|^2}{\psi_n - \psi_{n+1}} \\ &= \sum_{n=0}^{\infty} |x_n|^2 \left(\frac{1}{\psi_n - \psi_{n+1}} \cdot \sum_{m=0}^{n} e_m\psi_m\right) \leqslant \sum_{n\geqslant 0} |x_n|^2 d_n. \qquad \square\end{aligned}$$

In the case of the kernel (4.8) of Proposition 4.3, we have

$$d_n = (\alpha+2+2\beta)(-1)^n\binom{-\alpha-2}{n} = (\alpha+2+2\beta)\frac{(\alpha+2)_n}{n!}$$

and

$$e_n = \beta(-1)^n\binom{-\alpha}{n} = \beta\frac{(\alpha)_n}{n!},$$

where $(x)_n := x(x+1)\dots(x+n-1)$ is a standard Pochhammer's symbol. To prove Proposition 4.3, it is enough now to apply condition (iii) of Proposition 4.4 with the choice

$$\psi_n = \frac{(\alpha/2)_n}{(\alpha)_n}.$$

This accomplishes also the proof of Theorem 4.2.

One may observe that the spaces A^2_α where the operators C^b_φ or $C^{b,\gamma}_\varphi$ are contractive by theorems 4.1 and 4.2 have much bigger parameter α than the spaces where these operators are bounded by results of Section 2 (except for the trivial case $b \geqslant 1$). This leads to a natural question whether the bounds for contractivity $\alpha = 0$ for operators C^b_φ with $b \in (0,1)$ and $\alpha = 2|b|$ for $C^{b,\gamma}_\varphi$ with $b < 0$ are sharp. It turns out that this is the case.

The proof of this fact is based on the following

Lemma 4.5. *Let $\alpha > -1$ and $C \neq 0$. If a kernel*

$$X(u,v) = A\frac{(1-u)\overline{(1-v)}}{(1-\bar{v}u)^{\alpha+2}} + B\frac{(1-u)\overline{(1-v)}}{(1-\bar{v}u)^{\alpha+1}} + \frac{C}{(1-\bar{v}u)^{\alpha}}$$

is positive definite in $\mathbb{D}$, then $\alpha \geqslant 0$ and $A \geqslant \max\left(0, \frac{-4(\alpha+1)}{\alpha}C\right)$.

Proof. First, the inspection of diagonal coefficients $\hat{X}(n,n)$ (with big n) of the Taylor series expansion

$$X(u,v) = \sum_{n,k\geqslant 0} \hat{X}(n,k)u^n\bar{v}^k$$

shows that $A \geqslant 0$. Next, we may assume without loss of generality that $B \geqslant 0$ because otherwise we can replace X by bigger (in the sense of positive definiteness) kernel

$$\tilde{X}(u,v) = A\frac{(1-u)\overline{(1-v)}}{(1-\bar{v}u)^{\alpha+2}} + |B|\frac{(1-u)\overline{(1-v)}}{(1-\bar{v}u)^{\alpha+1}} + \frac{C}{(1-\bar{v}u)^{\alpha}}.$$

We consider first the case $C < 0$, $\alpha > 0$. Then positive definiteness of $X(u,v)$ takes the form

$$\frac{1}{(1-u)\overline{(1-v)}(1-\bar{v}u)^{\alpha}} \ll \frac{A/|C|}{(1-\bar{v}u)^{\alpha+2}} + \frac{B/|C|}{(1-\bar{v}u)^{\alpha+1}}$$

which is equivalent to the property that the operator $M_{\frac{1}{1-u}}$ defined by (4.9) is a contraction from $l^2\left(\frac{1}{e_n}\right)$ to $l^2\left(\frac{1}{d_n}\right)$, where

$$e_n = (-1)^n\binom{-\alpha}{n} = \frac{(\alpha)_n}{n!} \quad \text{and} \quad d_n = \frac{A}{|C|}\frac{(\alpha+2)_n}{n!} + \frac{B}{|C|}\frac{(\alpha+1)_n}{n!}.$$

Testing this operator on functions $f(z) = (1-rz)^{-(\alpha/2+\varepsilon)}$ (using a standard asymptotics of binomial coefficients) and letting first $r \to 1-0$ and then $\varepsilon \to 0+0$, we arrive at the necessary condition

$$\frac{A}{|C|} \geqslant \frac{4(\alpha+1)}{\alpha}.$$

Assume now that X is positive definite and $C < 0$, $\alpha < 0$. Then the product

$$X(u,v)\cdot\frac{1}{(1-\bar{v}u)^{\varepsilon-\alpha}}$$

is also positive definite for $\varepsilon > 0$. But

$$X(u,v)\cdot\frac{1}{(1-\bar{v}u)^{\varepsilon-\alpha}} = A\frac{(1-u)\overline{(1-v)}}{(1-\bar{v}u)^{\varepsilon+2}} + B\frac{(1-u)\overline{(1-v)}}{(1-\bar{v}u)^{\varepsilon+1}} + \frac{C}{(1-\bar{v}u)^{\varepsilon}}$$

and by the preceding case we have

$$\frac{A}{|C|} \geqslant \frac{4(\varepsilon+1)}{\varepsilon}.$$

Letting $\varepsilon \to 0+0$, we get a contradiction.

It remains to rule out the possibility $\alpha < 0$, $C > 0$. In this case we write positive definiteness of X in the following form

$$\sum_{n=1}^{\infty}|\alpha|\frac{(1-|\alpha|)_{n-1}}{n!}(\bar{v}u)^n \ll 1 + (1-u)\overline{(1-v)}\left(\frac{A/C}{(1-\bar{v}u)^{\alpha+2}} + \frac{B/C}{(1-\bar{v}u)^{\alpha+1}}\right). \tag{4.10}$$

Let

$$e_n = |\alpha|\frac{(1-|\alpha|)_{n-1}}{n!} \quad \text{and} \quad d_n = \frac{A}{C}\frac{(\alpha+2)_n}{n!} + \frac{B}{C}\frac{(\alpha+1)_n}{n!}.$$

By the classical theory of reproducing kernels, (4.10) is equivalent to the following property: for any function $f \in l^2\left(\frac{1}{e_n}\right)$ there is a constant $x \in \mathbb{C}$ and a function $g \in l^2\left(\frac{1}{d_n}\right)$ such that

$$f(u) = x + (1-u)g(u) \qquad \text{and} \qquad |x|^2 + \|g\|^2_{1/d_n} \leqslant \|f\|^2_{1/e_n}. \tag{4.11}$$

Since

$$\frac{1}{1-u} \notin l^2\left(\frac{1}{d_n}\right)$$

for $\alpha < 0$, the only possibility for the constant x in (4.11) is that $x = f(1)$ and hence (4.11) takes the form

$$|f(1)|^2 + \left\|\frac{f(u)-f(1)}{u-1}\right\|^2_{1/d_n} \leqslant \|f\|^2_{1/e_n}.$$

Choosing now

$$f(u) = \sum_{n \geqslant 1} e_n u^n = 1 - (1-u)^{|\alpha|},$$

we get a contradiction. □

Corollary 4.6. *Assume that for some fixed* α, b *and* γ *all operators* $C_\varphi^{b,\gamma}$ *with* $\varphi \in \mathcal{S}_1$ *are contractions in* A^2_α. *Then* $\alpha \geqslant \max(0, -2b)$.

Proof. We know already that positive definiteness of the kernel $\mathbf{X}(u,v)$ given by (4.7) is a necessary condition for contractivity of all operators $C_\varphi^{b,\gamma}$ in A^2_α. It remains to apply Lemma 4.5. □

The same method as we used for the proof of Theorem 4.2 can be applied to the study of contractivity of operators $C_\varphi^{b,\gamma}$ in general spaces $l^2(w_n)$. We have

Theorem 4.7. *Let* $(w_n)_{n\geqslant 0}$ *be a positive sequence. Let also* $c_n = w_n^{-1}$. *Let* b *and* γ *be fixed. Then two following conditions are equivalent*

(i) *All operators* $C_\varphi^{b,\gamma}$ *with* $\varphi \in \mathcal{S}_1$ *are contractive in* $l^2(w_n)$*;*
(ii) *The following kernel is positive definite in* $\mathbb{D}$

$$\mathbf{Y}(u,v) := (1-u)\overline{(1-v)} \sum_{n\geqslant 0} d_n (u\bar{v})^n - \sum_{n \geqslant 0} e_n (u\bar{v})^n, \tag{4.12}$$

where

$$d_n = -2bc_{n-1} + (n+\gamma)c_n - (n+\gamma-1)c_{n-1}, \quad n \geqslant 0$$

and

$$e_n = -b(c_n - 2c_{n-1} + c_{n-2}), \quad n \geqslant 0$$

(as usual, we use the convention that $c_{-1} = c_{-2} = 0$*).*

Proof. As in the proof of Theorem 4.2, we realize that contractivity of $C_\varphi^{b,\gamma}$ is equivalent to positive definiteness of the derivative

$$-\frac{\partial}{\partial t}K_{\varphi_t}(z,\lambda),$$

where now

$$K_\varphi(z,\lambda) := \left(\varphi'(z)\overline{\varphi'(\lambda)}\right)^b \left(\frac{\varphi(z)}{z}\frac{\overline{\varphi(\lambda)}}{\lambda}\right)^\gamma K(\varphi(z),\varphi(\lambda))$$

and

$$K(z,\lambda) = \sum_{n\geqslant 0} c_n (z\bar{\lambda})^n$$

is the reproducing kernel for $l^2(w_n)$. An explicit calculation based on the Löwner equation (4.3) shows that

$$-\frac{\partial}{\partial t}K_{\varphi_t}(z,\lambda) = \frac{2\left(\varphi_t'(z)\overline{\varphi_t'(\lambda)}\right)^b \left(\frac{\varphi_t(z)}{z}\frac{\overline{\varphi_t(\lambda)}}{\lambda}\right)^\gamma}{(1-k_t\varphi_t(z))^2\overline{(1-k_t\varphi_t(\lambda))^2}} \cdot \mathbf{Y}(u,v),$$

where $\mathbf{Y}(u,v)$ is given by (4.12), $u = k_t\varphi_t(z)$ and $v = k_t\varphi_t(\lambda)$. This proves the Theorem. □

The condition of positive definiteness of the kernel $\mathbf{Y}(u,v)$ defined by (4.12) can be reformulated as positive definiteness of certain Jacobi matrix. Namely, the matrix $J = (J_{kl})_{k,l\geqslant 0}$ of Taylor coefficients in the decomposition

$$Y(u,v) = \sum_{k,l\geqslant 0} J_{k,l}u^k\bar{v}^l$$

has the following entries:

$$\begin{aligned} J_{nn} &= d_n + d_{n-1} - e_n; \\ J_{n,n+1} &= J_{n+1,n} = -d_n; \\ J_{kl} &= 0 \quad \text{if} \quad |k-l| \geqslant 2 \end{aligned}$$

and clearly positive definiteness of $\mathbf{Y}(u,v)$ is equivalent to positive definiteness of J. In the case where sequences $(d_n)_{n\geqslant 0}$ and $(e_n)_{n\geqslant 0}$ are positive, one can apply Proposition 4.4 to obtain a reformulation and a sufficient condition for positive definiteness of $\mathbf{Y}(u,v)$.

Several computational experiments performed with different sequences (w_n) led me to the following conjecture. Assume that (w_n) is sufficiently regular in appropriate sense. In the case $b \in (0,1)$, if γ is fixed, then contractivity of all operators $C_\varphi^{b,\gamma}$ with $\varphi \in \mathcal{S}_1$ in $l^2(w_n)$ implies that w_n decay as n^{-1} or faster. In the case $b < 0$ and fixed γ, if all operators $C_\varphi^{b,\gamma}$ are contractive in $l^2(w_n)$, then w_n decay as n^{-1+2b} or faster. In other words, contractivity bounds given by Theorems 4.1(ii) and 4.2 are sharp even in asymptotic sense.

References

[1] L. Ahlfors, *Conformal invariants: topics in geometric function theory.* McGraw-Hill Series in Higher Mathematics. McGraw-Hill Book Co., New York-Düsseldorf-Johannesburg, 1973.

[2] N. Aronszajn, *Theory of reproducing kernels.* Trans. Amer. Math. Soc. **68** (1950), 337–404.

[3] D. Bertilsson, *On Brennan's conjecture in conformal mapping.* Dissertation, Royal Institute of Technology, Stockholm, 1999.

[4] L. Carleson, N. Makarov, *Some results connected with Brennan's conjecture.* Ark. Mat. **32** (1994), no. 1, 33–62.

[5] G.M. Golusin, *Geometrische Funktionentheorie.* Hochschulbücher für Mathematik, Bd. **31**. VEB Deutscher Verlag der Wissenschaften, Berlin, 1957.

[6] H. Hedenmalm, S. Shimorin, *Weighted Bergman spaces and the integral means spectrum of conformal mappings*, to appear in Duke Math. Journal.

[7] B. MacCluer, *Compact composition operators on* $H^p(B_N)$. Michigan Math. J. **32** (1985), no. 2, 237–248.

[8] B. MacCluer, J. Shapiro, *Angular derivatives and compact composition operators on the Hardy and Bergman spaces.* Canad. J. Math. **38** (1986), no. 4, 878–906.

[9] N.K. Nikolski, *Operators, functions, and systems: an easy reading. Vol. 1. Hardy, Hankel, and Toeplitz.* Mathematical Surveys and Monographs, **92**. American Mathematical Society, Providence, RI, 2002.

[10] Ch. Pommerenke, *Univalent functions.* Studia Mathematica/Mathematische Lehrbücher, Band **XXV**. Vandenhoeck & Ruprecht, Göttingen, 1975.

[11] Ch. Pommerenke, *Boundary behaviour of conformal maps.* Grundlehren der Mathematischen Wissenschaften [Fundamental Principles of Mathematical Sciences], **299**. Springer-Verlag, Berlin, 1992.

[12] Ch. Pommerenke, *The integral means spectrum of univalent functions.* Zap. Nauchn. Sem. S.-Peterburg. Otdel. Mat. Inst. Steklov. (POMI) **237** (1997), Anal. Teor. Chisel i Teor. Funkts. 14, 119–128; translation in J. Math. Sci. (New York)**95** (1999), no. 3, 2249–2255

[13] J. Shapiro, *Composition operators and classical function theory.* Universitext: Tracts in Mathematics. Springer-Verlag, New York, 1993.

[14] S. Shimorin, *A multiplier estimate of the Schwarzian derivative of univalent functions.* Int. Math. Res. Not. 2003, no. **30**, 1623–1633.

[15] K. Zhu, *Operator theory in function spaces.* Monographs and Textbooks in Pure and Applied Mathematics, **139**. Marcel Dekker, Inc., New York, 1990.

Serguei Shimorin
Matematik
Royal Institute of Technology
S-100 44 Stockholm, Sweden
e-mail: shimorin@math.kth.se

Operator Theory:
Advances and Applications, Vol. 156, 239–255

Quadrature Identities and Deformation of Quadrature Domains

Tomas Sjödin

Abstract. We study the possibility of deforming quadrature domains into each other, and also discuss the possibility of changing the distribution in a quadrature identity from complex to real and from real to positive. The last question is in a sense also studied without the assumption that we have a quadrature domain.

Mathematics Subject Classification (2000). Primary 31B05; Secondary 31A05, 31C35.

Keywords. Quadrature domains, quadrature identities, deformation of quadrature domains.

1. Introduction

In this paper we will be interested in two things. We start by studying a question considered in the paper [4]. In that paper the class of domains which are quadrature domains for analytic functions with respect to one and the same real measure were under consideration, and for this case it was possible to prove that two domains in such a class can always be deformed into each other within the class. Here we will study the same question but for complex distributions with compact support in $\mathbb{R}^2$. This will be done in Section 3 by a direct argument similar to the one used in [4].

Another approach to this would be to show that a complex distribution always may be replaced by a real measure in a quadrature identity for analytic functions. That this is possible for a single quadrature domain turns out to be rather easy to prove. However, it is not obvious that for two domains which are quadrature domains with respect to the same complex distribution it is possible to change to the same real measure for both domains. We do not have a complete answer to this, but at-least obtain some necessary and sufficient conditions on when this is possible.

Since we proved that a complex distribution may always be replaced by a real a similar question of whether a real measure may always be replaced by a positive one is natural to pose. This question is just as natural to study in any number of dimensions, and it has already been studied in [6]. There the authors were able to prove that it is in-fact always possible in two dimensions, and gave some sufficient conditions to guarantee it in the higher-dimensional case. They only treated the case for harmonic quadrature domains. Here we aim at relaxing the conditions in the higher-dimensional case, and also briefly discuss the case of analytic quadrature domains in two dimensions.

During the work with the latter questions above it became clear that it was more natural to study these without the assumption that we have a quadrature domain. What we mean here is simply that if we are given any open set Ω in $\mathbb{R}^N$ and a set of functions $V \subset C(\Omega)$ (real or complex-valued) and two classes A, B of Radon measures with compact support in Ω, then we would like to get necessary and sufficient conditions on the measures μ in A that guarantees the existence of a measure η in B such that

$$\int f d\mu = \int f d\eta \qquad \forall f \in V.$$

The two cases we are interested in are first in dimension two when V is a class of analytic functions, A is the class of complex Radon measures and B the class of real (signed) Radon measures. This is treated in Section 4. The second is in any number of dimensions and V is a class of harmonic functions, A the class of real (signed) Radon measures and B positive Radon measures, which is treated in Section 5.

I would also like to take the opportunity to thank first of all my supervisor Björn Gustafsson for his support, and also professor Makoto Sakai for several constructive comments which led to improvements of the paper. In particular professor Sakai pointed out to me the possibility of using a theorem of Bers (Theorem 2.2 part 1) in Lemma 4.4, which in turn made it possible to get rid of an unnecessary assumption in Theorem 4.6.

2. Basic definitions and notation

Let $\Omega \subset \mathbb{R}^N$ be open. We now define the following classes in Ω:

$H(\Omega) :=$ {harmonic functions in Ω},
$L^1(\Omega) :=$ {Lebesgue-Integrable functions in Ω}, $m =$ Lebesgue-measure,
$P(\Omega) :=$ {positive functions in Ω},

If $N = 2$ we let $A(\Omega) :=$ {holomorphic functions in Ω},

$\mathcal{D}'(\Omega, \mathbb{R}) :=$ {real-valued distributions in Ω},
$\mathcal{D}'(\Omega, \mathbb{C}) :=$ {complex-valued distributions in Ω},
$\mathcal{E}'(\Omega, \mathbb{R}) :=$ {real-valued distributions with compact support in Ω},
$\mathcal{E}'(\Omega, \mathbb{C}) :=$ {complex-valued distributions with compact support in Ω}.

There are natural identifications of $\mathcal{D}'(\Omega, \mathbb{R})$ and $\mathcal{E}'(\Omega, \mathbb{R})$ with subspaces of $\mathcal{D}'(\Omega, \mathbb{C})$ resp. $\mathcal{E}'(\Omega, \mathbb{C})$ which we will use.

We also order $\mathcal{D}'(\Omega,\mathbb{R})$ by saying that $u \leq v$ if $\langle u,\phi\rangle \leq \langle v,\phi\rangle$ for every $\phi \geq 0$ in $C_c^\infty(\Omega,\mathbb{R})$.

$$h_\varepsilon(x) := \begin{cases} \frac{1}{\varepsilon^N} C \exp(\frac{\varepsilon^2}{|x|^2-\varepsilon^2}) & |x| < \varepsilon \\ 0 & |x| \geq \varepsilon \end{cases},$$

where C is s.t. $\int h_\varepsilon dm = 1$, h_ε are the standard smooth radial mollifiers.

$$\Phi(x) = \{\text{Newtonian kernel}\} := \begin{cases} C_2 \log|x| \text{ if } N = 2 \\ C_N|x|^{2-N} \text{ if } N \geq 3 \end{cases}$$

where C_N is s.t. $-\Delta\Phi = \delta$, the Dirac measure at the origin.

We now define $U^\mu = \{\text{Newtonian potential of } \mu\} = \Phi * \mu$.

An open set $\Omega \subset \mathbb{R}^N$ is called Greenian if it has a Green's function, or what amounts to the same thing, the function $\Phi(\bullet - y)$ has a subharmonic minorant for each $y \in \Omega$. Then

$$G(x,y) := \Phi(x-y) - h(x,y)$$

where $h(\bullet, y)$ is the largest subharmonic minorant to $\Phi(\bullet - y)$, and G is called the Green's function for Ω. Also if G is the Green's function for some domain we denote the Green-potential of μ by G^μ (for $\mu \in \mathcal{E}'(\mathbb{R}^N,\mathbb{R})$).

$$A(z) = \{\text{Cauchy kernel}\} := \frac{1}{\pi z}, \text{ s.t. } \frac{\partial}{\partial \bar{z}} A = \delta.$$

The Cauchy transform of μ will be denoted by $A^\mu := A * \mu$ (for $\mu \in \mathcal{E}'(\mathbb{R}^2,\mathbb{C})$).

$\mathrm{Bal}(\mu,\partial\Omega)$ = Balayage of μ onto $\partial\Omega$ (where μ is a real signed Radon measure with compact support in Ω). We will only use balayage when $\partial\Omega$ is fairly smooth, and then it may be defined as the unique Radon measure η with support on $\partial\Omega$ such that

$$\int h d\mu = \int h d\eta \qquad \forall h \in H(\Omega) \cap C(\overline{\Omega}).$$

Furthermore we use the super-positioning principle such that for instance

$$HL^1(\Omega) = H(\Omega) \cap L^1(\Omega).$$

We also set the following conventions regarding measures. By a real measure we mean a real (possibly) signed measure always indicating when we mean positive. If we write something like $\mu = \alpha + i\beta$ is a complex measure it is to be understood that α and β are real measures. Also, by support we always mean closed support.

Definition 2.1. *If $\mu \in \mathcal{E}'(\mathbb{R}^N,\mathbb{R})$, then we define the class $Q(\mu, HL^1)$ to consist of those open subsets $\Omega \subset\subset \mathbb{R}^N$ (i.e., compactly contained in) such that $\mu \in \mathcal{E}'(\Omega,\mathbb{R})$ and $\langle \mu, f\rangle = \int_\Omega f dm$ $\forall f \in HL^1(\Omega)$.*

If $N = 2$ we make a similar definition of $Q(\mu, AL^1)$ for $\mu \in \mathcal{E}'(\mathbb{R}^2,\mathbb{C})$ to consist of those open subsets $\Omega \subset\subset \mathbb{R}^2$ such that $\mu \in \mathcal{E}'(\Omega,\mathbb{C})$ and $\langle \mu, f\rangle = \int_\Omega f dm$ $\forall f \in AL^1(\Omega)$.

Elements in $Q(\mu, HL^1)$ are called quadrature domains for harmonic functions with respect to μ, and elements in $Q(\mu, AL^1)$ quadrature domains for analytic functions with respect to μ.

Remark. Note that we do not require quadrature domains to be connected, but since μ has compact support in Ω it follows that a finite number of components covers μ and there can't be any component that doesn't contain some part of μ, because the characteristic function of a component lies in the test class. Hence a quadrature domain has a finite number of components. The definition above is more restrictive than the one used in [4] where the measure is not required to have compact support in the domain.

Example. The most basic example of a quadrature domain is a ball which is a quadrature domain with respect to a point-mass at its center. In two dimensions we may use conformal mappings to get other examples. So let us look at $B_1(0)$ in $\mathbb{R}^2$, which belongs to $Q(\pi\delta, AL^1)$. Now let $f(z) = z + iz^2/2$ which is a conformal map of $B_1(0)$ onto some domain which we denote Ω. We have $f(0) = 0$ and $f'(z) = 1 + iz$ so $f'(0) = 1$. If $g \in AL^1(\Omega)$ then $g \circ f \in AL^1(B_1(0))$ and

$$\int_\Omega g dm = \int_{B_1(0)} (g \circ f)(z)|f'(z)|^2 dm(z).$$

We have $|f'(z)|^2 = (1-y)^2 + x^2$ so for $n \geq 0$:

$$\int_{B_1(0)} z^n |f'(z)|^2 dm = \int_{\theta=0}^{2\pi} \int_{r=0}^{1} r^n e^{in\theta}(1 - 2r\sin\theta + r^2) r dr d\theta =$$

$$\begin{cases} 3\pi/2 & \text{if } n = 0 \\ -i\pi/2 & \text{if } n = 1 \\ 0 & \text{if } n \geq 2. \end{cases}$$

Hence we get for $g \in AL^1(\Omega)$

$$\int_\Omega g dm = \int_{B_1(0)} (g \circ f)(z)|f'(z)|^2 dm(z) =$$

$$\frac{3\pi}{2}(g \circ f)(0) - \frac{i\pi}{2}(g \circ f)'(0) = \frac{3\pi}{2} g(0) - \frac{i\pi}{2} g'(0).$$

Therefore $\Omega \in Q(3\pi\delta/2 - i\pi\delta'/2, AL^1)$.

(This is a standard example, and many others may be produced by analytic maps. See [8] for instance).

The following theorem, where the first part is due to L. Bers [2] and the second to M. Sakai [8], will be of fundamental importance for us.

Theorem 2.2.

1. *If $\Omega \subset\subset \mathbb{R}^2$, then the linear span of the Cauchy kernels with poles in Ω^c is dense in $AL^1(\Omega)$.*
2. *If $\Omega \subset\subset \mathbb{R}^N$ then the linear span of the Newtonian kernels and its first order partial derivatives with poles in Ω^c are dense in $HL^1(\Omega)$.*

Corollary 2.3.

1. $\Omega \in Q(\mu, HL^1)$ *if and only if* $U^\Omega = U^\mu$ *and* $\nabla U^\Omega = \nabla U^\mu$ *on* Ω^c.
2. $\Omega \in Q(\mu, AL^1)$ *if and only if* $A^\Omega = A^\mu$ *on* Ω^c.

For the proof see [4]. (Notice that if μ is real then $A^\mu = A^\Omega$ is equivalent to $\nabla U^\mu = \nabla U^\Omega$.)

3. Deformation of quadrature domains

Here we will study quadrature domains in $\mathbb{R}^2$ with respect to the class AL^1. We start by introducing the following partial order on the subsets of $\mathbb{R}^N$: $\Omega_1 \prec \Omega_2$ means that $U^{\Omega_1} \geq U^{\Omega_2}$ in $\mathbb{R}^N$. Before we state and prove the main theorem of this section we need a few preliminaries.

Lemma 3.1. *For* $\varepsilon > 0$, *let* $B_\varepsilon = B_\varepsilon(0)$ *and let* $l_\varepsilon(x) := \frac{1}{m(B_\varepsilon)}\chi_{B_\varepsilon}(x)$.

1. *A l.s.c. function* u *on* $\mathbb{R}^N$ *is superharmonic if and only if* $u * l_\varepsilon \leq u \ \forall \varepsilon > 0$. *Furthermore* $u * h_\varepsilon \leq u$ *for any superharmonic function* u *on* $\mathbb{R}^N$.
2. *If* $u \in \mathcal{D}'(\mathbb{R}^N, \mathbb{R})$ *and* $u \leq w$ *where* w *is superharmonic on* $\mathbb{R}^N$, *then* $u * l_\varepsilon \leq w$ *and* $u * h_\varepsilon \leq w$.

Proof. 1. Since $u * l_\varepsilon(x) = \int u(y) l_\varepsilon(x-y) dm(y) = \frac{1}{m(B_\varepsilon)} \int_{B_\varepsilon(x)} u(y) dm(y)$ this is simply a reformulation of the super-mean-value inequality. The second statement is a direct consequence of the inequality $u(x) \leq \frac{1}{S(\partial B_\varepsilon)} \int_{\partial B_\varepsilon(x)} u dS$ where S is surface-area, and the argument may be found in most books on potential theory. It is actually true that u l.s.c. and $u * h_\varepsilon \leq u$ implies that u is superharmonic, but we will not need it.

2. Since $u \leq w \Rightarrow u * l_\varepsilon \leq w * l_\varepsilon$ this follows from (1). The proof for h_ε is identical. □

The following is our main result in this section.

Theorem 3.2. *If* $\mu \subset \mathcal{E}'(\mathbb{R}^2, \mathbb{C})$ *and* $\Omega_0, \Omega_1 \in Q(\mu, AL^1)$, *then there is a one-parameter family* $\Omega(t) \in Q(\mu, AL^1)$ *s.t.* $\Omega(0) = \Omega_0$, $\Omega(1) = \Omega_1$ *and* $U^{\Omega(t)}$ *depends continuously on* t. *Furthermore there is a least upper bound* $\Omega_0 \vee \Omega_1$ *in* $Q(\mu, AL^1)$ *of* Ω_0 *and* Ω_1 *w.r.t.* $\prec$. *In other words we have* $\Omega_j \prec \Omega_0 \vee \Omega_1$ $(j = 0, 1)$ *and if* $D \in Q(\mu, AL^1)$ *with* $\Omega_j \prec D$ $(j = 0, 1)$ *then* $\Omega_0 \vee \Omega_1 \prec D$.

Proof. For $(t_0, t_1) \in \mathbb{R}^2$ define

$$F(t_0, t_1) := \{u \in \mathcal{D}'(\mathbb{R}^2, \mathbb{R}) : u \leq \min\{U^{\Omega_0} + t_0, U^{\Omega_1} + t_1\}, -\Delta u \leq 1\}.$$

To keep the notation less cumbersome we omit (t_0, t_1) in the first part of the proof since it is considered as fixed for the moment. The function

$$w := \min\{U^{\Omega_0} + t_0, U^{\Omega_1} + t_1\}$$

is superharmonic and continuous. If $t_0 \geq t_1$, then $w = U^{\Omega_1} + t_1$ far away since $U^{\Omega_0} = U^{\Omega_1}$ there, and the other way around if $t_0 \leq t_1$. In either case

$$w = U^{\nu} + \min\{t_0, t_1\}$$

where $\nu = -\Delta w$ is a positive Radon measure with compact support.

First of all F is nonempty, because if we mollify w with h_ε, then $w_\varepsilon = w * h_\varepsilon \leq w$ (and actually $w_\varepsilon = w$ far away) and $-\Delta w_\varepsilon \leq 1$ if ε is large enough. So $w_\varepsilon \in F$ for large ε.

We now prove that F has a largest element by a Perron-type of argument (see [1] [3] or [7] for instance for the facts needed). By adding $|x|^2/4$ to each element of F they become subharmonic, so $F \subset L^1_{loc}(\mathbb{R}^2)$ and each member has a unique representation as an upper semi-continuous (u.s.c.) function. Also, if $u_1, u_2 \in F$, then so does $\sup\{u_1, u_2\}$, therefore we may put

$$v := \sup\{\phi : \phi \in F\}$$

and

$$V := \inf\{f : f \text{ is u.s.c. and } f \geq v\}.$$

Then $V \in F$, $V = v$ a.e. and hence is the largest element of F.

We now start by proving that V is in-fact superharmonic. By Lemma 3.1 we have that $V_\varepsilon := V * l_\varepsilon \in F$ for each $\varepsilon > 0$, because $V \leq w$ implies $V_\varepsilon \leq w * l_\varepsilon \leq w$ and $-\Delta V_\varepsilon = (-\Delta V) * l_\varepsilon \leq 1$. (Note that $V_\varepsilon \in C(\mathbb{R}^2)$). Since $V \in L^1_{loc}(\mathbb{R}^2)$, $V_\varepsilon \to V$ in L^1_{loc} and also point-wise a.e. by Lebesgue's differentiation theorem. So if we look at $u := \sup_{\varepsilon > 0}\{V_\varepsilon\}$ point-wise, then u is l.s.c. by construction ($\{x \in \mathbb{R}^2 : u(x) > a\} = \cup_{\varepsilon > 0}\{x \in \mathbb{R}^2 : V_\varepsilon(x) > a\}$ is open $\forall a \in \mathbb{R}$).

We also have $u = V$ a.e. on $\mathbb{R}^2$, hence $u * l_\varepsilon = V * l_\varepsilon = V_\varepsilon \leq u$ for every $\varepsilon > 0$. But this implies that u is superharmonic and belongs to F. That is $0 \leq -\Delta u \leq 1$, and the first inequality implies that $-\Delta u$ may be represented as a positive Radon measure, whereas the second implies that this measure is absolutely continuous with respect to m. Also note that $-\Delta u = 0$ far away. So if we put $f = -\Delta u$ as a function in $L^1(\mathbb{R}^2) \cap L^\infty(\mathbb{R}^2)$ we may choose a representative that is identically zero outside some compact set and always fulfills $0 \leq f \leq 1$. All this implies that u is in-fact continuous and therefore we have $u = V$ everywhere by the choice of V (because $u + |x|^2/4 = V + |x|^2/4$ a.e. and both are subharmonic). Now $V = U^f + \min\{t_0, t_1\}$ on $\mathbb{R}^2$, so $V \in W^{2,p}_{loc}(\mathbb{R}^2)$ for $1 \leq p \leq \infty$.

We now introduce the sets

$$\begin{aligned}
&\omega = \omega(t_0, t_1) := \{x \in \mathbb{R}^2 : V(x) < w(x)\},\\
&\Omega = \Omega(t_0, t_1) := \mathbb{R}^2 \setminus \operatorname{supp}(1 + \Delta V),\\
&D_0 := \{x \in \mathbb{R}^2 : U^{\Omega_0} + t_0 < U^{\Omega_1} + t_1\},\\
&D_1 := \{x \in \mathbb{R}^2 : U^{\Omega_0} + t_0 > U^{\Omega_1} + t_1\},\\
&S := \{x \in \mathbb{R}^2 : U^{\Omega_0} + t_0 = U^{\Omega_1} + t_1\}.
\end{aligned}$$

Note that Ω, ω, D_0 and D_1 are open and S is closed since all functions involved are continuous. Also note that $\omega \subset \Omega$, because if $\omega \setminus \Omega$ contained an open ball B, then on this ball neither the bound $V = w$ nor the bound $-\Delta V = 1$ are attained, which obviously contradicts the maximality of V. We next aim at proving that

$$V = U^{\Omega} + \min\{t_0, t_1\}.$$

First of all,

$$-\Delta V = 1 \text{ in } \Omega$$

by definition. We also have

$$V = U^{\Omega_j} + t_j \text{ on } \Omega^c \cap (D_j \cup S),$$

and therefore, since $V, U^{\Omega_j} \in W^{2,p}_{loc}$, we get

$$\nabla V = \nabla U^{\Omega_j} \text{ a.e. on } \Omega^c \cap (D_j \cup S)$$

$$-\Delta V = -\Delta U^{\Omega_j} = \chi_{\Omega_j} \text{ a.e. on } \Omega^c \cap (D_j \cup S).$$

Also $\Omega_j \cap D_j \subset \Omega$ $(j = 0, 1)$ because $-\Delta w = 1$ here.

This implies that $\Omega^c \cap D_j \cap \Omega_j = \emptyset$. On $\Omega_0 \cap \Omega_1$ we have $-\Delta w \geq 1$, because on $\Omega_0 \cap \Omega_1$,

$$\min\{U^{\Omega_0} + t_0, U^{\Omega_1} + t_1\} = U^{\Omega_0 \cap \Omega_1} + \min\{U^{\Omega_0 \setminus \Omega_1} + t_0, U^{\Omega_1 \setminus \Omega_0} + t_1\}.$$

Therefore $\Omega_0 \cap \Omega_1 \subset \Omega$, hence $\Omega^c \cap S \cap \Omega_0 \cap \Omega_1 = \emptyset$.

Summing up

$$\begin{aligned}\Omega^c &= (\Omega^c \cap D_0) \cup (\Omega^c \cap D_1) \cup (\Omega^c \cap S)\\ &= (\Omega^c \cap D_0 \cap \Omega_0^c) \cup (\Omega^c \cap D_1 \cap \Omega_1^c) \cup (\Omega^c \cap S \cap (\Omega_0^c \cup \Omega_1^c)).\end{aligned}$$

Therefore

$$-\Delta V = 0 \text{ a.e. on } \Omega^c.$$

All together this implies that $-\Delta V = \chi_\Omega$, and hence $V = U^\Omega + \min\{t_0, t_1\}$ because this holds on the unbounded component.
Also, since

$$A^{\Omega_j} = A^\mu \text{ on } \Omega_j^c$$

we have

$$A^{\Omega} = A^\mu \text{ on } \Omega^c.$$

So by Corollary 2.3

$$\Omega \in Q(\mu, AL^1).$$

The least upper bound is easily seen to be $\Omega(0,0)$ in the above construction. As for the one-parameter family we note that two variables were introduced mainly for symmetry reasons, since obviously

$$\Omega(t_0 + \alpha, t_1 + \alpha) = \Omega(t_0, t_1) \; \forall \alpha \in \mathbb{R},$$

one dimension collapses automatically. If we now look at $\Omega(0,t)$, then for large t $\Omega(0,t) = \Omega_0$ and for small t $\Omega(0,t) = \Omega_1$ of-course. By translating and rescaling the parameter t this gives us our one-parameter family.

The only thing that remains is the continuity of $U^{\Omega(0,t)}$ with respect to t. We start by noting that

$$\bigcup_{(t_0,t_1)\in\mathbb{R}^2} \Omega(t_0,t_1) \subset\subset \mathbb{R}^2.$$

This is proved by choosing a ball B so large that

$$\Omega_0 \cup \Omega_1 \subset\subset B$$

and

$$\gamma := \mathrm{Bal}(\mu, \partial B) = \mathrm{Bal}(\chi_{\Omega_j}, \partial B)$$

may be mollified to a measure γ_ε s.t. $\gamma_\varepsilon \leq 1$ and $\mathrm{supp}(\gamma_\varepsilon) \cap \overline{(\Omega_0 \cup \Omega_1)} = \emptyset$. This implies that $U^{\gamma_\varepsilon} + \min\{t_0, t_1\} \in F(t_0, t_1)$, and therefore that

$$\Omega(t_0, t_1) \subset B \cup \mathrm{supp}(\gamma_\varepsilon),$$

independent of (t_0, t_1). We only prove that $U^{\Omega(0,t+\epsilon)} \searrow U^{\Omega(0,t)}$ as $\epsilon \searrow 0$ since the other cases are similar to handle. So given $\epsilon > 0$ then obviously

$$\min\{U^{\Omega_0}, U^{\Omega_1} + t + \epsilon\} \geq \min\{U^{\Omega_0}, U^{\Omega_1} + t\}.$$

Hence if we let V_ϵ denote the maximum of $F(0, t+\epsilon)$ (not to be confused with the mollification V_ε of V) and $V = V_0$, then $V_\epsilon > V$ and V_ϵ is decreasing with ϵ. So we may define $V' := \lim_{\epsilon\to 0} V_\epsilon$. Now since

$$V_\epsilon \leq \min\{U^{\Omega_0}, U^{\Omega_1} + t + \epsilon\} \text{ for every } \epsilon$$

we have

$$V' \leq \min\{U^{\Omega_0}, U^{\Omega_1} + t\},$$

and also $-\Delta V' \leq 1$, so $V' \in F(0,t)$ and therefore we have $V' \leq V \leq V'$, i.e., $V' = V$. But this immediately implies that $U^{\Omega(0,t+\epsilon)} \to U^{\Omega(0,t)}$ as $\epsilon \to 0$, which finishes the proof. □

Remark. In the construction above we used μ very little, and in-fact if there are two different distributions μ_1, μ_2 such that $\Omega_0, \Omega_1 \in Q(\mu_i, AL^1)$ for $i = 1, 2$ then the family $\Omega(t) \in Q(\mu_i, AL^1)$ for $i = 1, 2$.

Also, the first part of the proof is not dimension-dependent, and it gives an alternative to the obstacle-problem approach used in [4]. Except for this difference the proof above is essentially based on the one for real measures in [4].

Examples. There are simple examples of quadrature domains

$$\Omega_0, \Omega_1 \in Q(\mu, AL^1)$$

where one is connected and the other one is not.
We may for instance take $\mu = \pi\delta_0 + 2S\lfloor_{\partial B_2(0)}$ in $\mathbb{R}^2$, where S is arc-length. Now we may take

$$\Omega_0 = B_3(0)$$

and

$$\Omega_1 = B_1(0) \cup (B_{\sqrt{41}/2}(0) \setminus \overline{B_{3/2}(0)}).$$

This leads to a natural question that we do not have an answer for:

If Ω_0 and Ω_1 are connected must $\Omega(t_0, t_1)$ constructed as above consist of only connected domains?

If for instance $\Omega_0 \cap \Omega_1$ is connected this is trivially so, but in general we do not know. It is mainly for this reason that we do not require quadrature domains to be connected.

4. Complex to real identities

In this section it will be natural to use complex notation, and we will write $z = x + iy$ for a point in the complex plane $\mathbb{C}$. The following elementary lemma which is an immediate consequence of the Cauchy-Riemann equations will also be used.

Lemma 4.1. *Let $u + iv$ be analytic in Ω, where Ω is a domain (i.e., open and connected) in $\mathbb{C}$. Let $a \in \Omega$ be fixed. In that case we have for any $z \in \Omega$*

$$(*)\ v(z) = v(a) + \int_\gamma (-\frac{\partial u}{\partial y}dx + \frac{\partial u}{\partial x}dy)$$

for any piecewise C^1 Jordan arc γ in Ω between a and z.

Theorem 4.2. *Let $\Omega \subset \mathbb{C}$ be open. Suppose $\mu = \alpha + i\beta$ (α,β real Radon measures) has compact support in Ω.*
Then there is a real Radon measure η with compact support in Ω such that

$$\int f d\mu = \int f d\eta \qquad \forall f \in A(\Omega)$$

if and only if $\beta(U) = 0$ for each component U of Ω.

Proof. We start by assuming that Ω is connected. Fix $a \in \Omega$ and for each $z \in \operatorname{supp}(\mu)$ a piecewise C^1 Jordan arc γ_z between a and z in such a way that all γ_z are contained in a fixed compact $K \subset \Omega$ and the lengths of all γ_z are bounded by a constant $l < \infty$. (This is possible since Ω is connected and $\operatorname{supp}(\mu)$ is compact.)

Now define $T : C_c^\infty(\Omega, \mathbb{R}) \to \mathbb{R}$ by

$$\langle T, \phi \rangle := \int \phi d\alpha - \int \int_{\gamma_z} (-\frac{\partial \phi}{\partial y}dx + \frac{\partial \phi}{\partial x}dy)d\beta.$$

Then

$$|\langle T, \phi \rangle| \leq \sup_K |\phi| ||\alpha|| + (\sup_K |\frac{\partial \phi}{\partial y}| + \sup_K |\frac{\partial \phi}{\partial x}|) l ||\beta||.$$

T is obviously linear, so by the above $T \in \mathcal{E}'(\Omega, \mathbb{R})$, and it may now be mollified to a measure η with compact support in Ω. Let $f = u + iv \in A(\Omega)$. Then

$$\begin{aligned}\int f d\eta &= \int u d\eta + i \int v d\eta = \langle T, u\rangle + i\langle T, v\rangle \\ &= \int u d\alpha - \int\int_{\gamma_z} (-\frac{\partial u}{\partial y} dx + \frac{\partial u}{\partial x} dy) d\beta + i\int v d\alpha - i\int\int_{\gamma_z} (-\frac{\partial v}{\partial y} dx + \frac{\partial v}{\partial x} dy) d\beta \\ &= \int u d\alpha - \int (v(z) - v(a)) d\beta(z) + i\int v d\alpha - i\int (u(a) - u(z)) d\beta(z) \\ &= \int u d\alpha + i\int u d\beta + i\int v d\alpha - \int v d\beta = \int f d\mu.\end{aligned}$$

Here we used that $\int v(a) d\beta(z) = v(a) \int d\beta = 0$ and similarly for u.

If Ω is not connected we first note that by compactness the support of μ must be contained in a finite number of components of Ω. Let these be denoted $U_1, \ldots, U_n$. By the above there are for each $j = 1, \ldots, n$ a real measure η_j with compact support in U_j such that

$$\int f d\mu \lfloor U_j = \int f d\eta_j \qquad \forall f \in A(U_j).$$

If we put $\eta = \sum_{j=1}^n \eta_j$ we get the required measure. In the other direction we get

$$\int \chi_U d\mu = \int \chi_U d\alpha + i \int \chi_U d\beta = \int \chi_U d\eta$$

which obviously implies $\int d\beta \lfloor U = 0$. □

Remark. What we actually proved above was the following. If $V \subset A(\Omega)$ and μ is as in the theorem, then the condition that $\beta(U) = 0$ for each component U of Ω is sufficient to guarantee the existence of a real measure η with compact support in Ω such that

$$\int f d\mu = \int f d\eta \qquad \forall f \in V$$

and if $\chi_U \in V$ for each component it is also necessary.

Corollary 4.3. *Suppose $\mu \in \mathcal{E}'(\mathbb{R}^2, \mathbb{C})$ and $\Omega \in Q(\mu, AL^1)$. Then there is a real Radon measure η with compact support in Ω such that $\Omega \in Q(\eta, AL^1)$.*

Proof. Since we may mollify μ to a measure we may without loss of generality assume that it already is a complex Radon measure $\mu = \alpha + i\beta$ with α,β real Radon measures with compact support in Ω. By definition we have

$$\int_\Omega \chi_U dm = \int \chi_U d\mu = \int d\alpha \lfloor U + i \int d\beta \lfloor U$$

for every component U of Ω. Hence $\int d\beta \lfloor U = 0$, and we may apply Theorem 4.2. □

In connection with the deformation of quadrature domains it would be good to have conditions on when it is possible to change a complex measure to the same real measure for two different domains simultaneously. We begin with an approximation lemma that is a simple consequence of Runge's theorem (Bers theorem for the integrable case). For simplicity we consider only bounded domains.

Lemma 4.4. *Suppose $\Omega_1, \Omega_2 \subset\subset \mathbb{C}$ are open and $f \in A(\Omega_1 \cap \Omega_2)$. For any compact set $K \subset \Omega_1 \cap \Omega_2$ and $\varepsilon > 0$ there are rational functions $r_i \in A(\Omega_i)$ $(i = 1, 2)$ such that $|f - (r_1 + r_2)| < \varepsilon$ on K. If furthermore $\Omega_i = \operatorname{int}(\overline{\Omega}_i)$ $(i = 1, 2)$ we may choose each $r_i \in C(\overline{\Omega}_i)$. Finally if $f \in AL^1(\Omega_1 \cap \Omega_2)$ then we may take $r_i \in AL^1(\Omega_i)$.*

Proof. Let $\{U_j\}_{j=1}^{\infty}$ denote the components of K^c. In each U_j we may assume that there is a point a_j in $(\Omega_1 \cap \Omega_2)^c$, because otherwise we may add U_j to K. This a_j then also belongs to some Ω_i^c trivially. By Runge's theorem we have

$$|f(z) - \sum_{k=1}^{m} \frac{p_k(z)}{(z - a_{j_k})^{n_k}}| < \varepsilon \qquad \text{on } K$$

for some choice of the polynomials p_k and constants m, j_k, n_k. Defining

$$A_1 := \{k \in \{1, \dots, m\} : a_{j_k} \in \Omega_1^c\}$$

$$A_2 := \{1, \dots, m\} \setminus A_1,$$

and

$$r_i(z) := \sum_{k \in A_i} \frac{p_k}{(z - a_{j_k})^{n_k}} \qquad (i = 1, 2),$$

then the first part easily follows. As for the second part we only need to note that with the notation as above, if $a_j \in \Omega_i^c$, then there is some $a_j' \in (\overline{\Omega}_i)^c \cap U_j$ by the topological assumption. So we may assume that in the r_i's $a_{j_k} \in (\overline{\Omega}_i)^c$ when $k \in A_i$ $(i = 1, 2)$. Finally, if $f \in AL^1(\Omega_1 \cap \Omega_2)$, if we refer to Bers theorem (Theorem 2.2 part 1) then we see that we may take p_k constant, and $n_k = 1$ for every k above, and this finishes the proof. □

Theorem 4.5. *Suppose $\Omega_1, \Omega_2 \subset\subset \mathbb{C}$ and $\mu = \alpha + i\beta$ is a complex Radon measure with compact support in $\Omega_1 \cap \Omega_2$. Then a necessary and sufficient condition for the existence of a real Radon measure η with compact support in $\Omega_1 \cap \Omega_2$ such that*

$$(1) \quad \int f d\mu = \int f d\eta \qquad \forall f \in A(\Omega_1) \cup A(\Omega_2)$$

is that $\beta(U) = 0$ for each component U of $\Omega_1 \cap \Omega_2$. In that case we also have automatically

$$(2) \quad \int f d\mu = \int f d\eta \qquad \forall f \in A(\Omega_1 \cap \Omega_2).$$

Proof. That $\beta(U) = 0$ for each component U of $\Omega_1 \cap \Omega_2$ is sufficient is obvious by Theorem 4.2. We start by proving that (1) $\Rightarrow$ (2) (which holds regardless of whether $\beta(U) = 0$ for each component U of $\Omega_1 \cap \Omega_2$ or not). Let $f \in A(\Omega_1 \cap \Omega_2)$ and $\varepsilon > 0$. Then we may by Lemma 4.4 choose $r_i \in A(\Omega_i)$ $(i = 1, 2)$ with

$$|f - (r_1 + r_2)| < \varepsilon \qquad \text{on } \operatorname{supp}(\mu) \cup \operatorname{supp}(\eta),$$

so

$$|\int f d\mu - \int f d\eta| = |\int f d\mu - \int (r_1 + r_2) d\mu + \int (r_1 + r_2) d\eta - \int f d\eta|$$
$$\leq |\int (f - (r_1 + r_2)) d\mu| + |\int (f - (r_1 + r_2)) d\eta| \leq \varepsilon(||\mu|| + ||\eta||).$$

Hence

$$\int f d\mu = \int f d\eta.$$

Now if U is a component of $\Omega_1 \cap \Omega_2$, then $\chi_U \in A(\Omega_1 \cap \Omega_2)$, and the proof is done. □

Theorem 4.6. *If $\Omega_1, \Omega_2 \in Q(\mu, AL^1)$ where $\mu = \alpha + i\beta$ is a complex Radon measure with compact support, then a necessary and sufficient condition for the existence of some real Radon measure η with compact support such that $\Omega_1, \Omega_2 \in Q(\eta, AL^1)$ is that $\beta(U) = 0$ for each component U of $\Omega_1 \cap \Omega_2$.*

Proof. The proof is equivalent to the previous one of theorem 4.5. The only thing we need to note is that we may take $r_i \in AL^1(\Omega_i)$ by Lemma 4.4 in the proof above. □

We remark that if either $\Omega_1 \cap \Omega_2$ is connected or Ω_1 and Ω_2 differ only on a set of Lebesgue measure zero we may replace μ with the same real measure η. (In the first case this follows immediately from theorem 4.2 and the following remark and in the second by Corollary 4.3 because $\Omega_1 \cap \Omega_2 \in Q(\mu, AL^1)$).

Theorem 4.7. *Suppose $\mu = \alpha + i\beta$ is a complex Radon measure with compact support in $\mathbb{C}$ and $\Omega_1, \Omega_2 \in Q(\mu, AL^1)$. Also assume that $\Omega_1 \cap \Omega_2$ consists of a finite number of components $U_1, \ldots, U_n$ where $\overline{U}_j \cap \overline{U}_k = \emptyset$ if $j \neq k$ and each U_j fulfills that $\partial D \cap \partial U_j \neq \emptyset$ where D is the unbounded component of $(\overline{\Omega_1 \cap \Omega_2})^c$. In that case we have $\beta(U_j) = 0$ for $j = 1, \ldots, n$.*

Proof. By the topological assumptions made we have that if $j \neq k$ then $\overline{U}_k$ lies in the unbounded component of $\mathbb{C} \setminus \overline{U}_j$. Let us now define V_j to be the union of U_j and all the bounded components of $\mathbb{C} \setminus U_j$. Then choose open $W_j's$ with $V_j \subset\subset W_j$ and $W_k \cap W_j = \emptyset$ if $j \neq k$. It is easy to see that it is enough to prove that $\beta(V_j) = 0$ for each j. Let

$$V = \begin{cases} U^{\Omega_1} & \text{on } D \setminus \Omega_1 \\ U^{\Omega_2} & \text{on } D \setminus \Omega_2 \\ 0 & \text{otherwise} \end{cases}$$
$$\nu := -\Delta V$$

so that

$$\nu \in \mathcal{E}'(\mathbb{C}, \mathbb{R})$$

with support on

$$\bigcup_{j=1}^{n} \partial V_j.$$

It follows that $A^\nu = A^\mu$ on D, and hence if we put

$$f_j := \begin{cases} 1 & \text{on } W_j \\ 0 & \text{otherwise} \end{cases}$$

we get

$$\langle \alpha + i\beta, f_j \rangle = \alpha(V_j) + i\beta(V_j) = \langle \nu, f_j \rangle \in \mathbb{R},$$

so

$$\beta(V_j) = 0. \qquad \square$$

5. Real to positive identities

In this section we look at similar questions as the one in the previous section, but instead we are mainly interested in changing a real measure to a positive one, and for test classes of harmonic functions instead of analytic ones. This question is natural to study in any number of dimensions, and that is what we intend to do. The question is also natural to consider for quadrature domains for analytic test functions, and we end by proving that we may find another real measure making these into quadrature domains for harmonic test functions, reducing the problem to the above one. In this section we will assume that Ω is connected because it is no essential loss, and makes some statements less cumbersome.

Theorem 5.1. *Suppose $\Omega \subset \mathbb{R}^N$ is a Greenian open domain and μ is a real (possibly signed) Radon measure with compact support in Ω, then the following are equivalent.*

(1) *$G^\mu > 0$ in $\Omega \setminus K$ for some compact $K \subset \Omega$, where G is the Green's function of Ω.*

(2) *There is a positive Radon measure $\eta \neq 0$ with compact support in Ω such that*

$$\int h d\mu = \int h d\eta \qquad \forall h \in H(\Omega).$$

(3) *$\int d\mu > 0$ and there is a compact $K \subset \Omega$ such that*

$$\int h d\mu \leq \Big(\int d\mu\Big) \sup_K |h| \qquad \forall h \in H(\Omega).$$

(4)
$$\int h d\mu > 0 \qquad \forall h \in HP(\Omega) \setminus \{0\}$$

Proof. (3) $\Rightarrow$ (2) because if we consider $H(\Omega)$ as a subspace of $C(\Omega)$, then the linear functional $\Lambda(h) := \frac{1}{\int d\mu} \int h d\mu \leq \sup_K |h|$ may be extended to the whole

$C(\Omega)$ s.t. $\Lambda(f) \leq \sup_K |f|\ \forall f \in C(\Omega)$ by the Hahn-Banach theorem. But for $f \in C(\Omega)$ and $f \geq 0$ on K we then have

$$\sup_K f - \Lambda(f) = \Lambda(\sup_K f - f) \leq \sup_K(\sup_K f - f) \leq \sup_K f,$$

so $\Lambda(f) \geq 0$ which proves (3) $\Rightarrow$ (2) (by Riesz's theorem).

(2) $\Rightarrow$ (1) Suppose that (2) holds but not (1). Then there is a sequence

$$\{x_n\} \subset \{x \in \Omega : G^\mu(x) \leq 0\}$$

which has no limit point in Ω. We now let

$$M(x,y) := \frac{G(x,y)}{G(a,y)} \qquad (a \in \Omega \text{ fixed }, y \in \Omega).$$

Then

$$\int M(t,x_n)d\mu(t) = \frac{1}{G(a,x_n)}\int G(t,x_n)d\mu(t) \leq 0.$$

It is well known by Martin boundary theory (or seen directly from Harnack's convergence theorem and a diagonal argument) that some subsequence of $\{M(t,x_n)\}$ converges u.c. to a function $h \in HP(\Omega)$ with $h(a) = 1$. But then we have

$$0 \geq \int h d\mu = \int h d\eta > 0$$

which gives a contradiction.
(1) $\Rightarrow$ (4). We choose a compact K such that $G^\mu > 0$ on $\Omega \setminus K$ and $\text{supp}(\mu) \subset \text{int}(K)$ (this may even be needed to guarantee that G^μ is well defined on $\Omega \setminus K$). Now let $K \subset \Omega_1 \subset\subset \Omega$ be open, then it is well known that any $h \in HP(\Omega)$ may be represented in Ω_1 as a potential G^ν for some positive measure with support on $\partial\Omega_1$. For $h \not\equiv 0$ we get

$$\int h d\mu = \int G^\nu d\mu = \int G^\mu d\nu > 0.$$

That is $\int h d\mu > 0\ \forall h \in HP(\Omega), h \not\equiv 0$.

(4) $\Rightarrow$ (3) Suppose that(4) holds but not (3). Then if we let

$$\Omega_n \nearrow \Omega \; (\Omega_1 \subset\subset \Omega_2 \subset\subset \cdots \quad \text{and} \quad \Omega = \cup_{n=1}^\infty \Omega_n)$$

there is by assumption for each n a function $k_n \in H(\Omega)$ such that $\int k_n d\mu > (\int d\mu) \sup_{\Omega_n} |k_n|$. If we let

$$h_n := (\sup_{\Omega_n} |k_n| - k_n)/(\sup_{\Omega_n} |k_n| - k_n(a))$$

where a is a fixed point in Ω_1, then $h_n > 0$ on Ω_n, $h_n(a) = 1$ and $\int h_n d\mu < 0$ for every n. Now applying Harnack's convergence theorem together with Cantor's diagonal argument gives us a subsequence that converges uniformly on compact subsets of Ω to some $h \in HP(\Omega)$ with $h(a) = 1$. But then (if we still denote this subsequence by h_n) we get

$$\int h_n d\mu \to \int h d\mu \text{ so } \int h d\mu \leq 0$$

which gives a contradiction. □

Remark. In analogy with Theorem 4.2 we may note that we actually have proved the following: If $V \subset H(\Omega)$ and μ is as in the theorem, then that $G^\mu \geq 0$ on $\Omega \setminus K$ for some compact $K \subset \Omega$ is a sufficient condition for the existence of a positive η such that

$$\int h d\mu = \int h d\eta \qquad \forall h \in V.$$

If furthermore $HP(\Omega) \subset V$ it is also necessary.

Corollary 5.2. *If μ is a real (possibly signed) Radon measure with compact support in $\mathbb{R}^N$, and $\Omega \in Q(\mu, HL^1)$ (and Ω connected). Then a sufficient condition for the existence of a positive Radon measure $\eta \neq 0$ with compact support in $\mathbb{R}^N$ such that $\Omega \in Q(\eta, HL^1)$ is that $G^\mu > 0$ on $\Omega \setminus K$ for some compact $K \subset \Omega$. (Here G is the Green's function of Ω). If $\Omega = \operatorname{int}(\overline{\Omega})$ it is also necessary.*

Proof. The first part is obvious by Theorem 5.1. As for the second part let us assume that η exists. We may now choose a compact $K \subset \Omega$ such that $\operatorname{supp}(\mu) \cup \operatorname{supp}(\eta) \subset K$ and each component of K^c intersects Ω^c. If now $\Omega = \operatorname{int}(\overline{\Omega})$ it follows that if O is a component of K^c then there is some open ball B in $O \cap \Omega^c$. Since $U^\mu = U^\eta$ on B it follows by harmonic continuation that this holds on all of O. But this implies that $G^\mu = G^\eta > 0$ on $O \cap \Omega$, hence $G^\mu > 0$ on $\Omega \setminus K$. □

We may also note that since both $U^\Omega = U^\mu$ and $\nabla U^\Omega = \nabla U^\mu$ on $\partial\Omega$ for $\Omega \in Q(\mu, HL^1)$ it seems highly likely that this implies that $G^\mu \geq 0$ in a neighborhood of $\partial\Omega$. One difference between this condition and the ones given in [6] is that although this is a stronger result to guarantee the existence of a positive measure, [6] gives conditions to guarantee $HP(\Omega) \subset L^1(\Omega)$ for quadrature domains, and the above does not. In two dimensions it is proved in [6] that η always exists by the characterization of quadrature domains known only for $N = 2$ (see [8] and [5]).

A natural question to ask is now if the same is true for the class AL^1 in two dimensions ([6] deals with HL^1). We end this article with a proof of this fact using the powerful characterization of quadrature domains in two dimensions due to Sakai.

Theorem 5.3. *Let $\mu \in \mathcal{E}'(\mathbb{C}, \mathbb{C})$, $\Omega \in Q(\mu, AL^1)$. Then there is a $\mu' \in \mathcal{E}'(\mathbb{C}, \mathbb{R})$ such that $\Omega \in Q(\mu', HL^1)$.*

Proof. By Corollary 4.3 we may assume without loss of generality that $\mu \in \mathcal{E}'(\mathbb{C}, \mathbb{R})$. It is known (see [9] Theorem 1.7 and [6]) that Ω contains a finitely connected quadrature domain $\Omega' \in Q(\mu, AL^1)$. Call the components of $\mathbb{C} \setminus \Omega'$ $S_1, \ldots, S_n$. We start by proving that $U^\Omega - U^\mu$ is constant on each component S_j.

For this we must refer to [9] again for the result that $\partial\Omega'$ has finite length and is in-fact the subset of a real-analytic variety, which contains at most a finite set of singularities. The only result we need from this is that it implies that each pair of points $z_1, z_2 \in S_j$ may be connected by a piecewise C^1 Jordan arc $\gamma \subset S_j$. Since by assumption

$$\nabla U^{\Omega} = \nabla U^{\mu} \text{ on } S_j$$

and $U^{\Omega} - U^{\mu}$ is C^1 in some neighborhood of $\partial\Omega'$ it follows by a simple line integral that

$$U^{\Omega}(z_1) - U^{\mu}(z_1) = U^{\Omega}(z_2) - U^{\mu}(z_2)$$

which proves the statement that $U^{\Omega} - U^{\mu}$ is constant on S_j. We let c_j denote the constant value of $U^{\Omega} - U^{\mu}$ on S_j, then we choose a set $\Omega'' \subset\subset \Omega'$ such that $\mathbb{C} \setminus \Omega''$ consists of n components $W_1, \ldots, W_n$ such that each $S_j \subset W_j$. To finish the proof we define

$$T := \sum_{j=1}^{n} c_j \Delta \chi_{W_j}.$$

Then $T \in \mathcal{E}'(\Omega', \mathbb{R})$ and

$$U^T = \begin{cases} -c_j & \text{on } W_j \\ 0 & \text{in } \Omega'' \end{cases}$$

(We may note that $c_j = 0$ for the unbounded component S_j, but we do not need it here). Now we just need to put

$$\mu' = \mu - T.$$

Then it follows that

$$U^{\mu'} = U^{\Omega'} = U^{\Omega} \text{ on } \Omega'^c$$

and

$$\nabla U^{\mu'} = \nabla U^{\mu} = \nabla U^{\Omega'} = \nabla U^{\Omega} \text{ on } \Omega'^c$$

and the proof is done. □

Remark 1. In connection with the earlier results we may note that here we cannot expect to get μ' as in the theorem above for two domains $\Omega_1, \Omega_2 \in Q(\mu, AL^1)$ simultaneously such that $\Omega_1, \Omega_2 \in Q(\mu', HL^1)$. This is because this would imply that $U^{\Omega_1} = U^{\Omega_2}$ on $(\Omega_1 \cup \Omega_2)^c$. But this is not true even for two concentric annuli Ω_1, Ω_2 with $\Omega_1 \cap \Omega_2 \neq \emptyset$ and $m(\Omega_1) = m(\Omega_2)$ which are in $Q(\mu, AL^1)$ for some measure μ (μ can be chosen as some constant times arc-length on some common circle in $\Omega_1 \cap \Omega_2$). Then in the bounded component of $(\Omega_1 \cup \Omega_2)^c$ both U^{Ω_1} and U^{Ω_2} are constant but different unless $\Omega_1 = \Omega_2$.

Remark 2. By the results of [6] we may even take $\mu' \geq 0$ in the theorem above.

References

[1] D.H. Armitage, S.J. Gardiner, *Classical Potential Theory,* Springer-Verlag (2001)

[2] L. Bers, *An Approximation theorem,* J. Analyse Math. **14** (1965).

[3] J.L. Doob, *Classical Potential Theory and its Probabilistic Counterpart,* Springer-Verlag (1983).

[4] B. Gustafsson, *On Quadrature Domains and an Inverse Problem in Potential Theory,* J. Analyse Math. **55** (1990).

[5] B. Gustafsson, M. Putinar, *An Exponential Transform and Regularity of Free Boundaries in Two Dimensions,* Scoula Normale Superiore, Pisa, 1997.

[6] B. Gustafsson, M. Sakai, H. Shapiro, *On Domains in which Harmonic Functions Satisfy Generalized Mean Value Inequalities,* Potential Analysis 7, Kluwer Academic Publishers (1997).

[7] L.L. Helms, *Introduction to Potential Theory,* Wiley, 1969.

[8] M. Sakai, *Quadrature Domains,* Lecture Notes in Mathematics, Vol. 934, Springer-Verlag, 1982.

[9] M. Sakai, *Regularity of boundaries of quadrature domains in two dimensions,* SIAM J. Math. Anal. **24** (1993).

Tomas Sjödin
Department of Mathematics
Royal Institute of Technology
S-100 44 Stockholm, Sweden
e-mail: tomas@math.kth.se

Operator Theory:
Advances and Applications, Vol. 156, 257–277

Subharmonicity of Higher Dimensional Exponential Transforms

Vladimir G. Tkachev

To Harold Shapiro on his 75th Anniversary, with admiration.

Abstract. Our main result states that the function $(1-E_\rho)^{(n-2)/n}$ is subharmonic, where $0 \leq \rho \leq 1$ is a density function in $\mathbb{R}^n$, $n \geq 3$, and $E_\rho(x) = \exp\left(-\frac{2}{n}\int \frac{\rho(\zeta)d\zeta}{|\zeta - x|^n}\right)$, is the exponential transform of ρ. This answers in affirmative the recent question posed by B. Gustafsson and M. Putinar in [6].

Mathematics Subject Classification (2000). Primary 31C05; Secondary 44A12, 53C65.

Keywords. Riesz potential, exponential transform, subharmonic function.

1. Introduction

The exponential transform can be viewed as a potential depending on a domain in $\mathbb{R}^n$, or, more generally, on a measure having a *density* function $\rho(x)$ (with compact support) in the range $0 \leq \rho \leq 1$. The two-dimensional version

$$E_\rho(z,w) = \exp\left[-\frac{1}{\pi}\int \frac{\rho(\zeta)\, dA(\zeta)}{(\zeta - z)(\bar{\zeta} - \bar{w})}\right] \tag{1.1}$$

has appeared in operator theory, as a determinantal-characteristic function of certain close to normal operators [4], [10], and has previously been studied and proved to be useful within operator theory, moment problems and other problems of domain identification, and for proving regularity of free boundaries (see [6], [11] for further references). A corresponding exponential transform on the real axis was already known and used by A.A. Markov (in the 19th century) and later by N.I. Akhiezer and M.G. Krein in their studies of one-dimensional moment problems [1], [2] (see, also [8]).

The author was supported by grant RFBR no. 03-01-00304 and by Göran Gustafsson Foundation.

In [6] the diagonal version of (1.1),

$$E_\rho(x) = \exp\left[-\frac{2}{\omega_n}\int \frac{\rho(\zeta)d\zeta}{|x-\zeta|^n}\right],$$

is studied in higher dimensional case $n \geq 3$. Here ω_n denotes the $(n-1)$-dimensional Lebesgue measure of the unit sphere in $\mathbb{R}^n$.

Clearly, $0 < E_\rho(x) < 1$ for all $x \notin \operatorname{supp}\rho$. In particular, it was shown in [6] that E_ρ is a subharmonic function. In two dimensions it is also known that the function $\ln(1-E_\rho)$ is subharmonic, which is a stronger statement. Here we extend the mentioned sub/superharmonicity in dimension $n \geq 3$, thereby answering in affirmative a recent question [6, p. 566]:

Theorem 1.1. *Let $E_\rho(x)$ be the exponential transform of a density $\rho \not\equiv 0$. Then the function*

$$\begin{cases} \ln(1-E_\rho), & \text{if} \quad n=2, \\ \frac{1}{n-2}(1-E_\rho)^{(n-2)/n}, & \text{if} \quad n \geq 3, \end{cases} \tag{1.2}$$

is subharmonic outside $\operatorname{supp}\rho$.

In fact, we show that a stronger version holds. To formulate it we need some notation. Given an integer $n \geq 1$, we define $\mathcal{M}_n(t)$ as the solution of the following ODE:

$$\mathcal{M}_n'(t) = 1 - \mathcal{M}_n^{2/n}(t), \qquad \mathcal{M}(0) = 0. \tag{1.3}$$

We call $\mathcal{M}_n(t)$ the *profile* function.

Theorem 1.2. *For $n \geq 2$ let ρ be a density function and*

$$V_\rho(x) = -\frac{n}{2}\ln E_\rho(x) \equiv \frac{n}{\omega_n}\int \frac{\rho(\zeta)d\zeta}{|x-\zeta|^n}. \tag{1.4}$$

Then the function

$$\begin{cases} \log \mathcal{M}_2(V_\rho(x)), & \text{if} \quad n=2 \\ [\mathcal{M}_n(V_\rho(x))]^{(n-2)/n}, & \text{if} \quad n \neq 2 \end{cases} \tag{1.5}$$

is subharmonic outside the support of ρ. Moreover, this function is harmonic in $\mathbb{R}^n \setminus B$, if B is an arbitrary Euclidean ball and $\rho = \chi_B$ is its characteristic function.

We discuss properties of the profile function in more detail in Section 4. In particular we show that $1 - \mathcal{M}_n(x)$ is a completely monotonic function in $\mathbb{R}^+$.

2. The main inequality

2.1. Variational problem

Let $x = (x_1, y) \in \mathbb{R}^n$, $y = (x_2, \ldots, x_n)$, and

$$\mathbb{R}^n_\pm = \{x = (x_1, y) : \pm x_1 > 0\}.$$

Given a measurable function $h(x)$ we denote by $\mathcal{J}(h)$ the integral

$$\mathcal{J}(h) = \fint_{\mathbb{R}^n} h(x)\, dx = \frac{n}{\omega_n}\int_{\mathbb{R}^n} h(x)\, dx$$

where $dx = dx_1 dy$ denotes the n-dimensional Lebesgue measure in $\mathbb{R}^n$. In what follows we fix the following notations:

$$f(x) = \frac{1}{|x|^{n-2}}, \qquad g(x) = \frac{1}{|x|^n}, \qquad \varphi(x) = \frac{x_1}{|x|^n},$$

and suppose that $\rho(x)$ is a density function such that

$$0 \le \rho(x) \le 1.$$

If $n = 2$ we assume that $f(x) \equiv 1$. Throughout this section, unless otherwise stated, we will assume that $\rho \neq 0$ on a non-null set and the support of ρ does not contain a neighborhood of the origin. We write

$$\rho \in \mathcal{H}(w) \qquad \Leftrightarrow \qquad \mathcal{J}(\rho g) \equiv \fint_{\mathbb{R}^n} \frac{\rho}{|x|^n}\, dx = w \ge 0. \tag{2.1}$$

Our main subject is the ratio

$$\Phi(\rho) = \frac{\mathcal{J}^2(\varphi\rho)}{\mathcal{J}(f\rho)}.$$

Theorem 2.1. *Let ρ be a density function, $0 \notin \operatorname{supp}\rho$. Then*

$$\max_{\rho\in\mathcal{H}(w)} \Phi(\rho) = \mathcal{M}_n(w). \tag{2.2}$$

For any $w > 0$ the maximum is attained when $\rho(x)$ is the characteristic function of the ball centered at $(\tau, \mathbf{0})$ of radius $\tau\mathcal{M}_n(w)^{1/n}$, with $\tau > 0$.

We mention two limit cases of the last assertion. Namely, the boundedness of maximum in (2.2) easily follows from $\varphi^2 \le fg$ and the Cauchy-Schwarz inequality:

$$\frac{\mathcal{J}^2(\varphi\rho)}{\mathcal{J}(f\rho)} \le \mathcal{J}(g\rho) = w. \tag{2.3}$$

On the other hand, it was shown by Gustafsson and Putinar in [6, p. 563] that

$$\frac{\mathcal{J}^2(\varphi\rho)}{\mathcal{J}(f\rho)} < 1 \tag{2.4}$$

does hold. The last means that inequality (2.4) considerably refines (2.3) when $w > 1$ while the first estimate becomes to be sharper when w is a small value.

Corollary 2.2. *For any density function $\rho(x)$, $0 \notin \operatorname{supp}\rho$, the following sharp inequality holds*

$$\left(\fint_{\mathbb{R}^n} \frac{x_1\rho(x)}{|x|^n}\, dx\right)^2 \le \mathcal{M}_n \left(\fint_{\mathbb{R}^n} \frac{\rho(x)}{|x|^n}\, dx\right) \fint_{\mathbb{R}^n} \frac{\rho(x)}{|x|^{n-2}}\, dx \tag{2.5}$$

The inversion $x \to x/|x|^2$ gives another equivalent form of the preceding property.

Corollary 2.3. *For any density function $\rho(x)$, $0 \notin \operatorname{supp}\rho$, the following sharp inequality holds:*

$$\left(\fint_{\mathbb{R}^n} \frac{x_1\rho(x)}{|x|^{n+2}}\,dx\right)^2 \le \mathcal{M}_n\left(\fint_{\mathbb{R}^n} \frac{\rho(x)}{|x|^n}\,dx\right)\fint_{\mathbb{R}^n} \frac{\rho(x)}{|x|^{n+2}}\,dx \tag{2.6}$$

Remark 2.4. We note that for $n \ge 3$ the above inequality (2.5) can be interpreted as a pointwise estimate on the Coulomb potential

$$U_\rho(x) = \fint \frac{\rho(\zeta)dx}{|x-\zeta|^{n-2}}$$

with an *bounded* density function ρ, $0 \le \rho \le 1$. Indeed, using the inversion in $\mathbb{R}^n$ we see that (2.5) is equivalent to

$$|\nabla U_\rho(x)|^2 \le \mathcal{M}_n[V_\rho(x)]U_\rho(x), \qquad x \notin \operatorname{supp}\rho,$$

where $V_\rho(x)$ is defined by (1.4). In particularly, $\mathcal{M}_n(w) < 1$ gives us the inequality due to Gustafsson and Putinar [6]:

$$|\nabla U_\rho(x)|^2 < U_\rho(x), \qquad x \notin \operatorname{supp}\rho.$$

2.2. Auxiliary integrals

In order to prove Theorem 2.1, we need to evaluate the integrals in (2.5) for a specific choice of the density function. Namely, let $\tau > \alpha > 0$ and consider the following density function

$$\widehat{\rho}(x) = \chi_{\mathbb{D}}(x),$$

where

$$\mathbb{D} \equiv \mathbb{D}(\alpha,\tau) := \left\{x = (x_1, y): \quad (x_1-\tau)^2 + |y|^2 < \tau^2 - \alpha^2\right\}. \tag{2.7}$$

First, we note that the function $f(x) = |x|^{2-n}$ is harmonic in $\overline{\mathbb{D}}$. Using the fact that the ball $\mathbb{D}$ is of radius $\sqrt{\tau^2-\alpha^2}$ and centered at $x = (\tau, \mathbf{0})$, we have by the mean value theorem

$$\mathcal{J}(\widetilde{\rho}f) = \fint_{\mathbb{D}} \frac{dx}{|x|^{n-2}} = \frac{(\tau^2-\alpha^2)^{n/2}}{\tau^{n-2}} = \alpha^2\frac{\sinh^n\xi}{\cosh^{n-2}\xi} \tag{2.8}$$

where

$$\cosh\xi = \frac{\tau}{\alpha}. \tag{2.9}$$

Similarly, harmonicity of $\varphi(x) = x_1|x|^{-n}$ implies

$$\mathcal{J}(\widetilde{\rho}\varphi) = \alpha\frac{\sinh^n\xi}{\cosh^{n-1}\xi}.$$

To evaluate $\mathcal{J}(\widetilde{\rho}g)$ we consider the following auxiliary function

$$\lambda(x) = \frac{|x|^2+\alpha^2}{2\tau x_1}.$$

Then $\lambda(x)$ is positive on $\mathbb{D}$ and ranges in

$$\frac{\alpha}{\tau} \leq \lambda(x) < 1, \quad x \in \mathbb{D}.$$

Moreover, it is easy to see that

$$\lambda(x) \equiv z, \qquad x \in S(z) = \partial\mathbb{D}(\alpha, \tau z). \tag{2.10}$$

Hence, the co-area formula yields

$$\mathcal{J}(\widetilde{\rho}g) = \fint_{\mathbb{D}} \frac{dx}{|x|^n} = \frac{n}{\omega_n} \int_{\alpha/\tau}^{1} dz \int_{S(z)} \frac{dS}{|x|^n |\nabla\lambda(x)|}. \tag{2.11}$$

Here dS is the $(n-1)$-dimensional surface measure of the level set $S(z)$.

On the other hand, we have for the gradient

$$|\nabla\lambda|^2 = \frac{|y|^2}{\tau^2 x_1^2} + \frac{(x_1^2 - \alpha^2 - |y|^2)^2}{4\tau^2 x_1^4},$$

which by virtue of (2.10) implies the corresponding value on the level set $S(z)$:

$$|\nabla\lambda|^2\Big|_{S(z)} = \frac{\tau^2 z^2 - \alpha^2}{\tau^2 x_1^2}.$$

Substitution of the last expression into (2.11) yields

$$\mathcal{J}(\widetilde{\rho}g) = \frac{n}{\omega_n} \int_{\alpha/\tau}^{1} \frac{\tau dz}{\sqrt{\tau^2 z^2 - \alpha^2}} \int_{E(z)} \frac{x_1}{|x|^n} dS.$$

Since $\varphi(x) = x_1|x|^{-n}$ in the inner integral is a harmonic function and $S(z)$ is a sphere, we have by the mean value theorem

$$\mathcal{J}(\widetilde{\rho}g) = \frac{n}{\omega_n} \int_{\alpha/\tau}^{1} \frac{\tau dz}{\sqrt{\tau^2 z^2 - \alpha^2}} \cdot \frac{(\tau^2 z^2 - \alpha^2)^{(n-1)/2}}{(\tau z)^{n-1}} = n \int_0^{\xi} \tanh^{n-1} t dt.$$

where ξ is defined by (2.9). Thus we obtain

$$\mathcal{J}(\widetilde{\rho}g) = T_n(\tau/\alpha) = T_n(\xi) := n \int_0^{\xi} \tanh^{n-1} t \, dt. \tag{2.12}$$

We point out that *the latter integral depends only on the ratio* τ/α. One can easy verify that

$$\mathcal{M}_n(T_n(\xi)) \equiv \tanh^n \xi = \left(\frac{\sqrt{\tau^2 - \alpha^2}}{\tau}\right)^n. \tag{2.13}$$

Remark 2.5. After a suitable shift in the x_1-direction, the last computation is equivalent to the following relation

$$\mathcal{M}_n\left(\fint_{\mathbb{B}(R)} \frac{d\zeta}{|x-\zeta|^n}\right) = \left(\frac{R}{|x|}\right)^n, \tag{2.14}$$

which holds for any ball $\mathbb{B}(R)$ of radius R centered at the origin.

2.3. Proof of Theorem 2.1

Let us denote

$$M_n^+(w) = \sup_{\rho\in\mathcal{R}} \Phi(\rho) \tag{2.15}$$

where $\mathcal{R}$ denotes the class of all density functions ρ such that $\operatorname{supp}\rho\cap\mathbb{R}^n_-$ has null measure. Then Theorem 2.1 follows from the following lemmas.

Lemma 2.6. $M_n^+(w) = \mathcal{M}_n(w)$.

Lemma 2.7. $\sup_\rho \Phi(\rho) = M_n^+(w)$.

Proof of Lemma 2.6. Our first step is to reduce the problem (2.15) to the following linear extremal problem with additional constraints:

$$N_n(w) := \sup_{\rho\in\mathcal{R}}\{\mathcal{J}(\rho\varphi): \quad \mathcal{J}(\rho f) = 1,\ \mathcal{J}(\rho g) = w\}. \tag{2.16}$$

Then we have

$$M_n^+(w) = N_n^2(w). \tag{2.17}$$

Indeed, in order to prove (2.17), let $\rho_a(x) = \rho(ax)$ be a homothety of $\rho(x)$ with positive coefficient a. Clearly, this transformation preserves the class $\mathcal{R}$. On the other hand, one can easily see that

$$\Phi(\rho_a) = \Phi(\rho)$$

by the virtue of homogeneity of Φ. Moreover,

$$\mathcal{J}(\rho_a\varphi) = \frac{1}{a}\mathcal{J}(\rho\varphi), \qquad \mathcal{J}(\rho_a f) = \frac{1}{a^2}\mathcal{J}(\rho f),$$

which proves (2.17).

Next, we claim that for any nonnegative w there exists an $\alpha > 0$ and $\tau > \alpha$ such that

$$\mathcal{J}(\widetilde{\rho} f) = 1, \qquad \mathcal{J}(\widetilde{\rho} g) = w, \tag{2.18}$$

where $\widetilde{\rho} = \chi_{\mathbb{D}(\alpha,\tau)}$ is the characteristic function of the ball $\mathbb{D}(\alpha,\tau)$ in (2.7). Indeed, using the definition of function $T_n(t)$ in (2.12) one can easily see that there exist a unique root $\xi > 0$ of the equation

$$T_n(\xi) = w. \tag{2.19}$$

Then we chose $\alpha > 0$ such that

$$\alpha^2 = \frac{\cosh^{n-2}\xi}{\sinh^n \xi},$$

and let $\tau = \alpha\cosh\xi$. Now (2.18) immediately follows from (2.8) and (2.12).

Thus, the function $\widetilde{\rho}(x)$ satisfies (2.18) and it follows that it is admissible for the problem (2.16). This implies

$$N_n(w) \geq \mathcal{J}(\widetilde{\rho}\varphi).$$

To prove that the inverse inequality holds, we fix any function $\rho \in \mathcal{R}$ which is admissible for (2.16). Then

$$\mathcal{J}(\widetilde{\rho}(f+\alpha^2 g)) = \mathcal{J}(\rho(f+\alpha^2 g)) = 1+\alpha^2 w.$$

The last property means that both the functions ρ and $\widetilde{\rho}$ are test functions for the following extremal problem

$$\sup_{\rho\in\mathcal{R}}\{\mathcal{J}(\rho\varphi): \quad \mathcal{J}(\rho(f+\alpha^2 g)) = 1+w\alpha^2\}. \tag{2.20}$$

Let us consider the ratio

$$h(x) := \frac{\varphi(x)}{f(x)+\alpha^2 g(x)} = \frac{x_1}{|x|^2+\alpha^2}.$$

Then,

$$\{x\in\mathbb{R}^n : h(x) > \frac{1}{2\tau}\} = \mathbb{D}(\alpha,\tau),$$

and it follows from the Bathtub Principle [9, p. 28] that $\widetilde{\rho}$ is the extremal density for (2.20). Thus, we have

$$\mathcal{J}(\rho\varphi) \le \mathcal{J}(\widetilde{\rho}\varphi),$$

and consequently

$$N_n(w) \le \mathcal{J}(\widetilde{\rho}\varphi).$$

Hence, we conclude that

$$N_n(w) = \mathcal{J}(\widetilde{\rho}\varphi) = \alpha\frac{\sinh^n\xi}{\cosh^{n-1}\xi}.$$

Now, it follows from (2.17) and our choice of α that

$$M_n^+(w) = N_n^2(w) = \alpha^2\frac{\sinh^{2n}\xi}{\cosh^{2n-2}\xi} = \tanh^n\xi,$$

and from (2.13), we find

$$M_n^+(w) = \mathcal{M}_n(T_n(\xi)) = \mathcal{M}_n(w),$$

and the lemma follows. □

Proof of Lemma 2.7. It suffices only to prove the one-side inequality

$$\sup_{\rho}\Phi(\rho) \le M_n^+(w). \tag{2.21}$$

Let ρ is an arbitrary admissible for (2.1) density function. Excluding the trivial case $\rho\in\mathcal{R}$ we distinguish two rest cases:

(i) the set supp $\rho\cap\mathbb{R}^n_+$ has the null measure;

(ii) the set supp ρ has non-zero counterpart in the both half-spaces.

Let ρ satisfy (i). Then the function

$$\rho^*(x_1,y) := \rho(-x_1,y)$$

belongs to $\mathcal{R}$, and it follows that

$$\sup_{\rho\in(\mathrm{i})} \Phi(\rho) = \sup_{\rho\in\mathcal{R}} \Phi(\rho) = M_n^+(w). \tag{2.22}$$

Now, let ρ satisfy (ii). We set $\rho^{\pm}(x) = \chi_{\mathbb{R}^n_{\pm}}(x)\rho(x)$. Then

$$\mathcal{J}(\rho\varphi) = \mathcal{J}(\rho^+\varphi) - \mathcal{J}((\rho^-)^*\varphi),$$
$$\mathcal{J}(\rho f) = \mathcal{J}(\rho^+ f) + \mathcal{J}((\rho^-)^* f),$$

where the last integrals are positive. Using the elementary inequality

$$\frac{(a-b)^2}{c+d} \le \max\left[\frac{a^2}{c}, \frac{b^2}{d}\right]$$

which holds for any set of positive numbers a, b, c, d, we conclude that

$$\Phi(\rho) = \frac{\mathcal{J}^2(\rho\varphi)}{\mathcal{J}(\rho f)} \le \max[\Phi(\rho^+), \Phi((\rho^-)^*)].$$

Hence, we have by Lemma 2.6

$$\Phi(\rho) \le \max[M_n^+(w_1), M_n^+(w_2)] = \max[\mathcal{M}_n(w_1), \mathcal{M}_n(w_2)],$$

where

$$w_1 = \mathcal{J}(\rho^+ g), \qquad w_2 = \mathcal{J}((\rho^+)^* g).$$

But

$$w = \mathcal{J}(\rho g) = w_1 + w_2,$$

whence $w_i \le w$, $i = 1, 2$. Since $\mathcal{M}_n$ is an increasing function we obtain $\Phi(\rho) \le \mathcal{M}_n(w)$, and consequently

$$\sup_{\rho\in(\mathrm{ii})} \Phi(\rho) \le \mathcal{M}_n(w) = M_n^+(w).$$

Combining the last inequality with (2.22) we obtain

$$\sup_{\rho} \Phi(\rho) = \sup_{\mathcal{R}\cup(\mathrm{i})\cup(\mathrm{ii})} \Phi(\rho) \le M_n^+(w)$$

which proves (2.21). □

3. Proof of the main results

Lemma 3.1. *For any $n \ge 1$ we have*

$$\mathcal{M}_n(w) \le Q_n(w) := \frac{e^{2w/n} - 1}{e^{2w/n} - \frac{n-2}{n}}. \tag{3.1}$$

Proof. Note that in the cases $n = 1, 2$, we have

$$\mathcal{M}_1(w) = \tanh w = \frac{e^{2w} - 1}{e^{2w} + 1},$$

$$\mathcal{M}_2(w) = 1 - e^{-w}$$

which turns (3.1) into equality.

Now, let $n \geq 3$. We have $M_n(0) = Q_n(0) = 0$ and by Definition (1.3) it suffices only to prove that

$$Q_n'(w) \geq 1 - Q_n^{2/n}, \quad w > 0. \tag{3.2}$$

We have

$$Q_n'(w) = (1 - \frac{n-2}{n} Q_n)(1 - Q_n)$$

and (3.2) becomes to be equivalent to the inequality

$$\frac{1 - t^{1-\gamma}}{1 - t} < 1 - \gamma t,$$

where $t = Q_n(w) \in (0, 1)$ and $\gamma = (n-2)/n$. To verify the last inequality we rewrite it in the form

$$\frac{1 - t^{\gamma}}{1 - t} > \gamma t^{\gamma}.$$

For $t \in (0, 1)$, the function in the left hand side is a decreasing function while the right hand side member is an increasing one. Since the both functions have the same limit value γ at $t = 1$, we have the desired inequality. □

Proof of Theorem 1.1. Let $f(x)$ denote the function in (1.2). Then we have for any $n \geq 2$ and $x \notin \operatorname{supp} \rho$

$$\nabla f(x) = -(1 - E_\rho)^{-2/n} \nabla E_\rho,$$

$$\Delta f(x) = -\frac{2}{n}(1 - E_\rho)^{-\frac{2+n}{n}} \left[\frac{n}{2}(1 - E_\rho)\Delta E_\rho + |\nabla E_\rho|^2\right].$$

Then the inequality $\Delta f(x) \geq 0$ to be proved becomes

$$\frac{n}{2}(1 - E_\rho)\Delta E_\rho + |\nabla E_\rho|^2 \leq 0. \tag{3.3}$$

On the other hand,

$$\nabla E_\rho(x) = 2E_\rho(x) \fint \frac{(x - \zeta)\rho(\zeta) d\zeta}{|x - \zeta|^{n+2}},$$

$$\Delta E_\rho(x) = 4E_\rho(x) \left(\left| \fint \frac{(x - \zeta)\rho(\zeta) d\zeta}{|x - \zeta|^{n+2}} \right|^2 - \fint \frac{\rho(\zeta) d\zeta}{|x - \zeta|^{n+2}} \right),$$

and (3.3) becomes

$$\left(1 - \frac{n-2}{n} E_\rho\right) \left| \fint \frac{(x - \zeta)\rho(\zeta) d\zeta}{|x - \zeta|^{n+2}} \right|^2 \leq (1 - E_\rho) \fint \frac{\rho(\zeta) d\zeta}{|x - \zeta|^{n+2}}. \tag{3.4}$$

In order to prove (3.4) we can assume without loss of generality that $x = 0$. In this case, after a suitable rotation we can write the vector integral as follows

$$\left| \fint \frac{\zeta\rho(\zeta)d\zeta}{|\zeta|^{n+2}} \right| = \fint \frac{\zeta_1\rho(\zeta)d\zeta}{|\zeta|^{n+2}}.$$

Thus, we arrive at the inequality to be proved

$$\left(1 - \frac{n-2}{n} e^{-\frac{2w}{n}}\right) \left(\fint \frac{\zeta_1\rho(\zeta)d\zeta}{|\zeta|^{n+2}} \right)^2 \leq (1 - e^{-\frac{2w}{n}}) \fint \frac{\rho(\zeta)d\zeta}{|\zeta|^{n+2}}, \tag{3.5}$$

with

$$w = \fint \frac{\rho(\zeta)d\zeta}{|\zeta|^n}.$$

But, it is easy to see that (3.5) follows from Corollary 2.3 and Lemma 3.1. The theorem follows. □

Proof of Theorem 1.2. Let $F(x)$ denote the function in (1.5) and $V(x) = V_\rho(x)$. Then the argument similar to that above yields for $n \geq 3$

$$\begin{aligned} \Delta F(x) &= \Delta(\mathcal{M}_n(V))^{(n-2)/n} \\ &= (1 - \mathcal{M}_n^{2/n}(V)) \left(\frac{n-2}{n} \mathcal{M}_n^{-\frac{2}{n}}(V)\Delta V - \frac{2(n-2)}{n^2} \mathcal{M}_n^{-\frac{2+n}{n}}(V)|\nabla V|^2 \right) \\ &= 2(n-2)(1 - \mathcal{M}_n^{2/n}(V)) \Big[\mathcal{M}_n(V)B - |A|^2 \Big], \end{aligned} \tag{3.6}$$

where

$$A = \fint \frac{(x-\zeta)\rho(\zeta)d\zeta}{|x-\zeta|^{n+2}}, \qquad B = \fint \frac{\rho(\zeta)d\zeta}{|x-\zeta|^{n+2}}.$$

Similarly, we have for $n = 2$

$$\Delta F(x) = \frac{1 - \mathcal{M}_2(V)}{\mathcal{M}_2^2(V)} \Big[\mathcal{M}_2(V)B - |A|^2 \Big].$$

Hence, for all integer $n \geq 2$, the sign of the Laplacian $\Delta F(x)$ coincides with the sign of $[M_n(V)B - |A|^2]$.

Let us fix an arbitrary point $x \notin \operatorname{supp} \rho$. Then after a suitable rotation we can reduce the vector integral A to the scalar one such that the value in last brackets in (3.6) becomes

$$\mathcal{M}_n \left(\fint \frac{\rho_1(\zeta)d\zeta}{|\zeta|^n} \right) \fint \frac{\rho_1(\zeta)d\zeta}{|\zeta|^{n+2}} - \left(\fint \frac{\zeta_1\rho_1(\zeta)d\zeta}{|\zeta|^{n+2}} \right)^2,$$

where $\rho_1(\zeta)$ is the correspondent transformed density. Then Corollary 2.3 again implies that the latter difference is nonnegative and subharmonicity of $\mathcal{E}_\rho$ easily follows.

Now, let us prove the second assertion of the theorem. Let $\mathbb{B}(R)$ be the ball of radius R with center at the origin and $\widehat{\rho}(x) = \chi_{\mathbb{B}(R)}(x)$ be the corresponding characteristic function. Then

$$V_{\widehat{\rho}}(x) := \int_{\mathbb{B}(R)} \frac{d\zeta}{|x-\zeta|^n},$$

and we have from (2.14) that in this case

$$\mathcal{M}_n(V_{\widehat{\rho}}(x)) = \left(\frac{R}{|x|}\right)^n,$$

which obviously yields harmonicity of

$$[\mathcal{M}_n(V_{\widehat{\rho}}(x))]^{(n-2)/n} = R^{n-2}|x|^{2-n}$$

for $n \geq 3$, and

$$\ln \mathcal{M}_2(V_{\widehat{\rho}}(x)) = 2\ln \frac{R}{|x|},$$

if $n = 2$. The theorem is completely proved. □

4. The profile function

Here we study the profile function $\mathcal{M}_n$ in more detail. This higher transcendental function, apart of its appearance in the above theorems, admits also number-theoretical applications (e.g., in connection with the Euler-Mascheroni constant γ, see Section 4.2). Our main result (Theorem 4.1 below) states that $1 - M_n(w)$ is a completely monotonic function. We also show (Theorem 4.5) that this function can be analytically extended across $w = +\infty$ by making use of a specific logarithmic transformation.

4.1. Complete monotonicity

It is convenient to consider the general case of (1.3). Namely, given a real $\alpha > 0$ we define $F_\alpha(x)$ as a solution to the following ODE

$$F_\alpha'(x) = 1 - F_\alpha^\alpha(x), \qquad F_\alpha(0) = 0. \tag{4.1}$$

Then for an integer n we have $\mathcal{M}_n(w) = F_{2/n}(w)$.

We recall that a function $f(x)$ defined on $[0; +\infty)$ is said to be *completely monotonic* if

$$(-1)^k f^{(k)}(x) \geq 0, \qquad x \in \mathbb{R}^+.$$

Theorem 4.1. *Let $\alpha > 0$. Then*

(i) *$F_\alpha(x)$ is an increasing function for $x \geq 0$ such that $F_\alpha(x) : \mathbb{R}^+ \to [0; 1)$;*
(ii) *for all $\alpha \in (0; 1]$ the function*

$$\widetilde{F}_\alpha(x) = 1 - F_\alpha(x)$$

is completely monotonic on $\mathbb{R}^+$.

It follows from the well-known Bernstein theorem [3] (see also [13, p. 161]) that $\widetilde{F}_\alpha(x)$ is a Laplace transform of a positive measure supported on $\mathbb{R}^+$.

Corollary 4.2. *For all $\alpha \in (0;1]$ the following Laplace-Stieltjes representation holds:*

$$\widetilde{F}_\alpha(x) = \int_0^{+\infty} e^{-xt}\, d\sigma_\alpha(t) \tag{4.2}$$

where $d\sigma_\alpha$ is a positive probability measure with finite variation,

$$\int_0^{+\infty} d\sigma_\alpha(t) = \widetilde{F}_\alpha(0) = 1. \tag{4.3}$$

The following subadditive property is a consequence of the general result due to Kimberling [7] and concerns complete monotonic functions satisfying (4.3).

Corollary 4.3. *For all $0 < \alpha \leq 1$ the function $\widetilde{F}_\alpha(x)$ is subadditive in the sense that*

$$\widetilde{F}_\alpha(x)\widetilde{F}_\alpha(y) \leq \widetilde{F}_\alpha(x+y). \tag{4.4}$$

Remark 4.4. It is easy to verify that for $\alpha > 1$ the third derivative of $F_\alpha'''(x)$ has no constant sign on $\mathbb{R}^+$. Thus our constraint is optimal for positive values of α. On the other hand, if $\alpha = 1$ then $F_1(x)$ can be derived as follows

$$F_1(x) = 1 - e^{-x}, \qquad \widetilde{F}_1(x) = e^{-x}$$

which implies the complete monotonicity immediately. Moreover, in the latter case $\widetilde{F}_1(x)$ satisfies a full additive property instead of (4.4). We notice also that in this case one can easily find that $d\sigma_1(t) = \delta_1(t)$ the delta-Dirac probability measure supported at $t = 1$. More precisely, we have

$$\sigma_1(t) = \chi_{[1,+\infty)}(t).$$

Proof of Theorem 4.1. The only non-trivial part of the theorem is (ii). We notice first that

$$\widetilde{F}_\alpha(x) \geq 0, \qquad \widetilde{F}_\alpha^{(k)}(x) = -F_\alpha^{(k)}(x), \quad k = 1, 2, \dots$$

and

$$F_\alpha''(x) = -\alpha(1 - F_\alpha^\alpha)\frac{F_\alpha^\alpha}{F_\alpha(x)}.$$

On the other hand, one can easily show by induction that the following property holds for all $k \geq 0$

$$F_\alpha^{(k+2)}(x) = \alpha t(1-t)\frac{H_k(t)}{F_\alpha(x)^{k+1}} \tag{4.5}$$

where

$$t = F_\alpha^\alpha(x)$$

and $H_j(t)$ is a polynomial of degree at most j. Moreover, we have the following recurrent relationship

$$H_{k+1}(t) = [(k+1-2\alpha)t - (k+1-\alpha)]H_k(t) + \alpha t(1-t)H_k'(t), \qquad k \geq 2 \quad (4.6)$$

with initial condition

$$H_0(t) = -1. \qquad (4.7)$$

Since $t = F_\alpha^\alpha(x)$ ranges in $[0;1)$ we have only to prove that the polynomials $(-1)^{k+1}H_k(t)$ are nonnegative in $\Delta = [0,1)$.

We will use the following Bernstein-type transformation

$$P^*(z) = (1+z)^n P\left(\frac{1}{1+z}\right), \qquad n \geq \deg P$$

which transforms a polynomial P to a polynomial of degree at most n.

Let

$$P(t) = a_0 + a_1 t + \ldots + a_n t^n$$

(here we use the assumption that $\deg P \leq n$ and some coefficients may vanish). Then we can write

$$P(t) = \sum_{j=0}^{n} b_j t^{n-j}(1-t)^j \qquad (4.8)$$

where

$$P^*(z) = b_0 + b_1 z + \ldots + b_n z^n, \qquad z = \frac{1-t}{t}.$$

We recall that (4.8) is the Bernstein-type expansion of P by the basis $t^j(1-t)^{n-j}$.

It follows then from (4.8) that if all (non-zero) coefficients of the associate polynomial $P^*(z)$ have the same sign: $\operatorname{sgn} b_j = \varepsilon$, then $P(t)$ changes no sign in Δ and its sign coincides with ε.

Let $H_k^*(z)$ be the associative polynomial for $H_k(t)$. Then

$$H_k(t) = t^k H_k^*\left(\frac{1-t}{t}\right)$$

and

$$H_k'(t) = kt^{k-1}H_k^*\left(\frac{1-t}{t}\right) - t^{k-2}H_k^{*\prime}\left(\frac{1-t}{t}\right).$$

It follows from (4.6) that

$$-H_{k+1}^*(z) = [\alpha + (k+1)(1-\alpha)z]H_k^*(z) + \alpha z(1+z)H_k^{*\prime}(z). \qquad (4.9)$$

We notice that by (4.7)

$$H_0^* = H_0 = -1.$$

On the other hand, since $0 \leq \alpha \leq 1$ the multipliers $(\alpha+(k+1)(1-\alpha)z)$ and $\alpha z(1+z)$ in (4.9) have positive coefficients with respect to z. Hence, it immediately follows from (4.9) by induction that all coefficients of $-H_{k+1}^*(z)$ have the same sign as

$H_k^*(z)$ does. Moreover, the sign of the coefficients of $H_k^*(z)$ is $(-1)^{k+1}$ which yields by the above remark that

$$(-1)^{k+1} H_k(t) \geq 0, \qquad t \in \Delta.$$

Clearly, the last property together with (4.5) yields the desired assertion. □

4.2. Exponential series for the profile function

Here we establish an explicit form of the above exponential representation for $\mathcal{M}_n(x)$. As above, it is convenient to consider a general $F_\alpha(x)$ instead of $\mathcal{M}_n(x)$ (see the definition (4.1)).

Let

$$\phi_\alpha(t) := 1 - F_\alpha\left(-\frac{1}{\alpha}\ln t\right).$$

According to its definition, $\phi_\alpha(t)$ is defined in $(0, 1]$. But it turns out that a stronger property holds

Theorem 4.5. *The following properties hold:*

(i) *For any $\alpha > 0$ the function $\phi_\alpha(t)$ admits an analytic continuation on $(-\epsilon, 1)$ with some $\epsilon > 0$ depending on α.*

(ii) *The corresponding Taylor series at $t = 0$ are*

$$\phi_\alpha(t) = \sum_{k=1}^{\infty} \sigma_k (\gamma_\alpha t)^k, \tag{4.10}$$

where

$$\gamma(\alpha) = \frac{1}{\alpha}\exp\left(-\int_0^1 \frac{1 - x^{\frac{1-\alpha}{\alpha}}}{1-x}\, dx\right),$$

and σ_k are the coefficients defined by the following recurrence

$$\sigma_1 := 1, \quad \sigma_k = \frac{1}{k(k-1)} \sum_{\nu=1}^{k-1} \sigma_\nu \sigma_{k-\nu} [(1+\alpha)\nu - \alpha k]\nu. \tag{4.11}$$

(iii) *If $\alpha \in (0, 1)$ then $\sigma_k > 0$ for all $k \geq 1$ and series (4.10) converges in $(-1, 1)$.*

(iv) *For all $0 < \alpha < 1$, $\phi_\alpha(t)$ is a strictly increasing convex function in $(-\infty, 1)$.*

Remark 4.6. The exact value of γ_α has the following form

$$\ln \gamma_\alpha = -\Psi(1/\alpha) - \gamma + \ln(1/\alpha), \tag{4.12}$$

where $\Psi(z)$ is the Digamma function: $\Psi(z) = \Gamma'(z)/\Gamma(z)$, and $\gamma = 0.5772156\ldots$ is the Euler-Mascheroni constant. The assertion of the theorem is still valid for $\alpha = 0$ which formally corresponds to $n = \infty$. In this case, $\phi_0(x)$ satisfies the following ODE:

$$\phi_0'(x) = -\frac{\ln(1 - \phi_0(x))}{x}, \qquad \phi_0(0) = 0.$$

It follows from (4.12) that in this case $\gamma_0 = e^\gamma$.

Corollary 4.7. *Let $n \geq 2$ be an integer. Then*

$$1 - \mathcal{M}_n(x) = \sum_{k=1}^{\infty} a_k e^{-2kx/n},$$

where $a_k = \sigma_k \gamma_{2/n}^k > 0$ and the series converges for all $x \geq 0$. In particular, the measure in (4.2) is an atomic measure supported at the set $\frac{2}{n}\mathbb{Z}^+$.

We are grateful to Björn Gustafsson for pointing out another useful consequence of the preceding property. Let us define an (n-dimensional) version of the exponential transform as follows

$$\mathbb{E}_\rho(x) = 1 - \mathcal{M}_n\left(\frac{n}{\omega_n}\int \frac{\rho(\zeta)d\zeta}{|x-\zeta|^n}\right).$$

where ρ is a density function.

Corollary 4.8. *The function $E_\rho(x)$ is analytic if and only $\mathbb{E}_\rho(x)$ is. Moreover, these functions are linked by the following identity*

$$\mathbb{E}_\rho(x) = \phi_{2/n} \circ E_\rho(x). \tag{4.13}$$

Proof. The cases $n = 1$ and $n = 2$ are trivial. For $n \geq 3$ we notice that the desired property follows from (4.13) and the fact that $\phi'_\alpha(0) \neq 0$ (see (4.18) below). □

Proof of Theorem 4.5. First we consider (i). The case $\alpha = 1$ is trivial. Let $\alpha > 0$, $\alpha \neq 1$ and $F_\alpha(x)$ be the solution to (4.1). We notice that this function is determined uniquely by virtue of the condition $F_\alpha(0) = 0$, and it is a real analytic function of x in $(0, +\infty)$. It follows that $\phi_\alpha(t)$ also is a real analytic function of t for $t \in (0,1)$ it is bounded there: $|\phi_\alpha(t)| < 1$. Moreover, $y = \phi_\alpha(t)$ satisfies the following differential equation

$$y'(t) = \frac{1-(1-y(t))^\alpha}{\alpha t}, \quad t \in (0,1), \tag{4.14}$$

and the initial condition has to be transformed to $\phi_\alpha(1) = 0$.

Now we prove that $\phi_\alpha(x)$ admits an analytic continuation in a small disk in the complex plane. Let us define the following auxiliary function

$$S(\zeta) := \exp\left(-\int_\zeta^1 \frac{\alpha d\xi}{1-(1-\xi)^\alpha}\right).$$

Here we fix the branch of $(1-\xi)^\alpha$ which assumes the value 1 at $\xi = 0$. Then $S(\zeta)$ is a single-valued holomorphic function in the unit disk $\mathbb{D}(1)$, where

$$\mathbb{D}(r) = \{\xi \in \mathbb{C} : |\zeta| < r\}.$$

Moreover, we have $S(\zeta) \neq 0$ and

$$S'(\zeta) = \frac{\alpha S(\zeta)}{1-(1-\zeta)^\alpha}. \tag{4.15}$$

On the other hand, we notice that the following renormalization of the above integrand

$$\frac{\alpha}{1-(1-\xi)^{\alpha}}-\frac{1}{\xi}=\frac{\alpha\xi-1+(1-\xi)^{\alpha}}{(1-(1-\xi)^{\alpha})\xi}=\frac{\alpha-1}{2}+\frac{(\alpha-1)(2\alpha-1)}{12}\xi+\dots$$

is an analytic function in $\mathbb{D}$ since it admits a regular Taylor expansion near $\xi=0$. This allows us to rewrite the above definition of $S(\zeta)$ as follows:

$$S(\zeta)=\zeta\,\exp\Big\{-\int_{\zeta}^{1}\Big(\frac{\alpha}{1-(1-\xi)^{\alpha}}-\frac{1}{\xi}\Big)d\xi\Big\}.$$

In particular, this implies

$$c_{\alpha}:=S'(0)=\exp\Big(-\int_{0}^{1}\Big(\frac{\alpha}{1-(1-x)^{\alpha}}-\frac{1}{x}\Big)dx\Big)\neq 0. \tag{4.16}$$

On the other hand,

$$\operatorname{Re}\frac{S'(\zeta)\zeta}{S(\zeta)}=\operatorname{Re}\frac{\zeta}{1-(1-\zeta)^{\alpha}}=\operatorname{Re}\frac{1}{\alpha}(1+\frac{\alpha-1}{2}\zeta+\dots).$$

Hence for $r>0$ sufficiently small,

$$\operatorname{Re}\frac{S'(\zeta)\zeta}{S(\zeta)}>0,\quad \zeta\in\mathbb{D}(r) \tag{4.17}$$

Taking into account (4.16), (4.17), and the well-known Alexander's property [5, p. 41] we conclude that the function $S(z)$ is starlike in $\mathbb{D}(r)$, and therefore univalent there.

Let $\psi(z)$ be the inverse function to $S(z)$. Clearly, it is defined in some small disk $\mathbb{D}(\epsilon)$ which is contained in the image $S(\mathbb{D}(r))$. Moreover, by its definition $\psi(z)$ assumes real values for real $z\in\mathbb{D}(\epsilon)$. We also have

$$\psi(0)=0,\qquad \psi'(0)=\frac{1}{S'(0)}=\frac{1}{c_{\alpha}}. \tag{4.18}$$

Furthermore, differentiation of the identity $S(\psi(z))=z$ together with (4.15) yields

$$1=S'(\psi(z))\psi'(z)=\frac{z\psi'(z)}{1-(1-\psi(z))^{\alpha}},$$

consequently, $y=\psi(z)$ is a solution of (4.14) in $\mathbb{D}(\epsilon)$.

Our next step is to prove that $\psi(z)$ is the desired analytic continuation. One suffices to show that $\psi(x)=\phi_{\alpha}(x)$ in some open subinterval of $(0,1)$, that in turn, is equivalent to establishing of the following identity

$$F_{\alpha}(x)=1-\psi(e^{-\alpha x}) \tag{4.19}$$

for all x in some interval $\Delta\subset(0,+\infty)$.

Taking into account the above remarks, we note that $g(x):=1-\psi(e^{-\alpha x})$ is a real-valued solution of (4.1) in $\Delta:=(-\frac{1}{\alpha}\ln\epsilon,+\infty)$. On the other hand, since $g(x)$

satisfies an obvious inequality $g(x) < 1$, and by virtue of the autonomic character of (4.1), we conclude that

$$g(x) = F_\alpha(x + c), \quad x \in \Delta,$$

for some constant $c \in \mathbb{R}$. Thus, we have only to check that $c = 0$.

To this aim, we note that

$$\alpha x = \int_0^{F_\alpha(x)} \frac{\alpha dt}{1 - t^\alpha} = -\int_0^{F_\alpha^\alpha(x)} \frac{1 - \tau^{\frac{1-\alpha}{\alpha}}}{1 - \tau} d\tau + \int_0^{F_\alpha^\alpha(x)} \frac{d\tau}{1 - \tau} d\tau.$$

Since $\lim_{\tau \to \infty} F_\alpha(\tau) = 1$, we arrive at

$$\lim_{x \to +\infty} (1 - F_\alpha^\alpha(x)) e^{x\alpha} = \exp\Big(- \int_0^1 \frac{1 - \tau^{\frac{1-\alpha}{\alpha}}}{1 - \tau} \, d\tau \Big) = \alpha\gamma_\alpha,$$

or

$$\lim_{x \to +\infty} (1 - F_\alpha(x)) e^{x\alpha} = \gamma_\alpha.$$

As a consequence we have,

$$\begin{aligned} \lim_{x \to +\infty} \psi(e^{-x\alpha}) e^{x\alpha} &= \lim_{x \to +\infty} (1 - g(x)) e^{x\alpha} \\ &= \lim_{x \to +\infty} [1 - F_\alpha(x + c)] e^{x\alpha} = e^{-c\alpha} \gamma_\alpha. \end{aligned} \tag{4.20}$$

On the other hand,

$$\lim_{x \to +\infty} \psi(e^{-x\alpha}) e^{x\alpha} = \lim_{t \to +0} \frac{\psi(t)}{t} = \psi'(0) = \frac{1}{c_\alpha}. \tag{4.21}$$

Finally, splitting the integral in the definition of c_α and making the change variables $\tau = (1 - t)^\alpha$, we obtain

$$\begin{aligned} \ln \frac{1}{c_\alpha} &= \lim_{s \to +0} \int_s^1 \Big(\frac{\alpha}{1 - (1 - x)^\alpha} - \frac{1}{x} \Big) dx = \lim_{s \to +0} \Big(\int_0^{(1-s)^\alpha} \frac{\tau^{\frac{1-\alpha}{\alpha}} d\tau}{1 - \tau} + \ln s \Big) \\ &= \lim_{s \to +0} \Big(- \int_0^{(1-s)^\alpha} \frac{1 - \tau^{\frac{1-\alpha}{\alpha}}}{1 - \tau} d\tau - \ln \frac{1 - (1 - s)^\alpha}{s} \Big) \\ &= -\ln \alpha - \int_0^1 \frac{1 - \tau^{\frac{1-\alpha}{\alpha}}}{1 - \tau} d\tau = \ln \gamma_\alpha. \end{aligned}$$

Thus, combining the latter identity with (4.20), and (4.21) we obtain $c = 0$, which yields (4.19) and the mentioned analytic continuation property follows.

Now we prove (ii) and (iii). We note that in view of (4.14)

$$(\alpha t \phi_\alpha')' = \alpha (1 - \phi_\alpha)^{\alpha - 1} \phi_\alpha' = \alpha \phi_\alpha' \frac{1 - \alpha t \phi_\alpha'}{1 - \phi_\alpha},$$

which implies

$$t\phi''_\alpha = \phi_\alpha(t\phi''_\alpha + \phi'_\alpha) - \alpha t\phi'^2_\alpha. \tag{4.22}$$

Setting

$$\phi_\alpha(t) = \sum_{k=1}^{\infty} a_k t^k \tag{4.23}$$

for the Taylor series of ϕ_α around $t = 0$ (we recall that $\phi_\alpha(0) = 0$) we obtain after comparison of the corresponding coefficients for all $k \geq 2$,

$$a_k = \frac{1}{k(k-1)} \sum_{\nu=1}^{k-1} a_\nu a_{k-\nu}[(1+\alpha)\nu - \alpha k]\nu = \sigma_k a_1^k.$$

Here $a_1 = \phi'_\alpha(0) = 1/c_\alpha = \gamma_\alpha$ and σ_k are defined as in (4.11). This yields the desired Taylor expansion. Moreover, we show that for $\alpha \in (0,1)$ the coefficients $\sigma_k > 0$ for all $k \geq 1$. Indeed,

$$\sigma_k = \frac{1}{2k(k-1)} \sum_{\nu=1}^{k-1} A_{\nu,k-\nu}\sigma_\nu\sigma_{k-\nu},$$

where

$$\begin{aligned} A_{\nu,k-\nu} &= [(1+\alpha)\nu - \alpha k]\nu + [(1+\alpha)(k-\nu) - \alpha k](k-\nu) \\ &= (1+\alpha)\left(\nu - \frac{k}{2}\right)^2 + \frac{1-\alpha}{2}k^2 > 0, \end{aligned}$$

unless $k = 2\nu$ when we also have

$$A_{\nu,\nu} = 2[(1+\alpha)\nu - 2\nu\alpha]\nu = 2(1-\alpha)\nu^2 > 0.$$

Since $\sigma_1 = 1$ and for $k \geq 1$ the coefficients $A_{\nu,k-\nu}$ before $\sigma_\nu\sigma_{k-\nu}$ are positive, the positiveness of σ_k follows now by induction.

Thus, $\phi_\alpha(t)$ has the Taylor expansion with positive coefficients. By standard facts of the power series theory we conclude that the radius R of convergence of (4.23) is at least $R = 1$ since $\phi_\alpha(t)$ is analytic along $t \in (-\epsilon, 1)$.

It remains only to prove (iv). We have $\phi'(0) > 0$ which yields $\phi_\alpha(t) < 0$ for sufficiently small $t < 0$. Then a standard analysis of (4.14) shows that these property holds for *all* negative t's where $\phi_\alpha(t)$ is defined. In view of (4.14), this proves the strictly increasing character of $\phi_\alpha(t)$.

In order to prove convexity, we note that (4.22) implies

$$\phi''_\alpha(t) = \phi'_\alpha(t)\frac{\phi_\alpha(t) - \alpha t\phi'_\alpha(t)}{t(1-\phi_\alpha(t))}, \quad \phi''(0) = (1-\alpha)\phi'^2_\alpha(0) > 0.$$

Clearly, it suffices to prove that $\phi''(t) \neq 0$. Assuming the contradictory, we have $\phi''(t) = 0$ an some point $t \neq 0$, and it follows that

$$\phi_\alpha(t) - \alpha t\phi'_\alpha(t) = 0,$$

which yields $(1-\phi_\alpha(t))^\alpha = 1 - \phi_\alpha(t)$. The contradiction is obtained.

Finally, since $\phi_\alpha(t)$ is convex and analytic in its region of definition, we conclude that it can be infinitely extended into the left side of $\mathbb{R}$. □

5. Final remark

Here we discuss in short an appearance of the profile function $\mathcal{M}_1(w)$ as interpretation of the exponential transform. We recall that the original result of A.A. Markov on the L-problem asserts that a sequence of reals $\{s_j\}_{j=0}^{\infty}$ is represented as the moments

$$s_k = \int x^k \rho(x) dx$$

of certain function $0 \le \rho(x) \le 1$, if and only if there is a positive measure $d\mu$ such that the following identity holds

$$1 - \exp(-\sum_{k=0}^{\infty} \frac{s_k}{z^{k+1}}) = \sum_{k=0}^{\infty} \frac{a_k}{z^{k+1}},$$

where

$$a_k = \int x^k d\mu(x).$$

For the detailed discussion of this theory see [2, p. 72]. The latter moment sequence, $\{a_j\}_{j=0}^{\infty}$, can be characterized as a standard *positive* sequence in the sense that the Hankel forms

$$(a_{i+j})_{i,j=0}^{m} \ge 0$$

are positive semi-definite for all $m \ge 0$. For simplicity reasons, we refer to (s_k) as an L-sequence.

Given a sequence $(a_k)_{k=0}^{\infty}$ we set

$$\widehat{a}(z) := \sum_{k=0}^{\infty} \frac{a_k}{z^{k+1}}$$

for the corresponding z-transform. Our first observation is as follows.

Proposition 5.1. *Let $c \in \mathbb{R}$, and $\{a_j\}_{j=0}^{\infty}$ and $\{b_j\}_{j=0}^{\infty}$ be two sequences such that their generating functions satisfy*

$$\frac{1}{\widehat{b}(z)} - \frac{1}{\widehat{a}(z)} = c. \tag{5.1}$$

Then $\{a_j\}_{j=0}^{\infty}$ is a positive sequence if and only $\{b_j\}_{j=0}^{\infty}$ is. Moreover, we have

$$\det(a_{i+j})_{i,j=0}^{m} = \det(b_{i+j})_{i,j=0}^{m}. \tag{5.2}$$

Proof. We prove only (5.2) since it immediately implies the desired positivity property. Let for definiteness, $\widehat{a}(z)$ satisfies the positivity condition, i.e., the corresponding sequence (a_k) is positive semi-definite.

Then the famous result of Stieltjes [12, Ch. XI] asserts that given a function $\widehat{a}(z)$ with power series as above, the following continued J-fraction (actually, Jacobi's type) decomposition holds

$$\widehat{a}(z) = \cfrac{\alpha_0}{\beta_1 + z - \cfrac{\alpha_1}{\beta_2 + z - \cfrac{\alpha_2}{\beta_3 + z - \dots}}}. \tag{5.3}$$

Moreover, in this case we have for the determinants

$$\det(a_{i+j})_{i,j=0}^{m} = \alpha_0^{m+1}\alpha_1^{m}\alpha_2^{m-1}\cdots\alpha_{m-1}^{2}\alpha_m. \tag{5.4}$$

Now, it follows from (5.1) and (5.3) that

$$\widehat{b}(z) = \cfrac{1}{c+\cfrac{1}{\widehat{a}(z)}} = \cfrac{\alpha_0}{c\alpha_0+\beta_1+z-\cfrac{\alpha_1}{\beta_2+z-\cfrac{\alpha_2}{\beta_3+z-\ldots}}}.$$

The latter continuous fraction is the Stieltjes' J-fraction for $\widehat{b}(z)$ and hence we have for its determinants the same expressions as those in (5.4), and (5.2) follows. □

Corollary 5.2. *The sequence* $\{s_j\}_{j=0}^{\infty}$ *is an* L*-sequence if and only if*

$$\mathcal{M}_1\left(\frac{1}{2}\widehat{s}(z)\right) = \widehat{b}(z) \tag{5.5}$$

for some positive sequence $\{b_j\}_{j=0}^{\infty}$.

Proof. Indeed, we have

$$\mathcal{M}_1(w) = \tanh w = \frac{e^{2w}-1}{e^{2w}+1},$$

therefore,

$$\widehat{b}(z) \equiv \mathcal{M}_1\left(\frac{1}{2}\widehat{s}(z)\right) = \frac{1-v(z)}{1+v(z)},$$

where $v(z) = \exp(-\widehat{s}(z))$ is the standard exponential transform of $\widehat{s}(z)$. Since $1-v(z)$ is the generating function of some positive sequence (a_k), we have

$$\widehat{b}(z) = \frac{\widehat{a}(z)}{2-\widehat{a}(z)},$$

or

$$\frac{1}{\widehat{b}(z)} = \frac{2}{\widehat{a}(z)} - 1,$$

and the required property follows from positivity of $\widehat{a}(z)/2$. □

Remark 5.3. The previous observation makes it possible to consider an analogue of the (n-dimensional) transform by letting

$$\mathbb{E}_\rho^n(x) := 1 - \mathcal{M}_n(V_\rho(x)).$$

In particular, $\mathbb{E}_\rho^2(x) = E_\rho(x)$, while for $n = 1$ we have

$$\mathbb{E}_\rho^1(x) = \frac{2E_\rho(x)}{1+E_\rho(x)}.$$

Acknowledgment

The author is grateful to Björn Gustafsson, Mihai Putinar and Serguei Shimorin for conversations crucial to the development of this paper.

References

[1] N.I. Akhiezer, *The Classical Moment Problem and Some Related Questions in Analysis*, Oliver and Boyd, Edinburgh/London, 1965.

[2] N.I. Akhiezer and M.G. Krein, *Some Questions in the Theory of Moments*, Amer. Math. Soc. Transl., Vol. 2, Providence, R.I., 1962.

[3] S.N. Bernstein, *Sur les fonctions absolument monotones.* Acta math. **52** (1928), 1–66

[4] R.W. Carey and J.D. Pincus, *An exponential formula for determining functions*, Indiana Univ. Math. J., **23** (1974), 1031–1042.

[5] Duren, P.L.: *Univalent functions*, Grundlehren der Mathematischen Wissenschaften, vol. 259, Springer-Verlag, New York (1983).

[6] B. Gustafsson and M. Putinar, *The exponential transform: a renormalized Riesz potential at critical exponent.* Ind. Univ. Math. J., **52** (2003), no. 3, 527–568.

[7] C.H. Kimberling, *A probabilistic interpretation of complete monotonicity.* Aequationes Math., **10** (1974), 152–164.

[8] M.G. Krein and A.A. Nudelman, *Markov Moment Problems and Extremal Problems*, Translations of Math. Monographs Volume 50, Amer. Math. Soc., Providence, RI, 1977.

[9] E.H. Lieb and M. Loss, *Analysis*, Graduate Stud. in Math., V. 14. AMS. 1997.

[10] J.D. Pincus and J. Rovnyak, *A representation formula for determining functions*, Proc. Amer. Math. Soc., **22** (1969), 498–502.

[11] M. Putinar, *A renormilized Riesz transform and applications* (to appear)

[12] H.S. Wall, *Continued Fractions*, University Series in Higher Mathematics, New-York, 1948.

[13] D.V. Widder, *The Laplace Transform.* Princeton, University Press, Princeton, 1946.

Vladimir G. Tkachev
Volgograd State University
Department of Mathematics
2-ya Prodolnaya 30
400062, Volgograd, Russia
e-mail: vladimir.tkatchev@volsu.ru

GPSR Compliance

The European Union's (EU) General Product Safety Regulation (GPSR) is a set of rules that requires consumer products to be safe and our obligations to ensure this.

If you have any concerns about our products, you can contact us on ProductSafety@springernature.com

In case Publisher is established outside the EU, the EU authorized representative is:

Springer Nature Customer Service Center GmbH
Europaplatz 3
69115 Heidelberg, Germany

Batch number: 10732489

Printed by Printforce, the Netherlands